Biology, History, and Natural Philosophy

Contributors

Francisco J. Ayala
Ludwig von Bertalanffy
Jay Boyd Best
Allen D. Breck
Ronald R. Cowden
Walter M. Elsasser
Constantine J. Falliers
Martin A. Garstens
Mirko D. Grmek
Stephen Körner
Arne Naess
Robert E. Roeder
Henryk Skolimowski
George Ledyard Stebbins
Albert Szent-Györgyi
René Taton
Håkan Törnebohm
Hayden V. White
John O. Wisdom
Harry Woolf
Wolfgang Yourgrau

Biology, History, and Natural Philosophy

Based on the Second International Colloquium held at the University of Denver

Edited by
Allen D. Breck
and
Wolfgang Yourgrau
University of Denver
Denver, Colorado

Ⴔ **PLENUM PRESS • NEW YORK-LONDON • 1972**

The Second International Colloquium
was held at the University of Denver
November 27 – December 2, 1967

Library of Congress Catalog Card Number 70-186262

ISBN 0-306-30573-9

© 1972 Plenum Press, New York
A Division of Plenum Publishing Corporation
227 West 17th Street, New York, N.Y. 10011

United Kingdom edition published by Plenum Press, London
A Division of Plenum Publishing Company, Ltd.
Davis House (4th Floor), 8 Scrubs Lane, Harlesden, London
NW10 6SE, England

This volume is dedicated
with esteem and admiration to
Albert Szent-Györgyi

Foreword

In a world that peers over the brink of disaster more often than not it is difficult to find specific assignments for the scholarly community. One speaks of peace and brotherhood only to realize that for many the only real hope of making a contribution may seem to be in a field of scientific specialization seemingly irrelevant to social causes and problems.

Yet the history of man since the beginnings of science in the days of the Greeks does not support this gloomy thesis. Time and again we have seen science precipitate social trends or changes in the humanistic beliefs that have a significant effect on the scientific community. Not infrequently the theoretical scientist, triggered by society's changing goals and understandings, finds ultimate satisfaction in the work of his colleagues in engineering and the other applied fields.

Thus the major debate in mid-nineteenth century in which the evidence of natural history and geology at variance with the Biblical feats provided not only courage to a timid Darwin but the kind of audience that was needed to fit his theories into the broad public dialogue on these topics. The impact of "Darwinism" was felt far beyond the scientific community. It affected social thought, upset religious certainties and greatly affected the teaching of science.

We could find many other such examples. They lend hope to the belief that there is great worth in bringing together scientists and scholars in other fields for colloquia such as the one whose proceedings make up the main body of this book. In the presence of the philosopher of science, the pure researcher, the historian—in the community of minds and skills that is the very essence of a present and future world—one must hope that the past will repeat itself and new understandings, adequate to our needs and our times, will be the result.

That this could happen at all, here in the foothills of the great Rocky Mountain range, is one of those minor miracles of determination, organization and support that seem to occur less frequently today than a decade or two ago. Thanks are very much in order, and should be addressed first to the scholars who came to the University of Denver from far places and always at the expense of other work and other demands on their time and skills.

We also owe a vote of thanks to the Martin-Marietta Corporation for their contribution of funds and enthusiasm, which made this and its predecessor Colloquium possible. The proximity of one of their major installations to the University of Denver and the common interest which pervades their research and ours have always provided a close bond in the development of solutions beneficial to the future.

In the deepening sea of troubles which surround us many pray for another Renaissance, a rediscovery "of man and of nature" with its roots, as before, in classical thought and humanistic involvement in scientific ways of thought. Just as the turmoil and new ideas of the Renaissance had their roots in all that had gone before, so may another turning point in history rest on the foundations of scientific discovery and social awareness of the century now nearing its close. The great hope that this might indeed come to pass may well be our willingness to keep the talk going, to share ideas, to take the time from our specialized fields to try to understand both the men and the ideas that exist in other private and little-known realms. To do this in the natural sciences seems one of the highest orders of priority, and the usefulness of the effort may be estimated after one has read the pages that follow.

Since I am already convinced that we have done a most useful thing in holding and publishing this Colloquium, I can only close by expressing the hope that many of us will meet again one day on another such occasion.

University of Denver   Maurice B. Mitchell
Denver, Colorado   Chancellor
November, 1971

Preface

The First International Colloquium on *Physics, Logic, and History* held at the University of Denver in May 1966, produced a scholarly volume of papers and comments; it was published in 1970 by the Plenum Press, New York. The success of that conference and the predictable response to the diverse papers and their discussions emboldened its organizers to gather another distinguished group of thinkers for a second Colloquium in the fall of 1967. The results of that conference are collected in this volume.

Again, it was our hope to contribute to some kind of understanding among academic disciplines only too often thought to be quite separated in purpose, formulation, and intrinsic technique. It is one of the most tragic difficulties of our time that scientists, historians, and philosophers have failed to listen to each other. The inevitable and dramatic advance in the various aspects of systematic knowledge has, unfortunately, led to an almost cavalier disregard for those aspects or elements essential to all pursuits of knowledge based on intelligible communication.

Thinkers like Nietzsche, whose anti-scientific attitude is proverbial, stated shortly before his mind deteriorated in the quicksand of mental disease, that he would like to study biology in a systematic manner and he even flirted with the idea of enrolling as a student of this subject. He recognized, as one among the first so-called non-scientific thinkers, that a deeper insight into the biological sciences may be imperative for his philosophy of life, which had been based entirely upon his intensive study of the Greek classics and some subjective philosophic ideas in general.

This second Symposium places the biological sciences at the center of our attention, flanked by some excursions into history and an occasional journey into natural philosophy. When we decided upon the issues which were supposed to be considered, we intended to explore topics like: "Is biology an autonomous science?"; " 'Organisms have been likened to machines'—is this assertion true?"; "Schrödinger and Teilhard de Chardin—two controversial credos in contemporary biological sciences"; "Historical systems considered from a biological point of view," etc.

The diverse papers and the ensuing discussions among biologists, historians and philosophers, historians and philosophers of science, are here presented because we were guided by the possibility that the search for a comprehensive understanding of nature may be pursued along a broader spectrum of knowledge by means of objective investigation, discussion, and hopefully inspiration.

It should be noted that the lectures contained in this volume do not correspond exactly with those given at the Conference. In a few cases authors requested permission to withdraw their papers rather than see them in print. All contributors were given a chance to make such revisions as would bring their conclusions up to date when new discoveries were made in their respective fields or further reflection altered some aspects of their views. But, the most fruitful contribution of the Colloquium resulted from informal discussions of the participants and interested observers.

In one important way, however, this volume differs from its predecessor. The editors thought best to omit the lengthy discussions which followed each paper. The difficulties of transcription, the incredibly vexing problem of editing remarks which were later amplified or discarded, the many *ad hominem* comments—all this made such inclusion less commendable for this specific collection of essays.

Some scholars who had been invited to participate were unfortunately unable to join us. They have been none the less gracious enough to send copies of their papers so they might become a part of this presentation. Special thanks of the editors go to Dr. Maurice B. Mitchell, Chancellor of the University of Denver, for presiding over some of the sessions and for participating from the beginning in making the Colloquium possible. Dr. Edward A. Lindell, Dean of the College of Arts and Sciences, was indispensable in helping to plan the format, complete arrangements for bringing participants to the campus, and carry out the multitudinous details without which the Symposium could never have materialized. The organizing committee consisted of Dr. Breck, Dr. Lindell, and Dr. Yourgrau.

This second Colloquium (as well as the first) was made possible by a grant from the Martin-Marietta Corporation Foundation and further financial support by the University of Denver. The entire manuscript was typed several times by Mrs. Betty Greenwell, secretary of the program in the history and philosophy of science. For all her tireless efforts we are indeed most grateful.

University of Denver   The Editors
Denver, Colorado
September, 1971

Participants

FRANCISCO J. AYALA, The Rockefeller University, New York, New York

SAUL BENISON, University of Cincinnati, Cincinnati, Ohio

LUDWIG VON BERTALANFFY, State University of New York at Buffalo, Buffalo, New York

JAY BOYD BEST, Colorado State University, Fort Collins, Colorado

ALLEN D. BRECK, University of Denver, Denver, Colorado

WILLIAM T. DRISCOLL University of Denver, Denver, Colorado

WALTER M. ELSASSER, University of Maryland, College Park, Maryland

CONSTANTINE J. FALLIERS, Children's Asthma Research Institute, Denver, Colorado

GEORGE GAMOW, University of Colorado, Boulder, Colorado

MARTIN A. GARSTENS, University of Maryland and the Office of Naval Research, College Park, Maryland

MIRKO D. GRMEK, Archives Internationales d'Historie des Sciences, Paris, France

STEPHAN KÖRNER, University of Bristol, England

EDWARD J. MACHLE, University of Colorado, Boulder, Colorado

ARNES NAESS, University of Oslo, Norway

JOHN R. OLIVE, American Institute of Biological Sciences, Washington, D.C.

THEODORE T. PUCK, University of Colorado School of Medicine, Denver, Colorado

ROBERT E. ROEDER, University of Denver, Denver, Colorado

HARRY ROSENBERG, Colorado State University, Fort Collins, Colorado

GEORGE G. SIMPSON University of Arizona, Tucson, Arizona

HENRYK SKOLIMOWSKI, University of Southern California, Los Angeles, California

ALBERT SZENT-GYÖRGYI, Laboratory of the Institute of Muscle Research, Marine Biological Laboratory, Woods Hole, Massachusetts

RENÉ TATON, École Pratique des Hautes Étude, Paris, France

HÅKAN TÖRNEBOHM, University of Göteborg, Sweden

STANISLAW M. ULAM, University of Colorado, Boulder, Colorado

HAYDEN V. WHITE University of California, Los Angeles, California

JOHN O. WISDOM, York University, Toronto, Canada

HARRY WOOLF, The Johns Hopkins University, Baltimore, Maryland

WOLFGANG YOURGRAU, University of Denver, Denver, Colorado

Contents

FOREWORD
Maurice B. Mitchell, Chancellor, *University of Denver*
PREFACE
The Editors
I. THE AUTONOMY OF BIOLOGY AS A NATURAL SCIENCE,
Francisco J. Ayala, *The Rockefeller University* 1
II. THE MODEL OF OPEN SYSTEMS: BEYOND MOLECULAR
BIOLOGY, Ludwig von Bertalanffy, *State University of
New York at Buffalo* 17
III. ELECTRONIC MOBILITY IN BIOLOGICAL PROCESSES,
Albert Szent-Györgyi, *Marine Biological Laboratory,
Woods Hole, Massachusetts* 31
IV. THE EVOLUTION AND ORGANIZATION OF SENTIENT
BIOLOGICAL BEHAVIOR SYSTEMS, Jay Boyd Best, *Colorado
State University, Fort Collins* 37
V. THE EVOLUTIONARY SIGNIFICANCE OF BIOLOGICAL
TEMPLATES, George Ledyard Stebbins, *University of
California, Davis* 79
VI. EVOLUTIONARY MODULATION OF RIBOSOMAL RNA
SYNTHESIS IN OOGENESIS AND EARLY EMBRYONIC
DEVELOPMENT, Ronald R. Cowden, *University of Denver* 103
VII. RESPIRATION AS INTERFACE BETWEEN SELF AND
NON-SELF: HISTORICO-BIOLOGICAL PERSPECTIVES,
Constantine J. Falliers, *Children's Asthma Research Institute
& Hospital, Denver* 111
VIII. MEASUREMENT THEORY AND BIOLOGY, Martin A. Garstens,
University of Maryland and the Office of Naval Research 123
IX. THE TRANSITION FROM THEORETICAL PHYSICS INTO
THEORETICAL BIOLOGY, Walter M. Elsasser, *University of
Maryland* 135
X. SCIENTIFIC ENTERPRISES FROM A BIOLOGICAL POINT
OF VIEW, Håkan Törnebohm, *University of Göteborg* 165

XI. HISTORICAL OBSERVATIONS CONCERNING THE
RELATIONSHIP BETWEEN BIOLOGY AND MATHEMATICS,
René Taton, *École Pratique des Hautes Étude, Paris* 171

XII. A SURVEY OF THE MECHANICAL INTERPRETATIONS OF
LIFE FROM GREEK ATOMISTS TO THE FOLLOWERS OF
DESCARTES, Mirko D. Grmek, *Archives Internationales
d'Historie des Sciences, Paris* 181

XIII. THE PLACE OF NORMATIVE ETHICS WITHIN A BIO-
LOGICAL FRAMEWORK, Arne Naess, *University of Oslo* 197

XIV. THE EVOLUTIONARY THOUGHT OF TEILHARD DE
CHARDIN, Francisco J. Ayala, *The Rockefeller University* 207

XV. THE USE OF BIOLOGICAL CONCEPTS IN IN THE WRITING
OF HISTORY, Allen D. Breck, *University of Denver* 217

XVI. WHAT IS A HISTORICAL SYSTEM?, Hayden V. White,
University of California, Los Angeles 233

XVII. ON A DIFFERENCE BETWEEN THE NATURAL SCIENCES
AND HISTORY, Stephan Körner, *University of Bristol* 243

XVIII. HISTORICAL TAXONOMY, Robert E. Roeder, *University
of Denver* 251

XIX. THEORIES OF THE UNIVERSE IN THE LATE EIGHTEENTH
CENTURY, Harry Woolf, *The Johns Hopkins University* 263

XX. MUST A MACHINE BE AN AUTOMATON?, John O. Wisdom,
York University 291

XXI. EPISTEMOLOGY, THE MIND AND THE COMPUTER,
Henryk Skolimowski, *University of Southern California* 299

XXII. MARGINAL NOTES ON SCHRÖDINGER, Wolfgang Yourgrau,
University of Denver 331

INDEX 345

CHAPTER I

The Autonomy of Biology as a Natural Science

Francisco J. Ayala
The Rockefeller University

The goal of science is the systematic organization of knowledge about the material universe on the basis of explanatory principles that are genuinely testable. The starting point of science is the formulation of statements about objectively observable phenomena. Common-sense knowledge also provides information about the material world. The distinction between science and common-sense knowledge is based upon the joint presence in science of at least three distinctive characteristics. First, science seeks to organize knowledge in a systematic way by exhibiting patterns of relations among statements concerning facts which may not obviously appear as mutually related.

The information obtained in the course of ordinary experience about the material universe is frequently accurate, but it seldom provides any explanation of why the facts are as alleged. It is the second distinctive characteristic of science that it strives to provide explanations of why the observed events do in fact occur. Science attempts to discover and to formulate the conditions under which the observed facts and their mutual relationships exist.

Thirdly, the explanatory hypotheses provided by science must be genuinely testable, and therefore subject to the possibility of rejection. It is sometimes asserted that scientific explanatory hypotheses should allow to formulate predictions about their subject matter which can be verified by further observation and experiment. However, in certain fields of scientific knowledge, like in those fields concerned with historical questions, prediction is considerably restricted by the nature of the subject matter itself. The criterion of testability can then be satisfied by requiring that scientific explanations have precise logical consequences which can be verified or falsified by observation and experiment. The word "precise" is essential in the previous sentence. To provide genuine verification the logical consequences of the proposed explanatory hypotheses must not be compatible with alternate hypotheses.

1

It is the concern of science to formulate theories, that is, to discover patterns of relations among vast kinds of phenomena in such a way that a small number of principles can explain a large number of propositions concerning these phenomena. In fact, science develops by discovering new relationships which show that observational statements and theories that had hitherto appeared as independent are in fact connected and can be integrated into a more comprehensive theory. Thus, the Mendelian principles of inheritance can explain, about many different kinds of organisms, observations which appear as *prima facie* unrelated; like the proportions in which characters are transmitted from parents to offspring, the discontinuous nature of many traits of organisms, and why in outbreeding sexual organisms not two individuals are likely to be genetically identical even when the number of individuals in the species is very large. Knowledge about the formation of the sex cells and about the behavior of chromosomes was eventually shown to be connected with the Mendelian principles, and contributed to explain additional facts; like why certain traits are inherited independently from each other while other traits are transmitted together more frequently than not. Additional discoveries have contributed to the formulation of a unified theory of inheritance which explains many other diverse observations, including the distinctiveness of natural species, the adaptive nature of organisms and their features, and paleontological observations concerning the evolution of organisms.

The connection among theories has sometimes been established by showing that the principles of a certain theory or branch of science can be explained by the principles of another theory or science shown to have greater generality. The less general branch of science, called the secondary science, is said to have been *reduced* to the more general or primary science. A typical example is the reduction of Thermodynamics to Statistical Mechanics.[1] The reduction of one branch of science to another simplifies and unifies science.

Reduction of one theory or branch of science to another has repeatedly occurred in the history of science. During the last hundred years several branches of Physics and Astronomy have been to a considerable extent unified by their reduction to a few theories of great generality like Quantum Mechanics and Relativity. A large sector of Chemistry has been reduced to Physics after it was discovered that the valence of an element bears a simple relation to the number of electrons in the outer orbit of the atom. The impressive success of these and other reductions has led in certain circles to the conviction that the ideal of science is to reduce all natural sciences, including biology, to a comprehensive theory that will provide a common set of principles of maximum generality capable of explaining all our observations about the material universe.

To evaluate the validity of such claims, I will briefly examine some of the necessary conditions for the reduction of one theory to another. I will, then, attempt to show that at the present stage of development of the two sciences, the reduction of biology to physics cannot be effected. I will further claim, in the second part of this paper, that there are patterns of explanation which are indispensable in biology while they do not occur in the physical sciences. These are teleological explanations which apply to organisms and only to them in the natural world, and that cannot be reformulated in nonteleological form without loss of explanatory content.

Conditions for Reduction

In general, reduction can be defined in the present context as "the explanation of a theory or a set of experimental laws established in an area of inquiry, by a theory usually though not invariably formulated for some other domain."[2] Nagel has stated the two formal conditions that must be satisfied to effect the reduction of one science to another. First, all the experimental laws and theories of the secondary science must be shown to be logical consequences of the theoretical constructs of the primary science. This has been called by Nagel the *condition of derivability*.

Generally, the experimental laws formulated in a certain branch of science will contain terms which are specific to that area of inquiry. If the laws of the secondary science contain some terms that do not occur in the primary science, logical derivation of its laws from the primary science will not be *prima facie* possible. No term can appear in the conclusion of a formal demonstration unless the term appears also in the premises. To make reduction possible it is then necessary to establish suitable connections between the terms of the secondary science and those used in the primary science. This may be called the *condition of connectability*. It can be satisfied by a redefinition of the terms of the secondary science using terms of the primary science. For example, to effect the reduction of genetics to physical science such concepts as gene, chromosome, etc., must be redefined in physicochemical terms such as atom, molecule, electrical charge, hydrogen bond, deoxyribonucleic acid, length, etc.

The problem of reduction is sometimes formulated as whether the properties of a certain kind of objects, for instance organisms, can be explained as a function of the properties of another such group of objects, like the organism's physical components organized in certain ways. This formulation of the question is spurious and cannot lead to a satisfactory answer. Indeed it is not clear what is meant by the "properties" of a certain object which enters as a part or component of some other object. If *all* the properties are included, it appears that reduction can always be accomplished, and it is in fact a trivial issue. Among the properties of a certain object one will list the properties which it has when it is a component of the larger whole. To use a simple example, one may list among the properties of hydrogen that of combining in a certain way with oxygen to form water, a substance which possesses certain specified properties. The properties of water will then be included among the properties of oxygen and hydrogen.

The reduction of one science to another is not a matter of deriving the *properties* of a kind of objects from the properties of some other group of objects. It is rather a matter of deriving a set of *propositions* from another such set. It is a question about the possibility of deriving the experimental laws of the secondary science as the logical consequences of the theoretical laws of the primary science. Scientific laws and theories consist of propositions about the material world, and the question of reduction can only be settled by the concrete investigation of the logical consequences of such propositions, and not by discussion of the properties or the natures of things.

From the previous observation it follows that the question of reduction can only be solved by a specific reference to the actual stage of development of the two disciplines involved. Certain parts of chemistry were reduced to physics after the modern theories of atomic structure was developed some fifty years ago, but the reduction could not have been accomplished before such development. If the reduction of one science to another is not possible at the present stage of development of the two disciplines, it is empirically meaningless to ask whether reduction will be possible at some further time, since the question can only be answered dogmatically or in terms of metaphysical preconceptions.

Simpson has suggested that the unification of the various natural sciences be sought not "through principles that apply to all phenomena but through phenomena to which all principles apply." Science, according to Simpson, can truly become unified in biology, since the principles of all natural sciences can be applied to the phenomena of life.[3] To be sure, the theoretical laws of physics and chemistry apply to the physicochemical phenomena occurring in organisms. Besides, there are biological theories that explain observations concerning the living world but have no application to nonliving matter. To conclude that therefore biology stands at the center of all science is true as far as it goes, but it is trivial and provides no progress in scientific understanding that I can discern.

The goal of the reductionistic program is not, as Simpson seems to believe, to establish a "body of theory that might ultimately be *completely* general in the sense of applying to *all* material phenomena," nor a "search for a least common denominator in science." It is rather a quest for a comprehensive theory that would *explain* all material phenomena—the living as well as the inanimate world—with an economy of stated laws and a corresponding increase in our understanding of the world. Whether such an ideal can be accomplished is a different issue that I shall consider presently.

The Reduction of Biology to Physical Science

The question of the reducibility of biology to physicochemistry has been raised again in the last decade particularly in connection with the spectacular successes accomplished in certain areas of biology. In genetics, research at the molecular level has contributed to establish the chemical structure of the hereditary material, to decipher the genetic code, and to provide some understanding of the mechanisms of gene action. Brilliant achievements have also been obtained in neurophysiology and other fields of biology. Some authors have claimed that the understanding of all biological phenomena in physicochemical terms is not only possible but the task of the immediate future. It is thus proclaimed that the only worthy and truly "scientific" biological research is what is called in recent jargon "molecular biology," that is the attempt to explain biological phenomena in terms of the underlying physicochemical components and processes.

I will without much ado dispose of two extreme positions which seem equally unprofitable. On one end of the spectrum there are substantive vitalists which defend the irreducibility of biology to physical science because living phenomena are the effect of a nonmaterial principle which is variously called vital force, entelechy, élan vital, radial energy, or the like. A nonmaterial principle cannot be subject to scientific observation nor lead to genuinely testable scientific hypotheses.[4]

At the other end of the spectrum stand those who claim that reduction of biology to physicochemistry is in fact possible at present. At the current stage of scientific development, a majority of biological concepts, like cell, organ, species, ecosystem, etc., cannot be formulated in physicochemical terms. Nor is there at present any class of statements belonging to physics and chemistry from which every biological law could be logically derived. In other words, neither the condition of connectability nor the condition of derivability—two necessary formal conditions of reduction—are satisfied at the present stage of development of physical and biological knowledge.

Two intermediate positions have also appeared in the recent literature. The reductionist position maintains that although the reduction of biology to physics cannot be effected at present, it is possible in principle. The factual reduction is made contingent upon further progress in the biological or in the physical sciences, or in both. Certain antireductionist authors claim that reduction is not possible in principle because organisms are not merely assemblages of atoms and molecules, nor even of organs and tissues standing in merely external relation to one another. Organisms are alleged to be "wholes" that must be studied as wholes and not as the "sum" of isolable parts.[5]

Although biological laws are not in general derivable from any available theory of physics and chemistry, the reductionists claim that such accomplishment will be possible in the future. Such proposition is frequently based on metaphysical preconceptions about the nature of the material world. It cannot be in any case convincingly argued empirically, since it is only a statement of faith about the possibility of some future event. It must be noted, however, that advances in various areas of molecular biology are continuously extending the as yet exiguous realm of biological phenomena that can be explained in terms of physicochemical concepts and laws.

As for the antireductionist position that maintains that organisms and their properties cannot be understood as mere "sums" of their parts, I have already stated that it rests on an unsatisfactory formulation of the problem. The question of reduction is whether *propositions* concerning organisms can be logically derived from physicochemical laws, and not whether the *properties* of organisms can be explained as the result of the properties of their physical components. It should perhaps be added that the phenomenon of so-called "emergent" properties occurs also in the nonliving world. Water is formed by the union of two atoms of hydrogen with one atom of oxygen, but water exhibits properties which are not the immediately apparent consequence of the properties of the two gases, hydrogen and oxygen. Another simple example can be taken from the field of thermodynamics. A gas has a temperature although the individual molecules of the gas cannot be said to possess a temperature.[6]

The reduction of biology to physicochemistry cannot be effected at the present stage of scientific knowledge. Whether the reduction will be possible in the future is an empirically meaningless question. A majority of biological problems cannot be as yet approached at the molecular level. Biological research must then continue at the different levels of integration of the living world, according to the laws and theories developed for each order of complexity. The study of the molecular structure of organisms must be accompanied by research at the levels of the cell, the organ, the individual, the population, the species, the community, and the ecosystem. These levels of integration are not independent of each other. Laws formulated at one level of complexity illuminate the other levels, both lower and higher, and suggest additional research strategies.[7] It is perhaps worth pointing out that in fact biological laws discovered at a higher level of organization have more frequently contributed to guide research at the lower level than vice versa. To mention but one example, the Mendelian theory of inheritance preceded the identification of the chemical composition and structure of the genetic material and made possible such discoveries.

The Notion of Teleology

I will now proceed to discuss the role of teleological explanations in biology. I shall attempt to show that teleological explanations constitute patterns of explanation that apply to organisms while they do not apply to any other kind of objects in the natural world. I shall further claim that although teleological explanations are compatible with causal accounts they cannot be reformulated in nonteleological form without loss of explanatory content. Consequently, I shall conclude that teleological explanations cannot be dispensed with in biology, and are therefore distinctive of biology as a natural science.

The concept of teleology is in general disrepute in modern science. More frequently than not it is considered to be a mark of superstition, or at least a vestige of the nonempirical, a prioristic approach to natural phenomena characteristic of the prescientific era. The main reason for this discredit is that the notion of teleology is equated with the belief that future events—the goals or end-products of processes—are active agents in their own realization. In evolutionary biology, teleological explanations are understood to imply the belief that there is a planning agent external to the world, or a force immanent to the organisms, directing the evolutionary process toward the production of specified kinds of organisms. The nature and diversity of organisms are, then, explained teleologically in such view as the goals or ends-in-view intended from the beginning by the Creator, or implicit in the nature of the first organisms.

Biological evolution can be explained without recourse to a Creator or planning agent external to the organisms themselves. There is no evidence either of any vital force or immanent energy directing the evolutionary process toward the production of specified kinds of organisms. The evidence of the fossil record is against any necessitating force, external or immanent, leading the evolutionary process toward specified goals. Teleology in the stated sense is, then, appropriately rejected in biology as a category of explanation.

In *The Origin of Species* Darwin accumulated an impressive number of observations supporting the evolutionary origin of living organisms. Moreover, and perhaps most importantly, he provided a causal explanation of evolutionary processes—the theory of natural selection. The principle of natural selection makes it possible to give a natural explanation of the adaptation of organisms to their environments. Darwin recognized, and accepted without reservation, that organisms are adapted to their environments, and that their parts are adapted to the functions they serve. Penguins are adapted to live in the cold, the wings of birds are made to fly, and the eye is made to see. Darwin accepted the facts of adaptation, and then provided a natural explanation for the facts. One of his greatest accomplishments was to bring the teleological aspects of nature into the realm of science. He substituted a scientific teleology for a theological one. The teleology of nature could now be explained, at least in principle, as the result of natural laws manifested in natural processes, without recourse to an external Creator or to spiritual or nonmaterial forces. At that point biology came into maturity as a science.

The concept of teleology can be defined without implying that future events are active agents in their own realization nor that the end-results of a process are consciously intended as goals. The notion of teleology arose most probably as a result of man's reflection on the circumstances connected with his own voluntary actions. The anticipated outcome of his actions can be envisaged by man as the goal or purpose toward which he directs his activity. Human actions can be said to be purposeful when they are intentionally directed toward the obtention of a goal.

The plan or purpose of the human agent may frequently be inferred from the actions he performs. That is, his actions can be seen to be purposefully or teleologically ordained toward the obtention of a goal. In this sense the concept of teleology can be extended, and has been extended, to describe actions, objects or processes which exhibit an orientation toward a certain goal or end-state. No requirement is necessarily implied that the objects or processes tend consciously toward their specified end-states, nor that there is any external agent directing the process or the object toward its end-state or goal. In this generic sense, teleological explanations are those explanations where the presence of an object or a process in a system is explained by exhibiting its connection with a specific state or property of the system to whose existence or maintenance the object or process contributes. Teleological explanations require that the object or process contribute to the existence of a certain state or property of the system. Moreover, and this is the essential component of the concept, teleological explanations imply that such contribution is the explanatory reason for the presence of the process or object in the system. Accordingly, it is appropriate to give a teleological explanation of the operation of the kidney in regulating the concentration of salt in the blood, or of the structure of the hand of man obviously adapted for grasping. But it makes no sense to explain teleologically the motions of a planet or a chemical reaction. In general, as it will be shown presently, teleological explanations are appropriate to account for the existence of adaptations in organisms while they are neither necessary nor appropriate in the realm of nonliving matter.

There are at least three categories of biological phenomena where teleological explanations are appropriate, although the distinction between the categories need not always be clearly defined. These three classes of teleological phenomena are established according to the mode of relationship between the structure or process and the property or end-state that accounts for its presence. Other classifications of teleological phenomena are possible according to other principles of distinction. A second classification will be suggested later.

(1) When the end-state or goal is consciously anticipated by the agent. This is purposeful activity and it occurs in man and probably, although in a lesser degree, in other animals. I am acting teleologically when I buy an airplane ticket to fly to Mexico City. A cheetah hunting a zebra has at least the appearance of purposeful behavior. However, as I have said above, there is no need to explain the existence of organisms and their adaptations as the result of the consciously intended activity of a Creator. There is purposeful activity in the living world, at least in man; but the existence of the living world, including man, need not be explained as the result of purposeful behavior. When some critics reject the notion of teleology from the natural sciences, they are considering exclusively this category of teleology.

(2) Self-regulating or teleonomic systems, when there exists a mechanism that enables the system to reach or to maintain a specific property in spite of environmental fluctuations. The regulation of body temperature in mammals is a teleological mechanism of this kind. In general, the homeostatic reactions of organisms belong to this category of teleological phenomena. Two types of homeostasis are usually distinguished by biologists—physiological and developmental homeostasis, although intermediate and additional types do exist.[8] Physiological homeostatic reactions enable the organism to maintain certain physiological steady states in spite of environmental shocks. The regulation of the composition of the blood by the kidneys, or the hypertrophy of muscle in case of strenuous use, are examples of this type of homeostasis.

Developmental homeostasis refers to the regulation of the different paths that an organism may follow in its progression from zygote to adult. The development of a chicken from an egg is a typical example of developmental homeostasis. The process can be influenced by the environment in various ways, but the characteristics of the adult individual, at least within a certain range, are largely predetermined in the fertilized egg. Aristotle, Saint Augustine, and other ancient and mediaeval philosophers, took developmental homeostasis as the paradigm of all teleological mechanisms. According to Saint Augustine, God did not create directly all living species of organisms, but these were implicit in the primeval forms created by God. The existing species arose by a natural "unfolding" of the potentialities implicit in the primeval forms or "seeds" created by God.

Self-regulating systems or servo-mechanisms built by man belong in this second category of teleological phenomena. A simple example of such servo-mechanisms is a thermostat unit that maintains a specified room temperature by turning on and off the source of heat. Self-regulating mechanisms of this kind, living or man-made, are controlled by a feed-back system of information.

(3) Structures anatomically and physiologically constituted to perform a certain function. The hand of man is made for grasping, and his eye for

vision. Tools and certain types of machines made by man are teleological in this third sense. A watch, for instance, is made to tell time, and a fawcet to draw water. The distinction between the (3) and (2) categories of teleological systems is sometimes blurred. Thus the human eye is able to regulate itself within a certain range to the conditions of brightness and distance so as to perform its function more effectively.

Teleology and Adaptation

Teleological mechanisms and structures in organisms are biological adaptations. They have arisen as a result of the process of natural selection. Natural selection is a mechanistic process defined in genetic and statistical terms as differential reproduction. Some genes and genetic combinations are transmitted to the following generations on the average more frequently than their alternates. Such genetic units will become more common, and their alternates less common, in every subsequent generation.

The genetic variants arise by the random processes of genetic mutation and recombination. Genetic variants increase in frequency and may eventually become fixed in the population if they happen to be advantageous as adaptations in the organisms which carry them, since such organisms are likely to leave more descendants than those lacking such variants. If a genetic variant is harmful or less adaptive than its alternates, it will be eliminated from the population. The biological adaptations of the organisms to their environments are, then, the result of natural selection, which is nevertheless a mechanistic and impersonal process.

The adaptations of organisms—whether organs, homeostatic mechanisms, or patterns of behavior—are explained teleologically in that their existence is ultimately accounted for in terms of their contribution to the reproductive fitness of the species. A feature of an organism that increases its reproductive fitness will be selectively favored. Given enough generations it will extend to all the members of the population.

Patterns of behavior, such as the migratory habits of certain birds or the web-spinning or spiders, have developed because they favored the reproductive success of their possessors in the environments where the population lived. Similarly, natural selection can account for the existence of homeostatic mechanisms. Some living processes can be operative only within a certain range of conditions. If the environmental conditions oscillate frequently beyond the functional range of the process, natural selection will favor self-regulating mechanisms that maintain the system within the functional range. In man death results if the body temperature is allowed to rise or fall by more than a few degrees above or below normal. Body temperature is regulated by dissipating heat in warm environments through perspiration and dilation of the blood vessels in the skin. In cool weather the loss of heat is minimized, and additional heat is produced by increased activity and shivering. Finally, the adaptation of an organ or structure to its function is also explained teleologically in that its presence is accounted for in terms of the contribution it makes to reproductive success in the population. The vertebrate eye arose because genetic mutations responsible for its development occurred which increased the reproductive fitness of their possessors.

There are in all organisms two levels of teleology that may be labelled *specific* and *generic*. There usually exists a specific and proximate end for every feature or an animal or plant. The existence of the feature is explained in terms of the function or property that it serves. This function or property can be said to be the specific or proximate end of the feature. There is also an ultimate goal to which all features contribute or have contributed in the past—reproductive success. The generic or ultimate end to which all features and their functions contribute is increased reproductive efficiency. The presence of the functions themselves—and therefore of the features which serve them—is ultimately explained by their contribution to the reproductive fitness of the organisms in which they exist. In this sense the ultimate source of explanation in biology is the principle of natural selection.

Natural selection can be said to be a teleological process in a *causal* sense. Natural selection is not an entity but a purely mechanistic process. But natural selection can be said to be teleological in the sense that it produces and maintains end-directed organs and mechanisms, when the functions served by them contribute to the reproductive efficiency of the organism.

The process of natural selection is not at all teleological in a different sense. Natural selection does not tend in any way toward the production of specific kinds of organisms or toward organisms having certain specific properties. The over-all process of evolution cannot be said to be teleological in the sense of proceeding toward certain specified goals, preconceived or not. The only nonrandom process in evolution is natural selection understood as differential reproduction. Natural selection is a purely mechanistic process and it is opportunistic.[9] The final result of natural selection for any species may be extinction, as shown by the fossil record, if the species fails to cope with environmental change.

The presence of organs, processes and patterns of behavior can be explained teleologically by exhibiting their contribution to the reproductive fitness of the organisms in which they occur. This need not imply that reproductive fitness is a consciously intended goal. Such intent must in fact be denied, except in the case of the voluntary behavior of man. In teleological explanations the end-state or goal is not to be understood as the efficient cause of the object or process that it explains. The end-state is causally—and in general temporarily also—posterior.

Internal and External Teleology

Three categories of teleological phenomena have been distinguished above, according to the nature of the relationship existing between the object or mechanism and the function or property that it serves. Another classification of teleology may be suggested attending to the process or agency giving origin to the teleological system. The end-directedness of living organisms and their features may be said to be *internal* teleology, while that of man-made tools and servo-mechanisms may be called *external* teleology. It might also be appropriate to refer to these two kinds of teleology as *natural* and *artificial*, but the other two terms, "internal" and "external," have already been used.[10]

Internal teleological systems are accounted for by natural selection which is a strictly mechanistic process. External teleological systems are the products of the human mind, or more generally, are the result of purposeful activity consciously intending specified ends. An automobile, a wrench and a thermostat are teleological systems in the external sense; their parts and mechanisms have been produced to serve certain functions intended by man. Organisms and their parts are teleological systems in the internal sense; their end-directedness is the result of the mechanistic process of natural selection. Organisms are the only kind of systems exhibiting internal teleology. In fact they are the only class of *natural* systems that exhibit teleology. Among the natural sciences, then, only biology, which is the study of organisms, requires teleology as a category of explanation.

Organisms do not in general possess external teleology. As I have said above, the existing kinds of organisms and their properties can be explained without recourse to a Creator or planning agent directing the evolutionary process toward the production of such organisms. The evidence from paleontology, genetics, and other evolutionary sciences is also against the existence of any immanent force or vital principle directing evolution toward the production of specified kinds of organisms.

Teleological Explanations in Biology

Teleological explanations are fully compatible with causal accounts. "Indeed, a teleological explanation can always be transformed into a causal one."[11] Consider a typical teleological statement in biology, "The function of gills in fishes is respiration." This statement is a telescoped argument the content of which can be unraveled approximately as follows: Fish respire; if fish have no gills, they do not respire; therefore fish have gills. According to Nagel, the difference between a teleological explanation and a nonteleological one is, then, one of emphasis rather than of asserted content. A teleological explanation directs our attention to "the *consequences* for a given system of a constituent part of process." The equivalent nonteleological formulation focuses attention on "some of the *conditions* . . . under which the system persists in its characteristic organization and activities."[12]

Although a teleological explanation can be reformulated in a nonteleological one, the teleological explanation connotes something more than the equivalent nonteleological one. In the first place, a teleological explanation implies that the system under consideration is directively organized. For that reason teleological explanations are appropriate in biology and in the domain of cybernetics but make no sense when used in the physical sciences to describe phenomena like the fall of a stone. Teleological explanations imply, while nonteleological ones do not, that there exists a means-to-end relationship in the systems under description.

Besides connoting that the system under consideration is directively organized, teleological explanations also account for the existence of specific functions in the system and more generally for the existence of the directive organization itself. The teleological explanation accounts for the presence in an organism of a certain feature, say the gills, because it contributes to the performance or maintenance of a certain function, respiration. In addition it implies that the function exists because it contributes to the reproductive fitness of the organism. In the nonteleological translation given above, the major premiss states that "fish respire." Such formulation assumes the presence of a specified function, respiration, but it does not account for its existence. The teleological explanation does in fact account for the presence of the function itself by implying or stating explicitly that the function in question contributes to the reproductive fitness of the organism in which it exists. Finally, the teleological explanation gives the reason why the system is directively organized. The apparent purposefulness of the ends-to-mean relationship existing in organisms is a result of the process of natural selection which favors the development of any organization that increases the reproductive fitness of the organisms.

If the above reasoning is correct, the use of teleological explanations in biology is not only acceptable but indeed indispensable. Organisms are systems directively organized. Parts of organisms serve specific functions that, generally, contribute to the ultimate end of reproductive survival. One question biologists ask about organic structures and activities is "What for?" That is, "What is the function or role of such a structure or such a process?" The answer to this question must be formulated in teleological language. Only teleological explanations connote the important fact that plants and animals are directively organized systems.

It has been argued by some authors that the distinction between systems that are goal directed and those which are not is highly vague. The classification of certain systems as teleological is allegedly rather arbitrary. A chemical buffer, an elastic solid or a pendulum at rest are examples of physical systems that appear to be goal directed. I suggest using the criterion of utility to determine whether an entity is teleological or not. The criterion of utility can be applied to both internal and external teleological systems. Utility in an organism is defined in reference to the survival and reproduction of the organism itself. A feature of a system will be teleological in the sense of internal teleology if the feature has utility for the system in which it exists and if such utility explains the presence of the feature in the system. Operationally, then, a structure or process of an organism is teleological if it can be shown to contribute to the reproductive efficiency of the organism itself, and if such contribution accounts for the existence of the structure or process.

In external teleology utility is defined in reference to the author of the system. Man-man tools or mechanisms are teleological with external teleology if they have been designed to serve a specified purpose, which therefore explains their existence and properties. If the criterion of utility cannot be applied, a system is not teleological. Chemical buffers, elastic solids and a pendulum at rest are not teleological systems.

The utility of features of organisms is with respect to the individual or the species in which they exist at any given time. It does not include usefulness to any other organisms. The elaborate plumage and display is a teleological feature of the peacock because it serves the peacock in its attempt to find a mate. The beautiful display is not teleologically directed toward pleasing man's aesthetic sense. That it pleases the human eye is accidental, because it does not contribute to the reproductive fitness of the peacock (except, of course, in the case of artificial selection by man).

The criterion of utility introduces needed objectivity in the determination of what biological mechanisms are end-directed. Provincial human interests should be avoided when using teleological explanations, as Nagel says. But he selects the wrong example when he observes that "the development of corn seeds into corn plants is sometimes said to be natural, while their transformation into the flesh of birds or men is asserted to be merely accidental."[13] The adaptations of corn seeds have developed to serve the function of corn reproduction, not to become a palatable food for birds or men. The role of wild corn as food is accidental, and cannot be considered a biological function of the corn seed in the teleological sense.

Some features of organisms are not useful by themselves. They have arisen as concomitant or incidental consequences of other features that are adaptive or useful. In some cases, features which are not adaptive in origin may become useful at a later time. For example, the sound produced by the beating of the heart has become adaptive for modern man since it helps the physician to diagnose the condition of health of the patient. The origin of such features is not explained teleologically, although their preservation might be so explained in certain cases.

Features of organisms may be present because they were useful to the organisms in the past, although they are no longer adaptive. Vestigial organs, like the vermiform appendix of man, are features of this kind. If they are neutral to reproductive fitness these features may remain in the population indefinitely. The origin of such organs and features, although not their preservation, is accounted for in teleological terms.

To conclude, I will summarize the second part of this paper. Teleological explanations are appropriate to describe, and account for the existence of, teleological systems and the directively organized structures, mechanisms and patterns of behavior which these systems exhibit. Organisms are the only natural systems exhibiting teleology; in fact they are the only class of systems possessing internal teleology. Teleological explanations are not appropriate in the physical sciences, while they are appropriate, and indeed indispensable, in biology which is the scientific study of organisms. Teleological explanations, then, are distinctive of biology among all the natural sciences.[14]

NOTES

[1] E. Nagel, *The Structure of Science*, New York: Harcourt, Brace and World, 1961, pp. 338-345.

[2] E. Nagel, *The Structure of Science*, p. 338; see also pp. 336-397.

[3] G. G. Simpson, *This View of Life*, New York: Harcourt, Brace and World, 1964, p. 107. According to J. G. Kemeny (*A Philosopher Looks at Science*, New York: Van Nostrand, 1959, pp. 215-216) the most likely solution of the question of the reduction of biology to physics is that a new theory will be found, covering both fields, in new terms. Inanimate nature will appear as the simplest extreme case of this theory. In that case, one would say that physics was reduced to biology and not biology to physics.

[4] Except for the general conclusion that biological phenomena will never be satisfactorily explained by mechanistic principles.

[5] E. S. Russell, *The Interpretation of Development and Heredity*, Oxford, 1930; see also E. Mayr, "Cause and Effect in Biology," *Cause and Effect*, D. Lerner (ed.), New York: Free Press, 1965, pp. 33-50.

[6] The temperature of the gas is identical by definition with the mean kinetic energy of the molecules.

[7] See Th. Dobzhansky, "Biology, Molecular and Organismic," *The Graduate Journal*, VII(1), 1965, pp. 11-25.

[8] For instance, the maintenance of a genetic polymorphism in a population due to heterosis can be considered a homeostatic mechanism acting at the population level.

[9] See F. J. Ayala, "Teleological explanation in evolutionary biology," *Philosophy of Science*, in press.

[10] T. A. Goudge, *The Ascent of Life*, Toronto: University of Toronto Press, 1961, p. 193.

[11] E. Nagel, "Types of causal explanation in science," *Cause and Effect,* D. Lerner (ed.), New York: Free Press, 1965, p. 25.

[12] E. Nagel, *The Structure of Science,* p. 405; see also F. J. Ayala, "Teleological explanations in evolutionary biology," *Philosophy of Science,* in press.

[13] E. Nagel, *The Structure of Science,* p. 424.

[14] It is a pleasure to thank Miss Mary C. Henderson who read the manuscript and made many valuable suggestions.

CHAPTER **II**

The Model of Open Systems: Beyond Molecular Biology

Ludwig von Bertalanffy
State University of New York at Buffalo

A New View on Scientific Practice and Theory

Two questions appear to have crystallized in our present Colloquium. First is the question of specialization and generalization. We are all aware that the enormous content, the complex techniques and sophisticated concepts of modern science require specialization. On the other hand, the question arises: Is there nothing common in the sciences—from physics to biology to the social sciences to history—so that the scientific enterprise must remain a bundle of isolated specialities, without connecting link and progressively leading to the type of learned idiot who is perfect in his small field but is ignorant and unaware of the basic problems we call philosophical, and which are of primary concern to man in one of the greatest crises of his history?

Secondly—we all feel that this is a time of scientific re-orientation. Whether this is expressed in the indeterminacy principle of physics and the gaps in the theory of elementary particles, in biological problems so that mystical views like those of Teilhard de Chardin are taken into serious considerations, or in the present dissatisfaction with psychological and sociological theories—it is the common feeling that something new is required, and that yesterday's mechanistic universe which has safely guided science through some 250 years, has come to an end.

One way to approach the problem is what the program of our colloquium calls "the historical vs. the logical approach." The logical approach, as we find it in innumerable philosophical writings, is, in broad outline and oversimplification, something like this: We are confronted by observation with sense data, facts, pointer readings, protocol sentences—whatever expression you prefer. From these we derive generalizations which, when properly formulated, are called laws of nature. These are fitted into conceptual schemes called theories, which in the well-known way of hypothetico-deductive system, allow for the explanation, prediction and control of nature. The logical operations involved could be carried through even better and neater with sufficiently capable computers.

However, personal experience of the scientist as well as the history of science shows that the actual development of science is nothing of this sort. Psychology has shown that cognition is an active process, not a passive mirroring of reality. For this reason, there are no facts as ultimate data; what we call facts has meaning only within a pre-existing conceptual system. The famous pointer-readings positivist philosophers were fond to speak of as being the basis of scientific knowledge simply make no sense without a conceptual scheme. Whether we read electric currents in a suitable apparatus, or read oxygen consumption of a tissue in a Warburg machine, or simply read a watch—the so-called facts of observation make no sense except in a conceptual construct we already possess and presuppose. In consequence, history of science does not appear as an approximation to truth, a progressively improved mirroring of an ultimate reality. Rather, it is a sequence of conceptual constructs which map, with more or less success, certain aspects of an unknown reality. For example, one of the first models was that of myth and magic, seeing nature animated by gods and demons who may be directed by appropriate practices. Another one was Aristotle's seeing the universe guided by purposeful agents or entelechies. Then there was the Newtonian universe of solid atoms and blind natural forces.

Nowadays we seem to be dedicated to still another model, epitomized by the term "system," of which we shall speak more in detail. Neither were the previous models and world views simply superstitious nonsense, nor were they completely eradicated by subsequent ones. The mythical world view served mankind admirably well through many millenia, and produced unique achievements, such as the array of domesticated plants and animals which modern science did not essentially increase. And there is still far too much demonology around, in science and particularly in the pseudoscience of politics. Aristotle's physics was a bad model, as was shown by Galileo; but problems posed by him, such as that of teleology, are still alive in the theory of evolution—see Teilhard de Chardin—and in the considerations of cybernetics. That our thinking still is much too Newtonian, is the common complaint of physicists, biologists and psychologists.

The modern reorientation of thought, the new models, appear to be centered in the *concept of system.* This would need much more elaboration than I can provide and I must refer to the literature cited. In a very aphoristic characterization: the procedure of classical science was to resolve observed phenomena into isolable elements; these, then, can be put together, practically or conceptually, to represent the observed phenomenon. Experience has shown that this isolation of parts and causal chains, and their summation and superposition works widely, but that now we are presented, in all sciences, with problems of a more difficult sort. We are confronted with wholes, organizations, mutual interactions of many elements and processes, systems—whichever expression you choose. They are essentially non-additive, and therefore cannot adequately be dealt with by analytical methods. You cannot split them into isolable elements and causal trains. Compared with the approach of classical science, they require new concepts, models, methods—whether the problem is that of an atomic nucleus, a living system or a business organization. Mutual interaction instead of linear causality; organized complexity instead of summation of undirected and statistical events—these, somewhat loosely, define the new problems.

One such model is that of the title of my talk—open system. Such discussion appears to present two advantages. First, we do not talk abstractly or philosophically about what biology should be, whether it should or ultimately will be reduced to physics, and so on; rather, we consider concrete problems of research which, with more or less success, elucidate basic phenomena of life. Secondly, in so doing, we can shed light on a number of questions which were asked of the present Colloquium—the question whether the organism is a machine, whether biology is an autonomous science, the question of the evolution of biological systems, of Schrödinger's and Teilhard's ideas, and so forth. When we quite straightforwardly look at the organism as open system, some philosophical questions take care of themselves!

The Living Machine and its Limitations

An introduction to the open system model can very well start with one of those trivial questions, which are often very difficult to answer scientifically. What is the difference between a normal living organism, a sick and a dead organism? From the standpoint of physics and chemistry, we must answer: None. For from the viewpoint of "ordinary" physics and chemistry, a living organism is an aggregate of an enormous number of processes which, sufficient work and knowledge presupposed, can be defined by means of chemical formulas, equations of physics and the laws of nature in general. Obviously, these processes are different in a living, sick or dead dog; but the laws of physics don't tell us a difference; they are not interested whether dogs are dead or alive. This remains unchanged even if we take into consideration the last results of molecular biology. One DNA-molecule, protein, enzyme, or hormone-controlled process is just as good as any other; everyone is determined by physical and chemical laws, none is better, healthier or more normal than another.

But there is, no doubt, a fundamental difference between a live and a dead organism; usually, we don't have any difficulties to distinguish between a living organism and a dead object. In a living being innumerable chemical and physical processes are so *ordered* as to allow the living system to persist, to grow, to develop, to reproduce, etc. What, however, does this notion of "order" mean, for which we would look in vain in textbooks of physics and chemistry? In order to be able to define and explain it we need a *model*, a conceptual construct. One such model was used since the beginnings of natural science. This was the model of the living machine. Depending on the state of the art, the model found different interpretations. When in the 17th century Descartes introduced the concept of the animal as a machine, only *mechanical machines* existed. Hence the animal was a sort of complicated clockwork. Thus Borelli, Harvey and other so-called iatrophysicists explained the function of muscles, of the heart, etc., by mechanical principles of levers, of a pump and the like. You can still see this in a musical opera, when in the *Tales of Hoffmann* the beautiful Olympia turns out to be an artful doll, a clockwork or automaton as it was then called. Later on the steam engine and thermodynamics were introduced, which led to the organism being conceived as a *heat engine,* a fact to which we owe the calculation of calories, for example. As it turned out, however, the organism is not a heat engine, transforming the energy of fuel first into heat and then into mechanical energy. It rather is a *chemico-dynamic* machine, directly transforming the chemical energy of fuel into effective work, a fact on which, for example, the theory of muscle-contraction is based. Lately, self-regulating machines came to the fore, such as the thermostat, missiles aiming at a target and the servomechanisms of modern technology. In parallel, the organism became a *cybernetic machine.* The most recent development are *molecular machines.* When we talk about the "mill" of the Krebs cycle of oxidation or about the mitochondria as the "power-plant" of the cell, this means that machine-like structures on the molecular level determine the order of enzyme reactions; similarly, it is a micromachine which transforms or translates the genetic code of the DNA of the chromosomes into proteins and eventually into a complex organism.

Nevertheless, the machine model of the organism has its difficulties. One is the problem of *the origin of the machine.* Old Descartes didn't have a problem because he explained the "animal machine" as the creation of a divine watchmaker. But how do machines come about in a universe of undirected physico-chemical events? Clocks, steam engines and transistors don't grow by themselves in nature. Where do the infinitely more complicated living machines come from?

Secondly, there is the problem of *regulation.* Self-repairing machines are conceivable in terms of modern automata theory. However, the problem arises when regulation and repair take place after arbitrary disturbances. Can a machine such as an embryo or a brain be programmed for regulation, not after a defined disturbance or a finite set of disturbances, but an indefinite number of disturbances? Here, it seems, is a limit for the logical automaton or the so-called Turing machine, which arises from "enormous" numbers. The Turing machine can, in principle, resolve even most complex operations into a finite number of steps. But the number may be neither finite and countable nor infinite and non-countable, but just "enormous," that is, of a higher order than, for example, the number of elementary particles in the universe. Such enormous numbers do appear, e.g., in the interactions of even a moderate number of elements, for example, genes or nerve cells.

Even more important is a third question. The living organism is maintained in a continuous *exchange of its components*; metabolism is the basic characteristic of living systems. It is, as it were, a machine composed of fuel, spending itself continually and yet maintaining itself. Such machines do not exist in present technology. In other words: *A machine-like structure of the organism cannot be the ultimate reason of order of life processes,* because the machine itself is maintained in an ordered flow of processes. The primary order must, therefore, lie in the process itself.

The Model of Open System: A Scientific Approach to Organismic Biology

We express this by saying that living systems are essentially open systems. An open system is defined by the fact that it exchanges matter with its environment, that it persists in import and export, building-up and breaking-down of its material components. But here we are immediately confronted with difficulties. When 30 years ago I proposed the model of the organism as open system, there was no theory of such systems. Rather, kinetics and thermodynamics, the fields of physics concerned, were by definition restricted to closed systems, that is, systems without exchange of matter. The theory of open systems is therefore a relatively recent development and leaves many problems unsolved. The development of a kinetic theory of open systems has two roots: first the biophysical problem of the organism; second, developments in industrial chemistry which increasingly applies continuous i.e. open reaction systems which have greater efficiency and other advantages compared to reactions in a closed system. The thermodynamic theory of open systems is called irreversible thermodynamics. It has become an important generalization of physical theory through the work of Meixner, Onsager, Prigogine and others.

As in any physical theory, we must start with simple models. Already simple open systems (Fig. 1a) show remarkable characteristics, apparent from the solution of the set of simultaneous equations which define them. Under certain conditions, open systems approach a time-independent state, a steady state or *Fliessgleichgewicht.* The steady state is maintained in distance from true equilibrium and is therefore capable of doing work—as is the case in living systems, in contrast to systems in equilibrium. Furthermore, the system remains constant in its composition, in spite of continuous irreversible processes taking place, i.e. import and export, building-up and breaking-down of components. If such steady state is reached in an open system, it is independent of the initial conditions, and determined only by the system parameters, that is, the rates of reaction and transport. We call this *equifinality*, and it is found in many organismic processes, for example in growth: The same final state or "goal," that is, the same species-characteristic, the final size may be reached from different initial sizes and after arbitrary disturbances of the growth process (Fig. 2).

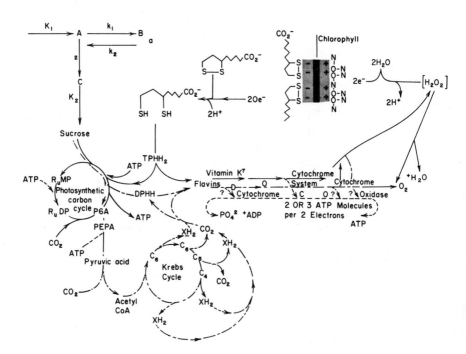

Figure 1a. Model of a simple open system, showing maintenance of constant concentrations in the steady state, equifinality, adaptation and stimulus-response, etc. The model can be interpreted as a simplified schema for protein synthesis (A: amino acids, B: protein, C: deamination products; k_1: polymerization of amino acids into protein, k_2: depolymerization, k_3: deamination; $k_2 \ll k_1$, energy supply for protein synthesis not indicated). In somewhat modified form, the model is Sprinson and Rittenberg's (1949) for calculation of protein turnover from isotope experiments. (After von Bertalanffy 1953.) 1b. The open system of reaction cycles of photosynthesis in algae. (After Bradley and Calvin 1957.)

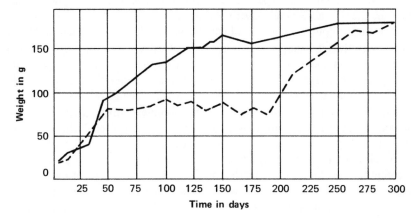

Figure 2. Equifinality of growth. Heavy curve: normal growth of rats. Broken curve: at the 50th day, growth was stopped by vitamin deficiency. After re-establishment of normal regime, the animals reached the normal final weight. (After Hober from von Bertalanffy 1960.)

From the viewpoint of thermodynamics, open systems can maintain themselves in a state of high statistical improbability, of order and organization. The reason is in the expanded entropy function of Prigogine. In closed systems, entropy and therefore disorder must increase owing to irreversible processes. In open systems, there is not only entropy production due to irreversible processes but also entropy transport due to import of matter as potential carrier of free energy or negative entropy. This is the basis of the negentropic trend in organismic systems and for Schrödinger's statement that the organism feeds on negative entropy. For this reason, open systems may even advance toward increasing differentiation and organization as is the case in the biological phenomena of development and evolution. In this way, an apparent contradiction between the inorganic and living universe is resolved. According to the second principle of thermodynamics, the general direction of physical events is toward increasing entropy, that is, toward states of increasing probability and decreasing differentiation. Organisms can evolve toward decreasing probability and increasing differentiation, because they represent open systems, exchanging matter with their environment.

The simple model I have shown is, so to speak, the granddaddy of more elaborate ones. One can, of course, mathematically set up much more complex models to study their properties, and find complex open systems both in nature and in industrial processes. I mention only two developments because of their theoretical implications: *computerization* and *compartment theory*. The solution of sets of simultaneous equations becomes most tiring with a larger number of reactants and equations; and if the equations are nonlinear, there is no general way to solving them.

Here the computer comes in, which can yield solutions for such problems. This is more than technique and facilitation of calculation; for nonlinearity is an obstacle in principle in solving more complex systems, and so the computer opens quite a new field. To give just one example, Benno Hess has analyzed the 14-step reaction chain of glycolysis in the living cell by way of a mathematical model of more than 100 nonlinear differential equations. Comparable analyses are routine in economics, market analysis, etc. This further makes possible *simulation*. Instead of investigating in the laboratory the complex interactions in a network of processes, one can do so easier on the computer, working out a suitable model and testing it against experiment. *Compartment theory* is concerned with the fact that reactions may take place, not in a homogenous space but in sub-systems which are partly permeable to the reactants. This obviously applies to many processes in the cell, and there are highly elaborate mathematical techniques to deal with such cases.

Compared to conventional closed systems, open systems show characteristics which seem to contradict the usual physical laws and were often considered as vitalistic characteristics of life, that is, as violations of physical laws explainable only by introducing soul-like or entelechial factors into the organic happenings. This is true of the equifinality of organic regulations, if for example the same "goal," i.e. a normal organism, is produced by a normal zygote, the half of a zygote, two zygotes fused, etc. This in fact was the most important, so-called proof of vitalism of Driesch. A similar consideration applies to the apparent contradiction between the tendency towards increasing entropy and disorder in inanimate nature, and negentropic processes in organic development and evolution. The apparent contradictions disappear with the expansion and generalization of physical theory to open systems.

In a discussion addressed to biologists, I would now have to show that the model of open system works, i.e. that it has power to explain phenomena previously not accounted for. We are here concerned with general and philosophical questions, so I must forego such discussion, and limit myself to quoting a few examples in the way of illustration.

There is, first, the wide field that has substantiated Goethe's *Stirb und Werde*, the continuous decay and regeneration, the dynamic structure of living systems at all levels of organization. Broadly we may say that this regeneration is taking place at far higher turnover rates than had been anticipated, Table 1. For example, it is certainly surprising that a calculation on the basis of a steady-state model revealed that the proteins of the human body have a turnover time of not much more than a hundred days. Essentially the same is true for cells and tissues. Many tissues of the adult organism are maintained in a steady state, cells being continuously lost by desquamation

TABLE 1

Turnover rates of intermediates of cellular metabolism (After Hess 1963)

structure	species	organ	turnover time in seconds
mitochondria	mouse	liver	1.3×10^6
hemoglobin	man	erythrocytes	1.5×10^7
aldolase	rabbit	muscle	1.7×10^6
pseudocholinesterase	man	serum	1.2×10^6
choloesterin	man	serum	9.5×10^5
fibrinogen	man	serum	4.8×10^4
glucose	rat	total organism	4.4×10^3
methionine	man	total organism	2.2×10^3
ATP glycolysis	man	crythrocytes	1.6×10^3
ATP glycolysis + respiration	man	thrombocytes	4.8×10^2
ATP glycolysis + respiration	mouse	ascites tumor	4.0×10^1
citrate cycle intermediates	rat	kidney	$1 - 10$
glycolytic intermediates	mouse	ascites tumor	$0.1 - 8.5$
flavoprotein$_{red.}$/flavoprotein$_{ox.}$	mouse	ascites tumor	4.6×10^{-2}
Fe^{2+}/Fe^{3+} – cytochrome a	grasshopper	wing muscle	10^{-2}
Fe^{2+}/Fe^{3+} – cytochrome a$_3$	mouse	ascites tumor	1.9×10^{-3}

and replaced by mitosis. Prior to such investigations, it was hardly expected that cells in the digestive tract or respiratory system have a life span of only a few days, Table 2.

After having extensively explored the paths of individual metabolic reactions in biochemistry, it has now become a pressing task to understand integrated metabolic systems as functional units. This is being done in the complex network and interplay of scores of reactions in functions such as photosynthesis (Fig. 1b), respiration and glycolysis. Here an important insight is implied. We begin to understand that beside visible morphologic organization, as we observe it macroscopically, with the ordinary or electron-microscope, there exists another, invisible organization, resulting from interplay of processes and defending itself against disturbances. With more time, I could show how the theory of open systems is applicable to many biological phenomena, such as phenomena of adaptation, action potentials, transport processes, pharmacodynamic action, growth, excitation, the energetics of the organism and many others.

Behind these facts we may trace the outline of an even further generalization. The theory of open systems is part of *general system theory*. This is a doctrine concerned with principles that apply to systems in general, irrespective of the nature of their components and the forces governing them. With general system theory we reach a level where we no longer talk about physical and chemical entities, but discuss wholes of a completely general

TABLE 2
Rate of mitosis in rat tissues (after F. D. Bertalanffy 1960)

	daily rate of mitosis (per cent)	renewal time (days)
Organs without mitosis nerve cells, neuroephithelium, neurilemma, retina, adrenal medulla	0	–
Organs with occasional mitosis but no cell renewal liver parenchyma, renal cortex and medulla, most glandular tissue, urethra, epididymis, vas deferens, muscle, vascular endothelium, cartilage, bone	less than 1	–
Organs with cell renewal upper digestive tract	7 – 24	4.3 – 14.7
large intestine and anus	10 – 23	4.3 – 10
stomach and pylorus	11 – 54	1.9 – 9.1
small intestine	64 – 79	1.3 – 1.6
trachea and bronchus	2 – 4	26.7 – 47.6
ureter and bladder	1.6 – 3	33 – 62.5
epidermis	3 – 5	19.1 – 34.5
sebaceous glands	13	8
cornea	14	6.9
lymph node	14	6.9
pulmonary alveolar cells	15	6.4
seminiferous epithelium	–	16

nature. Yet, certain principles of open systems still hold true and may successfully be applied at large, from ecology, the competition and equilibrium among species, to economics and other fields of sociology.

Unsolved Problems

At present, we do not have a thermodynamic criterion that would characterize the steady state in open systems in a similar way as maximum entropy defines equilibrium in closed systems. It was believed for some time that such criterion was provided by minimum entropy production, a statement known as "Prigogine's Theorem." However, Prigogine's Theorem, as was well known to its author, applies only under restrictive conditions. In particular, it does not define the steady state of chemical reaction systems.

Another unsolved problem of a fundamental nature originates in a basic paradox of thermodynamics. Eddington called entropy "the arrow of time." As a matter of fact, it is the irreversibility of physical events, expressed by the entropy function, which gives time its direction. Without entropy, that is, in a universe of completely reversible processes, there would be no difference between past and future. However, the entropy functions do not contain time explicitly. This is true of both the classical entropy function for closed systems by Clausius, and of the generalized function for open systems and irreversible thermodynamics by Prigogine.

Here we come into territory which is still largely unexplored. We can hardly doubt that in the living world there is a general phenomenon which can be termed anamorphosis, that is, increase in order and organization which is found in the development of an individual as well as in evolution. We have heard that this is not a violation of the second law of thermodynamics as was often assumed. But what, actually, determines the process? Thermodynamics presents no answer. We could further think of evolution as an accumulation of genetic information, i.e. the genetic code represented by the DNA of the chromosomes. But why does information accumulate in this special case of the genetic code, while in general, information is not only not preserved but is progressively dissipated into noise? The conventional answer given is the theory of selection. The so-called synthetic theory considers evolution as the result of chance mutations, that is, in a well-known simile, of "typing errors" which occasionally occur in the reduplication of the genetic code; the process being directed by selection, i.e. the survival of populations or genotypes with highest differential reproduction, that is, which produce the highest number of offspring under existing external conditions. Similarly, the origin of life is explained by a chance appearance of organic compounds (amino acids, nucleic acids, enzymes, ATP, etc.) in the primeval ocean which, by the way of selection, eventually formed reproducing units, virus-like forms, proto-organisms, cells, etc.

In contrast to this conventional view, it should be pointed out that selection, competition and "survival of the fittest" already *presuppose* the existence of self-preserving (and hence competing) systems; this can therefore not be the *result* of selection. At present we do not know of any physical law according to which, from a "soup" of organic compounds, self-preserving open systems are formed and maintained in a state of highest improbability. Likewise, even if such systems are accepted as "given," there is no physical law stating that evolution on the whole proceeds in the direction of increasing organization, i.e. improbability ("anamorphosis"). Selection of genotypes with maximum offspring doesn't help much in this respect; it is hard to understand why evolution ever should have gone beyond rabbits, herring or even bacteria which are unrivaled in their reproduction rate. Production of local conditions of higher order (and improbability) is physically only possible if "organizational forces" of some kind enter the scene; this is the case in the formation of crystals, where "organizational forces" are represented by valences, lattice forces and the like. Such organizational forces, however, are explicitly denied when the genome is considered as an accumulation of "typing errors."

Apparently, we need something more beyond conventional synthetic theory of evolution. One aspect is "organizational" forces and laws. As a matter of fact, we see with the electron-microscope structures far surpassing those encountered in ordinary physics and chemistry; insight into the laws of at least simpler supramolecular structures is slowly proceeding. "Biotonic" laws, as Elsasser calls them, at all levels of organization are apparently present, but are as yet known only in their visible manifestation.

The Russian biophysicist Trincher (1965) came to the conclusion that the entropy function is not applicable to living systems; he contrasts the entropy principle of physics with biological "principles of adaptation and evolution" expressing an increase of information. Here we have to consider, however, that the entropy principle has a physical basis in the Boltzmann derivation, in statistical mechanics and the transition to more probable distributions as it takes place in processes at random; while presently no physical explanation can be given for Trincher's phenomenological principles.

We can also speak of "internal" factors in evolution contrasted with "external," the outer-directedness of evolution expressed in the theory of selection. Following an ingeneous analysis by Herbert Simon, evolution appears to be connected with hierarchical order: Only by putting together subhierarchies, not by assembling complex systems from the start, it appears to be possible to have evolution in an acceptable time scale. However, what Simon calls stable hierarchic intermediates amounts to the same as what I called organizational principles—that is, supramolecular forces and laws that hold electron-microscopic, cell and supercellular structures together; and for this, Simon does not provide an explanation. These (and other) formulations mean essentially the same, and are different names for the same phenomenon and problem. We must admit that at present we do not have a theory. Presumably, irreversible thermodynamics, information theory, dynamical system theory, theory of structures higher than molecular ones, theory of hierarchic order and possibly others, will have to be united in a new way to account for anamorphosis. We have various hints in such direction, but nothing resembling a consistent theory.

The concept of the organismic model as an open system has proved to be very useful in the explanation and mathematical formulation of numerous life phenomena; it leads, as is to be expected in a scientific working hypothesis, to further problems, which are partly of a fundamental nature. This implies that it is not only of purely scientific but also of "meta-scientific" significance. The mechanistic concept of nature predominant so far emphasized the dissolution of happenings into linear causal chains, a conception of the world as a result of chance events or a physical and Darwinian "play of dice" (to quote Einstein's well-known saying) and the reduction of biological processes to laws presently known from inanimate nature. In contrast to this, in the theory of open systems (and its further generalization in general system theory) principles of multivariable interaction (e.g. reaction kinetics, flows and forces in irreversible thermodynamics, Onsager reciprocal relations) become apparent, a dynamic organization of processes and a possible expansion of physical laws under consideration of the biological realm. These developments therefore form part of a new formulation of the scientific world view.

REFERENCES

The following is a small selection of works quoted in the text or suggested for further reading on topics discussed.

Beier, W. *Biophysik*, 3rd edition. Jena: Fischer, 1968. English translation in preparation.

Bertalanffy, F. D., and C. Lau, "Cell Renewal," *Int. Rev. Cytol.* 13 (1962), 357-366.

von Bertalanffy, L., *Biophysik des Fliessgleichgewichts*. Translated by W. H. Westphal. Braunschweig: Vieweg, 1953. Revised edition with W. Beier and R. Laue in preparation.

von Bertalanffy, L., "Chance or Law," in *Beyond Reductionism*, edited by A. Koestler and J. R. Smythies. London: Hutchinson, 1969.

von Bertalanffy, L., *Robots, Men and Minds*. New York: Braziller, 1967.

von Bertalanffy, L., *General System Theory. Foundations, Development, Application*. New York: Braziller, 1968.

von Bertalanffy, L., "The History and Status of General System Theory," in *Trends in General Systems Theory*, edited by G. Klir. New York: Wiley, in press.

Denbigh, K. G., "Entropy Creation in Open Reaction Systems." *Trans. Faraday Soc.* 48 (1952), 389-394.

General Systems, L. von Bertalanffy and A. Rapport (eds.), Society for General Systems Research, Joseph Henry Building, Room 818, 2100 Pennsylvania Avenue, N.W., Washington, D.C. 20006, 12 vols. since 1956.

Hess, B., "Fliessgleichgewichte der Zellen," *Dt. Med. Wschr.*, 88 (1963), 668-676.

Hess, B., "Modelle enzymatischer Prozesse." *Nova Acta Leopoldina* (Halle, Germany), 1969.

Rescigno, A., and G. Segre, *Drug and Tracer Kinetics*, Waltham (Mass.): Blaisdell, 1966.

Rosen, R., *Dynamical System Theory in Biology*. Vol. I, Stability Theory and its Applications. New York: Wiley, 1970.

Simon, H. A., "The Architecture of Complexity," *General Systems* 10 (1965), 63-76.

Trincher, K. S., *Biology and Information: Elements of Biological Thermodynamics*. New York: Consultants Bureau, 1965.

Unity Through Diversity, edited by W. Gray and N. Rizzo, Vol. II: General and Open Systems. New York: London: Gordon and Breach, 1971.

Weiss, P. A., *Life, Order and Understanding*. The Graduate Journal, Vol. III, Supplement, University of Texas, 1970.

Whyte, L. L., *Internal Factors in Evolution*. New York: Braziller, 1965.

Whyte, L. L., A. G. Wilson and D. Wilson (eds.), *Hierarchical Structures*. New York: Elsevier, 1969.

Yourgrau, Wolfgang, A. Van der Merwe and G. Raw, *Treatise on Irreversible and Statistical Thermophysics*. New York: Macmillan, 1966.

CHAPTER **III**

Electronic Mobility in Biological Processes

Albert Szent-Györgyi
Institute for Muscle Research at the Marine Biological Laboratory
Woods Hole, Massachusetts

If you would ask a chemist to find out for you what an electric dynamo is and does, the first thing he would do is to dissolve it in hydrochloric acid. A molecular biologist would fare, perhaps, better. He might take the dynamo to pieces and describe the single pieces with the greatest care. However, if you would suggest to him that maybe there is an invisible fluid, electricity, flowing in that machine, and once he has taken it to pieces that fluid could not flow anymore, then he would scold you as a vitalist which is worse than to be called a communist by an FBI agent.

As you all know, biology at the moment stands entirely under the influence of the molecular concepts. Molecular biology tells us that the living organism is built up of very small closed units, molecules, and, consequently, what we have to know is the nature and composition and structure of these molecules and if we know all about them we will know all about life; the rest of it will take care of itself. What I want to do with you is to scrutinize this concept, that is, discuss with you the conceptual foundations of present biology, whether life processes can be fully explained and accounted for by molecules. I would add to all this that these molecules are macromolecules, that is, very big molecules which are very clumsy, very immobile. What gives special charm to biological research is the great subtlety of biological reactions and I have always found it difficult to believe that such a very subtle instrument as a living cell could be fully explained by the mutual relation of these clumsy, immobile units. I always felt that there must be something more in biology, something which engenders this great mobility and subtlety of the reactions. As you all know, the molecules are built up of atomic nuclei and electrons. The nuclei, practically, play no part in biology, they only act as charged points around which the electrons are arranged. All the biological phenomena thus must be due, in one way or another, to the reactions of these electrons. These electrons are very small, and very mobile, and I always have the feeling that the subtle biological reactions are due to the motion of electrons. Thus, for many years, I advocated an extension of biochemical research into the electronic dimension, and I even have written two little monographs about this subject, one under the title *Bioenergetics* and the other under the title *Introduction to a Submolecular Biology*. (Both are published by the Academic Press in New York.) It is now more than eight years ago that the second monograph has appeared and since then I have given very much time and thought to the study of electronic reactions.

The first and most basic question is now whether these electrons can have a more or less independent existence and mobility. Whether they can go from one molecule to the other producing those subtle reactions, or if they are chained to their units, the whole molecules which can participate in reactions only as one single whole closed unit.

Your program quotes a Chinese saying, according to which the longest journey of ten thousand leagues also begins with the first step. Therefore, if electrons do possess a mobility in biological systems the first question would be: how can they move within a molecule or go from one molecule to the other? Is there any possibility for such an electronic mobility?

I can give you the answer right away—it is affirmative.

Louis Brillouin has shown that if a molecule consists of regular repeat units, then the electrons within these single units disturb one another and their energy levels split up. If these electrons are not separated by very wide barriers, these levels conflow. And if such a big molecule has many units which contribute to the system there are many electron levels. Since each electron level splits up in two, there will be twice as many electronic levels as there are units, and in macromolecules this number can be very high. These many electron levels can more or less conflow to a continuous energy band, which can extend over the whole molecules. Whether the electrons will have free mobility in such an electron band depends on their number. According to the Pauli exclusion principle, no more than two electrons can be on the same energy level and this only if the two electrons have an opposite spin. So the maximum number of electrons which can be placed in such a band is the double of the number of the repeat units of that molecule which contributed to the band. If the number of the units is N and the number of the electrons in the band is 2N, then the band is saturated. This means there is no room for an extra electron and no room for mobility. The situation is similar to a very crowded cocktail party in which you can't move. To be able to move you have to wait till somebody leaves the room. So if, in this energy band, we should give a push to one electron in one direction, another electron will have to go the same length of the way in the opposite direction and there will be no net displacement and no electric conductivity.

In biology we are concerned, primarily, with protein molecules and so our first question is: do protein molecules have energy bands? There has been a great deal of discussion about this problem and as things stand today we have reason to believe that there are conduction bands in proteins. The question, only, is whether they are conductant—and this will depend on the number of electrons. It seems that in a protein molecule the number of electrons in such a band is 2N, that is, if there is such a band it is saturated, so in itself it could not conduct electricity and could not be the foundation of an electronic mobility. However, there are two possibilities of making such a system conductant. The single energy bands are separated from one another by "forbidden zones" in which no electron is allowed. Hence, the highest filled energy band has on top of it an empty band. Now the question is: how wide is this forbidden zone? If this forbidden zone is very narrow then even heat agitation could kick up an electron from the highest filled energy band to the lowest empty band. The question is thus: how wide is the forbidden zone in proteins? The calculations show that it is in the order of one to three electron volts and this is an enormous distance, so the electrons could not be moved up, let's say, by heat agitation from the filled band to the empty band. If the distance would be very small then even heat agitation could transfer electrons from the filled band to the empty band and the system would be what we call a "semiconductor" and could conduct electricity. This is, however, not the case in proteins.

Brillouin has shown that such a saturated band, which in itself could not conduct electricity and could not lead to an electronic mobility, could be rendered conductant if the system contains impurities and if these impure substances can pick up electrons from the saturated band. This would make the band desaturated and render it conductant and this could be a foundation for a great electronic mobility. My associate, Jane McLaughlin, has shown that proteins have, associated to them, amazingly big quantities of impurities, which can give up electrons. So if we attach these substances to a protein molecule we may develop an electronic mobility. If these impurities give electrons to an empty band then, on this empty band, the electrons will have a free mobility. Such conductivity is possible even without impurities because the protein molecule itself contains a great number of nitrogen and oxygen atoms, which are capable of giving off an electron as I have shown lately. They all have what we call a "lone pair of electrons," that is, electrons which are not involved in chemical bonding and thus are available for transmission.

To sum up what I have said till now is, that there is a possibility that electrons in proteins may have a great mobility and can therefore move from one end of a molecule to the other. This problem is still widely discussed and it will take a long time till it comes from the discussion table into the archives.

Another possibility of electronic mobility exists in molecules which contain double bonds in every second carbon to carbon link. This is what we call the "conjugated double bond" system which has an electronic conductivity similar to that of metals. This can be a very important factor in biological systems, in mobility of electrons, but the molecules which contain such a chain of conjugated double bonds are fairly short so they could not conduct on long distance but could transmit electrons merely from one molecule to another.

Now I am coming to the electronic mobility which has occupied me for the last years, I mean the question, how an electron can get from one molecule to the other. The Brillouin's bands explain mobility inside a molecule but the most important biological question is how an electron can proceed from one molecule to another and establish between them subtle relations. It was in 1942 that Joseph Weiss, in England, discovered that in certain molecule complexes formed of two molecules a high dipole moment is developed. He put together a strong oxidizing and reducing agent and found that they united to a molecular complex and this complex had its electronic charge distributed very asymmetrically. This he could explain only by supposing that an electron has gone over from one molecule to the other. The molecule which gave the electron he called the "donor" and the one which received the electron he called the "acceptor." This process he called the "charge transfer." This process was studied later from a quantum-mechanical point of view by R. S. Mulliken who did some wonderful work and systematized this whole group of reactions. Today, having electron spin resonance at our disposal, we would not look, in the first place, for a dipole moment but would look for an electron spin resonance signal which indicates that there is a molecule present with a free electron. In my opinion, this discovery of Joseph Weiss was one of the most momentous discoveries which may lead biochemistry into entirely new domains. However, this discovery did not meet with positive response in biochemistry at all and for a very simple reason. Joseph Weiss could observe such an electron transfer in molecules, one of which was a strong reducing agent, the other a strong oxidizing agent, and the presence of a strong oxidizing agent is incompatible with life; consequently, this reaction could play no role in biology. Later, it was shown that in many instances electrons could be moved over from one molecule to the other by the energy of light which excites the electron. This is so-called "charge transfer in the excited state." This charge transfer had not found access to biology because we have no light in our body, except in our eyes and our skin. So charge transfer has remained, for the biochemist, an item in the curiosity shop of nature. I could show, not so long ago, that such a charge transfer can take place also between molecules which are not strong oxidizing and reducing agents and do not need light for this electron transfer;

indeed I believe that this charge transfer is one of the most fundamental and most frequent biological reactions.

As you know, the electrons in a molecule are moving on certain paths which are strictly described by quantum mechanics. We call these paths "orbitals." If we approach a molecule from the outside, first we have to pass through orbitals which are not occupied, and then the occupied ones. Now, if the empty orbitals of one molecule touch the occupied orbitals of another molecule then there would be a chance for the electrons to go from the occupied orbital of the donor to the empty orbital of the acceptor molecules. Whether such a transfer will take place depends on the energy relation of the two orbitals but the electron will always have a tendency to go over to the empty orbital, because hereby it gains what is called the "resonance energy." So, as I see it today, this charge transfer is one of the most important biological reactions, and I could also show that certain very important hormones which are quite essential for life, like adrenal steroids, probably fulfill their function by taking up electrons and giving them off again, thus transmitting electrons and contribute in this way to the electronic mobility. I also think that the subtlety of biological reactions is due to these electronic interactions.

To sum up, I could thus say that I have been led by my extensive studies of the last ten years to believe that *the living machinery is essentially an electric device.* Of course, the molecules are excessively important as are the wires and the iron core in an electric dynamo. Without these, electricity could not move, but as the essential point about the dynamo is the motion of electricity which transforms electric energy into motion, analogously in biology the essential happening is the motion of this invisible "fluid" we call electricity, that is, the motion of electrons from one molecule to the other.

Such an electronic mobility has been entirely overlooked and I strongly believe that this is the reason why whole big chapters of biochemistry have remained, actually, blank pages. I should almost say that the most important chapters consist just of blank pages. What really interests us biologists is the nature of life. What is life? When we come to this most central question of biology we find a rather queer relation. I do not know what life is, but, all the same, I can tell life exactly from death. I know, for instance, beyond a doubt, when my dog is dead. If he does not move, has no reflexes, and leaves my carpet dry, he is dead. We know life, actually, by these functions, motions, reflexes and secretions. Actually, we never ask in science, what is something, because we have no means of understanding. What we ask in science only is, always: how does the thing behave? Is there any regularity in behavior and can we predict this behavior? So, life we know by its behavior, by its producing these symptoms I mention. If you analyze these symptoms, for instance, motion, you find that at its bottom is an energy transduction. We produce motion by the energy supplied by our foodstuffs. For instance, the muscle is actually a transformer which transforms chemical energy into mechanical energy. Similarly, a nervous function is a transduction which transduces chemical energy into electrical energy. A secretion is a transduction of chemical energy into osmotic work. In other words, all three big groups of symptoms of life are transductions, transformations of chemical energy into different forms of work. These transductions are the central problem of biology and we know nothing about them. I think that the reason for this ignorance is that we were thinking too much in terms of molecules and not in terms of electrons and electronic mobility.

I do not want to leave you with the impression that I depreciate molecular biology. It is one of the most wonderful achievements of the human mind and it was one of the most important strides in modern biology. However, I think this is not the last step and the next big stride will be the extension of biology to the electronic dimension and the study of the living cell as an electric appliance.

CHAPTER IV

The Evolution and Organization of Sentient Biological Behavior Systems

Jay Boyd Best
Colorado State University

Ideas about the evolution of psychological behavior systems are intertwined with notions about their organization and the design of "machine intelligence" systems to model them. Underlying all three is a "conventional wisdom" (Galbraith, 1958) seldom explicitly elaborated and without compelling theoretical or experimental basis. It has nevertheless exerted a profound influence on thinking in these areas. In this article I want to develop some notions which diverge substantially from this conventional wisdom and present a unifying conception of behavioral organization on a transphyletic basis.

To initially silhouette the target of discourse I shall express this conventional wisdom in explicit form. It goes as follows:

The psychological behaviors of sentient creatures came progressively into existence from simple stereotyped behaviors through the evolution of large complex brains from small brains. Stereotyped behaviors are simple and associated with simple neural systems.

Only complex neural or cybernetic systems produce psychological behaviors. The simple nervous systems of invertebrates show, accordingly, mostly stereotyped behaviors with psychological behaviors being exceptional. Higher vertebrates, conversely, show almost entirely plastic or psychological behaviors with little stereotyped behavior. The brain is somewhat similar to a large complex digital computer. The elements of both brain and computer are simple and stereotyped. Invertebrate nervous systems are similar to small computers with simple stereotyped programs.

This work was supported in part by Grant Number NsG 625 from the National Aeronautics and Space Administration and Grant Number RO1 MH07603 from the National Institute of Mental Health.

Artificial computers do not show sentient behavior because they are still too simple. Someday a computer named HAL will be constructed that will fill the basement of MIT and it will exhibit sentient behavior with perhaps incipient paranoia and megalomania.

What was the first brain to evolve like? Since we are one and a half billion years too late to directly verify our conjectures on this matter, probably the best we can do is to assume that the brain of an animal such as the fresh water planarian is fairly similar to that first dawn brain. What is the planarian brain like? Is its behavioral organization like a little piece of spinal cord with some stereotyped reflexes, or is it more like a miniature psychological system? To answer this I shall have to be more explicit about what I mean by a psychological system.

THE PROPERTIES OF PSYCHOLOGICAL SYSTEMS

By a psychological system I mean a behavioral system having the requisite properties to make it an object of interest to serious psychologists. If one asks an experimental psychologist what he studies he will say "behavior." Obviously one must take them at their word and observe the behavior of psychologists. If one does this one discovers that they do not study all behavior of all organisms but only a selected kind of behavior in a very selected set of organisms. They are not, for example, very interested in spinal cord aside from its necessity as a support system for the brain.

One discovers in this way that there are three minimal attributes which a behavioral system must have to capture the interest of a psychologist and, thereby, qualify as a psychological system. These attributes are:

1) It must have comething like drive or motivation.
2) It must exhibit habituation and/or learning.
3) It must exhibit something akin to emotion or emotional affect.

These attributes correspond closely to those which we, in a common sense way, would usually use to recognize a sentient creature in contrast to a nonsentient automaton. There are other useful attributes for making such a distinction but the three named certainly comprise the common denominator of all psychological systems.

MINIATURE MINDS AND BRAIN HOMOLOGIES

Does the planarian brain exhibit these minimal attributes of a psychological system? The answer is: "yes," it does, and more. In the limited space available I shall summarize evidence and arguments to show that:

1) It has behaviors reflecting such a psychological system.
2) It outlines the major anatomical and functional organization found in more highly evolved brains, e.g. those of vertebrates. This organizational

homology between the brains of planarians and vertebrates provides the basis of the behavioral homology between the two phyletic groups.

3) The sorts of functions and neurological systems which planarians (and probably also other bilaterally symmetric encephalized invertebrates) allocate to stereotyped behaviors correspond to the stereotyped systems of the vertebrates. Similarly, the sorts of functions and neurological systems which vertebrates allocate to psychological behaviors are comparable to those allocated to this purpose by planarians.

4) This organizational homology between the brains of planarians and vertebrates is a special aspect of a homology in the general organizations of the two kinds of animals.

5) This organizational homology between the brains of planarians and vertebrates is probably a special aspect of a basic organizational plan common to the central nervous systems of all bilaterally symmetric encephalized animals on the main evolutionary sequence.

THE PSYCHOLOGICAL SYSTEM OF PLANARIANS

Some similar kinds of experiments can be used to show motivation in both rats* and planarians. A rat, placed in some suitable test enclosure with food and fasted for some suitable length of time, sooner or later, goes to the food and eats. The elapsed time until the onset of feeding is called the feeding latency. This feeding latency, for a given test enclosure, decreases with length of fast, up to the point at which starvation deterioration sets in. Alternatively, one can use a fixed time interval of testing in the enclosure and simply score whether the rat does, or does not, eat. This will yield a relative frequency of feeding which increases with length of fast up to the point at which starvation deterioration begins to occur. If the rats are put into a new compartment or situation then they, for a given fast duration, tend to show an increased feeding latency, or decreased relative frequency of feeding (for a fixed test interval). It is generally agreed that this decreased tendency to feed is related to some kind of emotional or anxiety state (Munn, 1950; Hall, 1934). Planarians of the species *Cura foremanii* and *Dugesia dorotocephala* both show such reductions of feeding latency in a defined test situation as the duration of fast is increased (Best and Rubinstein, 1962; Best and Dunn, 1962). Results of such an experiment are shown in Figure 1.

---

*I shall refer to the rat repeatedly as a typical example of a higher vertebrate. My reason for choosing it resides in the fact that it has been extensively used for behavioral studies in experimental psychology and is not highly specialized for a particular ecological niche. Although there are important differences between laboratory and wild rats, these differences do not invalidate the cases in which I use rats as exemplary of the higher vertebrates. The evolutionary distance between rats and humans is a relative matter. It is small compared to those under consideration. Samarkand is close to Tulsa when one considers the distance of each from Tau Ceti.

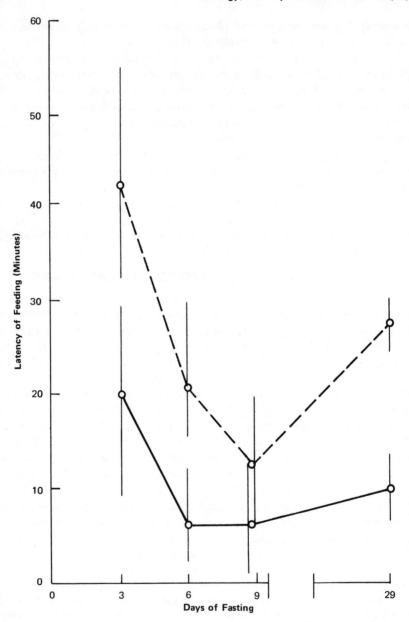

Figure 1. Average latency of feeding as a function of the duration of previous fast for planarians of the species *Cura foremanii*. ——○—— denotes those previously habituated to the test chamber. - -○- - denotes those not previously habituated to the test chamber. Vertical bars denote the standard deviations of the mean. (From Best and Rubinstein, 1962.)

Dugesia dorotocephala can exist in two different reproductive modes. In the asexual mode, gonads do not differentiate and the animal reproduces by simply tearing the tail portion away from the head portion. Following such fissioning, the anterior fragment regenerates a new tail and the tail fragment regenerates a new head to yield two complete, but smaller, planarians. In the sexual mode the planarian differentiates gonads (testes and ovaries), copulates, lays eggs which hatch into small planarians, but foregoes fissioning.

If asexual *Dugesia dorotocephala* are maintained at a higher population density for a period of time and then isolated, they exhibit an increased incidence of fissioning 2 to 5 days following such isolation. This effect is mediated through the brain (Best, Goodman and Pigon, 1969). In the presence of other planarians the brain exerts some kind of fission inhibiting influence. This influence evidently becomes attenuated with distance along the longitudinal axis of the worm since long planarians manifest a higher probability of fissioning than short ones following such isolation (Best, Goodman and Pigon, 1969). In keeping with this, tail fragments of a fissioning frequently undergo a subsequent secondary fissioning while the head piece never does. Although we do not know the stimulus cues and sensory modalities informing the brain of the presence of other planarians we know that it is not visual, or possesses water soluble chemical substances. Receptors in the anterior and lateral margins of the head are involved however (Pigon, Best, Howell and Riegel, 1970).

There appear to be circadian, circalunar and annual components of variation in this fissioning frequency for given values of population density (Best and Howell, 1970; Best, Goodman and Pigon, 1969).

Since the size of a planarian will depend upon its rate of growth and probability of fissioning per unit time for a given size, it is evident that mean size will be controlled by the brain through the mechanism described above. Prior studies on a closely related species indicate that transition to the sexual mode is also under control of the brain (Kenk, 1941). Size and sexual state are thus controlled by the brain and modulated by social interaction and "biological clock" phenomena in planarians as they are in vertebrates, annelids and arthropods (Florey, 1966; Wolfson, 1966; Cloudsley-Thompson, 1961). Such similarity in control of somatic functions by the brain, across phyletic boundaries, hints that the neural organization of these phyla may be homologous in other respects as well. We shall indeed find this to be true.

The situation in which a rat must cross an electric grid to obtain food belongs to the general class of situations called approach-avoidance paradigms. So does that in which a thirsty rat is administered an aversive shock through its metal drinking tube, or that in which a mouse is administered aversive shock for attempting to leave a small brightly lit space for a larger dark one (Jarvik and Kopp, 1967). Such experiments are particularly interesting in the present context because of their involvement of both emotional affect and learning. Suitable planarian analogs to such paradigms could be used to ascertain whether planarians show requisite attributes 2 and 3 for a psychological system. An additional point, which I shall subsequently stress, is that those situations yielding the most rapid and permanent learning in higher animals inevitably entail a controlled coupling of strong emotional affect, e.g. fear, to the situational contingencies. Single trial learning (obtained in rats with the metal drinking tube-shock paradigm or in mice with the Jarvik and Kopp paradigm) nearly always entails what can be regarded as emotional response conditioning. More specifically, they are conditioned fear or anxiety experiments. Another phenomenon is closely allied. Emotionally affective factors, e.g. those producing fear or anxiety, can be highly disruptive to learning when they enter into the situation as an extraneous variable. Such observations suggest an intimate link between the mechanisms of emotion and learning. They also suggest the possibility of obtaining much more rapid and stable learning in the lower invertebrates by harnessing such emotional factors into the paradigm where possible and minimizing the influence of those entering as extraneous variables. Such explicit consideration and control of these emotional factors are precluded if one assumes as *a priori* dictum that they do not exist. I shall therefore, without apology, deny this dictum its axiomatic position. The absence of any compelling rationale for the dictum and the real possibility that it may be counter productive, as well as arbitrary, are sufficient grounds for not making such an assumption. That such an assumption has been traditional is irrelevant.

In accord with these considerations learning experiments were performed with *D. dorotocephala*, originally collected in streams in the vicinity of Fort Collins, Colorado, and cultivated on fresh raw beef liver in the laboratory for at least a year in aged tap water in white dishpans. Each pan contained 500 to 1000 planarians. Several days prior to initiation of any training session the planarians of a particular experimental training group were selected and placed together in a preslimed glass bowl. This same group was subsequently maintained together in this same bowl and never picked up, handled, or transferred to another bowl during the entire period of the experiment (as long as 6 months in some instances). The usual number per such group was 20 (approximately threshold for suppression of fissioning). From their establishment such bowl groups were placed on an open shelf directly over the laboratory bench, where feeding and training sessions were conducted, and stored there for the entire period except these sessions. No special lights were employed to illuminate them or the bench top during a training session. Any jarring of bowls or bench during an experiment was carefully avoided. All of this was done to habituate the planarians to all extraneous stimuli and stimulus changes and thereby minimize any emotionality induced in them by these.

Two kinds of experiments were conducted. In both a metal washer (chromium plated steel) was placed gently on the floor of the bowl and a small piece of fresh beef liver put in the hole of the washer. The planarians, attracted by the scent of the liver, would crawl over the washer to feed upon the liver.

In one experiment, a localized electric shock* of brief duration was administered to the tip of the tail of any planarian attempting to crawl upon the washer or feed off the liver. Non-transgressors received no shock. At the end of such a session (30 minutes duration) the liver and washer were removed. The planarians were fed liver without washer or shock on different days than those in which they received such a training session. In this experiment, attempts to feed when the washer was present were always from the very first presentation of the washer, punished by a shock of the type described. The results of such an experiment are shown in Fig. 2.

In a second variety of experiment the planarians were always, from the very beginning, fed liver in the same sort of metal washer used in the previous experiment, but never shocked for such feeding. This regime was adhered to over a 3½ to 6 months period. Then these planarians, which had become thoroughly habituated to the washer, were administered a single 30 minute session in which each attempt to cross the washer and feed was "rewarded" by a shock to the tip of the tail of the transgressor. Nontransgressors were not shocked. They were then given a second session, under similar circumstances, five days to two weeks later. Although in the first session in which they received shocks for their attempts to feed they made significantly more attempts than normal, they exhibited a marked suppression in the number of

---

*The shock was administered by means of a small bipolar probe, insulated to the tip and held in the hand of the experimenter. The tip of the probe was held at a distance of about a millimeter and not allowed to touch the planarian. The shock was actuated for approximately a third of a second after the probe was positioned. The well localized field about the probe tip subjected no more than the posterior quarter of the planarian to direct stimulation and influenced the brain only by the normal avenues of nerve conduction. This accords well with the response, which was a vigorous, but normal, tail withdrawal (initiated about the time the shock was terminated) and escape behavior. In these important respects the present situation is quite different from the various versions (Corning and John, 1961; McConnell, Jacobson and Kimble, 1959; Baxter and Kimmel, 1963; Barnes and Katzung, 1963; Halas, Mulry and Deboer, 1962; James and Halas, 1964; Bennett and Calvin, 1964) of the Thompson and McConnell (1955) paradigm, all of which subject the entire central nervous system of the planarian to primary stimulation and are thus more comparable to a diffusely administered intracranial stimulation than the classical conditioning analogs they were intended to be. The responses obtained by the Thompson and McConnell paradigm are the consequence of selective excitation of certain neuronal size classes by the stimulus waveform fortuitously employed (Best and Elshtain, 1966; Best, Elshtain and Wilson, 1967; Best, 1967) and are unrelated to the normal behavioral repertoire of the animal; this is not the case with the shock used in the experiments described in the present paper.

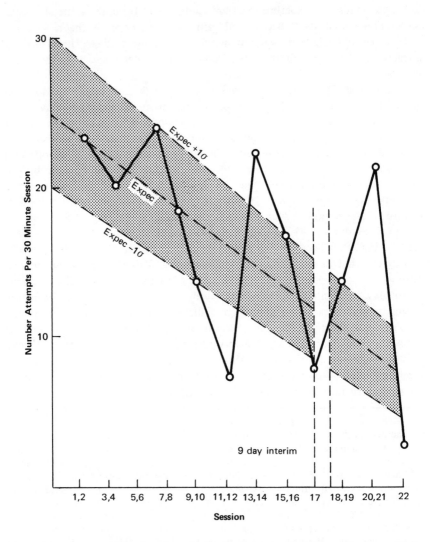

Figure 2. Average number of attempts to feed and shocks receiver per 30 minute training session by a group of 20 planarians of the species *Dugesia dorotocephala*. A respite of 9 days intervened between the 17th and 18th training sessions. The EXPEC line is the linear regression to the number of attempts. The EXPEC $\pm 1\,\sigma$ limits are computed on the basis of a Poisson distribution sampling model. It should be noted that the oscillations are much in excess of those to be anticipated from such statistical sampling fluctuations.

such attempts during the second session. The efficiency of the punishment in producing a subsequent suppression of feeding attempts was thus much greater in this experiment in which the planarians had previously been thoroughly habituated to positive reinforcement with the washer.

These two simple experiments are revealing and worth discussing. It is evident from Fig. 1 that the number of attempts to climb on the metal washer and feed, in the first experiment, shows a substantial decreasing trend with the number of training sessions. This trend downward does not occur uniformly, however, but rather in a series of fairly large fluctuations so that one finds planarians still attempting to feed even after a number of sessions. These fluctuations, which may be disturbing to some, are not ordinary statistical sampling fluctuations, but rather marks of authenticity. It is the way that higher animals behave in comparable situations (Arakilian, 1939). Intervals of acquiescence to the negative reinforcement are interrupted by bursts of what can be interpreted as "testing" behavior. In the course of its normal life there are many circumstances in which intensified effort, i.e. "trying harder," will result in achieving success though the immediate consequence is painful. I, the 'ominscient' experimenter, know that no amount of planarian ferocity, endurance, or stealth will succeed in the situation in question, but the planarian doesn't know it until it has "tried its bag of tricks."

The second experiment is comparable to the drinking tube shock paradigm which yields single trial learning in rats. It reveals not only a dimension of subtlety in planarian behavior but suggests a hitherto unrecognized* aspect of its rat counterpart. When an animal, such as a rat, has become trained through long habituation to expect that a metal drinking tube, for example, is always a source of positive reinforcement, e.g. water, and never a source of negative reinforcement, e.g. shock, the effect of a shock administered through the metal drinking tube is much more anxiety provoking than when administered in conjunction with a hitherto unfamiliar object, or an object known to have shock associated with it. This real significance of the paradigm's high conditioning efficiency has not been generally appreciated. The planarians had, in their counterpart paradigm, become habituated to positive reinforcement, liver, and to the absence of aversive stimuli, shock, in the context of climbing on the washer. When they were punished for doing so they were precipitated into a more or less specific "anxiety" state situation. Thus, in a situation with simple information structure and a high degree of emotional affect coupled to the paradigm, planarians can approach the maximum learning efficiency found in rats.

---

*Rats in the laboratory are nearly always maintained with water bottles having a metal drinking tube. Thus, unless otherwise stated, it is very likely that any rat used in experimental laboratory studies has been habituated to such metal drinking tubes from its time of weaning at about 25 days of age until its age at the time of the experiment.

These results contrast sharply with the large numbers of trials, drastic stimulus intensities, and/or marginal results reported in previous learning experiments on such lower invertebrates. With appropriate attention to the control of emotional factors one might expect a similarly rapid learning in many other invertebrates.

Are these behaviors due primarily to pheromone substances secreted into the water or onto the washer? The answer is no. The water and, usually, the washer are changed between sessions. The washer is boiled before use in a session. When the planarians are shocked they do deposit a thick gelatinous mucus onto the washer that is different from the thin viscid mucus that they normally deposit when not shocked. But it could be washed off readily. Nor does interchange of the washers cause any discernible alteration in the outcome. Is the effect due to some kind of effect of the tail shock on a chemotaxic or chemotrophic response, or to simply a change in locomotor activity? Again the answers are no. I base this on observations not reflected in graphical data. The planarians in the first washer-shock experiment tend to do several kinds of things at the point in the training when the number of attempts to feed on the liver in the washer begins to fall. They extend their heads onto the surface of the washer and then draw back. Another thing is that as many as half of them will be gathered at the perimeter of the washer with their pharynxes extended to suck up as much of the diffused juices as possible without actually going onto the surface of the washer. Yet another is that they try to "fly over" the liver and washer by moving on the air water interface. Those not at the washer perimeter are almost continuously in motion throughout the entire test training period. It is clear from such behavior that the planarians are attracted to the liver (or specific chemical factors from it) but are reluctant to cross beyond the perimeter of the washer to get it.

The planarians sometimes exhibit another behavior, much to the amusement of my laboratory colleagues who have observed it. At the point in training when the number of attempts per session has begun to fall, there will be a number of subjects which reduce speed at the edge of the washer and then move slowly, but unhesitatingly, toward the liver. Many of these will, when the tip of the probe, not electrified, breaks the water surface, veer away from the liver and crawl off the surface of the washer. Naive or control subjects seldom exhibit such behavior.

At the risk of offending orthodox mechanists I would say the most parsimonious interpretation of this aggregate of behaviors is that the planarians are still attracted to the liver after shock training but they are "afraid" to cross the surface of the washer to feed. Anticipating that there will be some skepticism that my reported observations may be "too subjective" I have documented these behaviors, and their changes in the experiment in motion pictures and can make demonstrations or the film available to serious investigators.

In a general way these results agree with earlier observations of subtle plastic behaviors in planarians that were termed "protopsychological" behaviors (Best and Rubinstein, 1962a, b; Best, 1963).

Many, myself included, are committed to the concept that structure begets function. Since the human brain (about 1200 grams) is about five million times, and the rat brain (about 1 gram) about four thousand times, the mass of planarian brain (about 250 micrograms) one might question whether the latter has the requisite neurological "hardware" to give rise to such a psychological system. In the next section I will turn attention to the structure of the planarian brain and the kind of behavioral substrate it appears to provide.

THE NEUROPHYSIOLOGICAL SUBSTRATE

The central nervous system of the planarian, as well as those of annelids, arthropods and vertebrates, can be divided into several major systems. There is the brain and segmental nervous system (cf. Figs. 3, 4 and 8). In vertebrates the spinal cord has a segmental structure and in its neural circuitry is contained the relatively stereotyped reflexive behavioral units of the type used repetitively as modular components of locomotor behavior and the maintenance of posture. A considerable volume of the spinal cord is devoted to relatively large longitudinal transmission nerve fibers. Since, in the vertebrates, these large nerve axonal fibers are sheathed with the whitish colored myelin, such longitudinal transmission systems give rise to the so call "white matter" of the spinal cord. The synaptic connections between neurons in this spinal cord are of the type that utilize actylcholine as the neurotransmitter substance. Such acetylcholine synapses contain aggregates of small diameter (DeRobertis, 1967) vesicles with a light colored core on the presynaptic side. This spinal cord also projects fibers into the sympathetic ganglia which regulate vegetative somatic functions, e.g. glandular and mucus secretaion, by release of adrenaline and noradrenaline. Among the somatic functions so regulated in both vertebrates and planarians is the release of mucus from a type of cell called "goblet cells" specialized for its manufacture and secretion (Best, Morita and Noel, 1968).

There is considerable resemblance in the organization of the segmental nervous systems of planarians and annelids to that described for the vertebrate spinal cord. That of planarians grossly resembles a ladder in which the "rungs" are comprised of bundles of nerve fibers linking the two mirror image sides in the regions of the segmental plexuses. A considerable portion of the volume of the segmental nervous systems of planarians and annelids is, as in the vertebrates, occupied by relatively large longitudinal transmission nerve fibers.

However, since these fibers are not myelinated in planarians or annelids, there is no white matter. The segmental nervous systems of these two invertebrate phyla contains, as does their vertebrate counterpart, the neural circuitry for the relatively stereotyped reflexive modular units of behavior involved in a repetitive manner in the locomotion of the animal. The synaptic connections between neurons in these invertebrate segmental systems predominantly are of the type utilizing the small light core synaptic vesicles of the acetyl choline variety. But the presence of some neurosecretory granules and dense core synaptic vesicles, together with some other observations to be mentioned shortly, suggest these invertebrate segmental nervous systems to be also the distribution apparatus of somatic control functions. They thus appear to evolutionarily anticipate or outline the organization of the vertebrate sympathetic nervous system and its relations to the spinal cord.

The systems of the brain which are closely linked to these segmental somatic control systems provide a good vantage point to turn our attention to a comparison of the organization of the brains of these different kinds of animals.

In planarians, annelids, arthropods, and vertebrates, control of size and sexual development is exercised by specialized structures of the brain (Best, *et al*, 1969; Florey, 1966, Ch. 21). The actual transmission and distribution of this control to the various relevant target tissues and organ systems is handled in varying degrees by the segmental somatotrophic system and by endocrine secretions carried in the circulatory system of the animal.

Two major sorts of external modulation of this master control of somatic function by the brain appear to be common, if not universal. One such modulation occurs by the daily photoperiod and those associated properties of it which have seasonal and/or lunar components, e.g. day to night length, shift in day to night length and spectral quality of the light (e.g. cf. Wolfson, 1966; Cloudsley-Thompson, 1961). Another sort of modulation involves the effect of the presence and state of other members of the animal's own species (Best, *et al*, 1969; Bruce and Parrott, 1960). The receptors usually involved in the first type of modulation are photic while those most commonly used by various species for the second sort appear to entail chemoreception. Occasionally one finds mechanoreception, usually auditory, and visual reception adapted to such purposes, but in such instances it appears as though the circuits for detection of such nonchemoreceptive social cues are "wired into" the old chemoreceptive structures of the brain.

The general organization of the planarian brain is shown in Fig. 3. The brain is comprised of two mirror image lobes with the plain of symmetry passing vertically through the longitudinal axis of the animal. The anterior ends of these two lobes are joined together by a large bundle of transverse fibers. The posterior end of each lobe joins onto the respective longitudinal

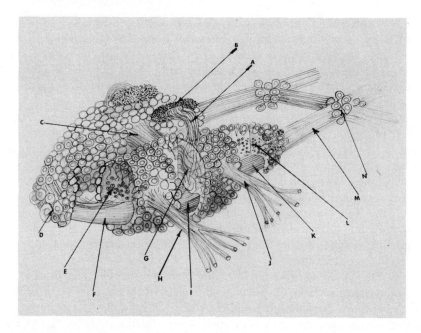

Figure 3. Drawing showing the gross anatomy of a planarian brain. Various selected portions are shown cut away to permit display of underlying structures. A is the optic tract on the left side, B is the eye on the left side, C is the major commisural tract which connects the two lobes in the vicinity of the point of entry of the optic tracts, D is the anterior motor system, E shows dense core synaptic vesicles in the neuropil, F is a longitudinal transmission tract which passes in the ventral region of the brain and almost certainly contains axons from the anterior motor system, G is some of the neuropil close to the point of entry of the optic tract, H is one of the chemoreceptive nerves, I denotes longitudinal transmission tract in ventral region of the brain close to the level of optic reception, J is an auricular nerve which probably carried information from hair cell mechano receptors to the brain, K is the ventral longitudinal transmission tract close to the posterior end of the brain and the region of merger into the ventral nerve cords, L is the region of the brain close to its juncture with the ventral nerve cords and the region containing the larger sized neurosecretory granules.

ventral nerve cord. The brain is comprised of three major types of cells: neurons, neurosecretory cells and neuroglia (Morita and Best, 1965, 1966). Nearly all of the cell bodies are arrayed into the cortical regions of these brain lobes. The central core of each lobe is almost entirely comprised of fiber processes from these three kinds of constituent cells which form this compact mass. Practically all, if not all, synaptic connections between these cells occur in the neuropil (Morita and Best, 1966).

Somatotrophic Control Systems. Neurosecretory cells are a type of cell that is specialized for release of hormonal substances in response to the appropriate patterns of nerve impulses. They characteristically resemble a kind of cross between a gland cell and a neuron and contain numbers of electron dense round granules with a sharp boundary of demarcation. These granules are called neurosecretory granules (cf. Fig. 5).

One finds a concentration of one type of such neurosecretory cells in the planarian brain toward its region of juncture with the segmental nervous system (cf. Fig. 3) suggesting a differentiation of this region for somatic control functions. A similar concentration in this region occurs in the annelid brain (Florey, 1966; Clark, 1959; Hubl, 1956). The medulla oblongata, the region of the vertebrate brain adjacent to its region of juncture with the spinal cord, the vertebrate segmental nervous system, is known to contain a concentration of somatic control centers. It is thus interesting that in planarians, annelids and vertebrates one finds this comparable region differentiated for somatic control.

In the vertebrate brain one also finds, however, another region, close to the optic tracts (cf. Fig. 4) containing the nuclei of the hypothalamus, known to be important somatotrophic control centers. As the optic tracts pass by the hypothalamus they deflect upward and connect in the optic tectum of the superior colliculus on the dorsal side of the brain stem. In this same general region is the pineal body which in most vertebrates appears to be both glandular and photosensitive and involved in the daily photoperiod modulation mentioned previously. In arthropods such as crustacea there is a concentration of neurosecretory cells about and on the optic stalk which exert a somatotrophic control and link it to the photoperiod (Florey, 1966). Polypeptide neurohormones appear as mediators of such control in both arthropods and vertebrates.

Anterior Motor System. There is an ensemble of oversized neurons in the anterior end of the planarian brain. It is fairly apparent from the role that such large neurons seem to play almost universally that this ensemble is the convergent pathway out of an anterior motor system. The nerve endings located in this same immediate neighborhood contain a number of fairly large dense core synaptic vesicles although there are also a number of synaptic endings with the conventional small light core synaptic vesicles of the acetylcholine type.

In the vertebrate brain there is a system of "basal motor ganglia" in the forebrain which includes the caudate nucleus, putamen and globus pallidus (cf. Chap. 12 of Ruch and Fulton, 1969). These form connections with structures such as the red nucleus and substantia nigra further down in the brain stem and are implicated in the integration of postural reflexes. In those vertebrates such as birds, reptiles, amphibia, fish and sharks which do not possess much

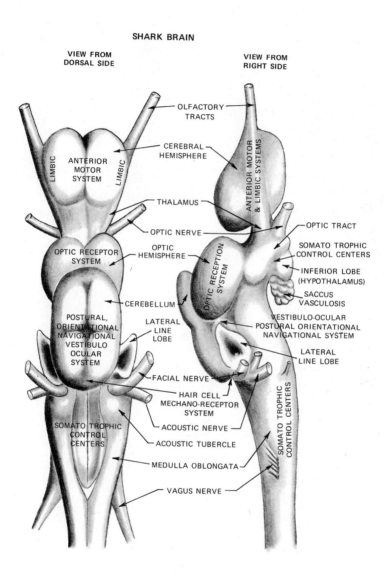

Figure 4. Drawing showing the gross anatomy of a shark brain.

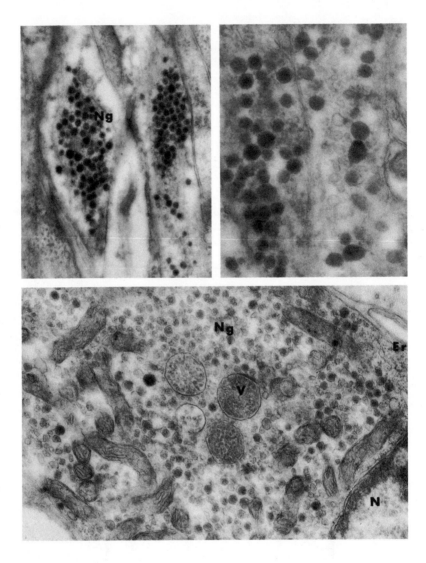

Figure 5. Electron micrographs showing neurosecretory granules in the brain of the planarian Dugesia dorotocephala. A and B show mature neurosecretory granules (Ng) in nerve fibers of the neuropil of the brain. C shows the cytoplasm of a neurosecretory cell in the planarian brain that has undergone transition into the mature storage phase.

cerebral cortex, these basal motor ganglia are the highest motor center of the animal and located toward the extreme anterior end of the neuraxis. In mammals this role of these basal motor ganglia seems to be superseded and transformed by the emergent motor area of the greatly enlarged overlaying cerebral cortex. Direct connections from the motor cortex to the segmental levels are formed in dextrous mammals such as primates and bears through the pyramidal tracts.

This basal ganglion system receives fibers containing dopamine and noradrenaline and, especially in the primitive vertebrates, e.g. sharks, lies close to the rhinencephalon ("smell brain").

The anterior motor system of the planarian brain appears to be a primordial homolog to this basal motor ganglion system of the vertebrates.

Vestibulo-ocular System. In primitive vertebrates such as sharks one finds a kind of mechano-receptor hair cell involved in three different roles which are, however, related. They appear in the lateral line organs which detect low frequency or nonperiodic movements of the aqueous environment relative to its body. They also appear in the cochlea and, in that role, detect sound frequencies as they do in the higher vertebrates. Finally, they occur in the ampullae of the semi-circular canals, or vestibular apparatus, which detects angular acceleration as it does in higher vertebrates. The nerves coming from all three of these receptor organ systems enter the brain in the same general region (cf. Fig. 4). The neural structures for initial reception and interpretation of the information from these three mechano-receptor systems are contained in the same general region of the brain stem or are mushroom-like outgrowths from it.

One very prominent mushroom-like outgrowth sprouts out of the dorsal side of the brain stem in between this region dedicated to mechano-receptor sensory information and the optic lobes situated further forward. This is the cerebellum. The cerebellum, together with a system of nuclei, such as the vestibular mucleus and red nucleus, located in the brain stem between these mechano and visual reception regions, integrates this information to provide the animal with a spatial reference frame and orient it appropriately in regard to it.

The eye has probably served longer as a sextant than it has as a television camera. But to use a light source as a navigational star to steer its course by, a mobile animal must know whether the apparent changes in direction of the reference arise from its own movements or a real movement of the reference point. Even using eyes for pattern vision as we do, it is necessary to compensate for bodily or head movements by suitable rotation of the eye ball in its socket in order to fixate on an object or point. The mechanism for achieving this entails an exquisitely precise, but highly stereotyped, linkage between the muscles which rotate the eyeball and the vestibular apparatus. If one starts the fluid circulating in one of the semicircular canals and thus deflects the ampulla of that canal, there will occur a tracking and sweep back movement of the eyes in the plane of that canal, and in the same way as though the subject was producing such vestibular circulation by actual rotation relative to the environment. Accompanying this is an inescapable sensation of vertigo as though the world were spinning about an axis perpendicular to the plain of the semicircular canal undergoing stimulation. This phenomenon, called nystagmus, is a consequence of this vestibulo-ocular integrating system.

Flying insects have a tiny vibrating weight on the end of a little rod on each side of their body close to the region of attachment of the wings. Like a tiny pendulum these "halteres" tend to maintain their original plane of vibration when the longitudinal axis of the thorax of the animal changes orientation. They thus supply the insect with information very similar to that provided by our own semicircular canals. Such insects commonly use the infinite light sources provided by the sun, moon or stars for visual navigational reference points to set a course by* (cf. Chap. 9 of Fraenkel and Gunn, 1961). The compound eye of the insect can provide a fairly good estimate of the angle of the infinite light source to the axis of the animal. But this information is of only limited value unless the animal has some way of integrating the information from the halteres with that provided by the eye. Although I shall not discuss the problem, it is also evident that to use the sun, moon or stars as navigational references for flights of any duration, the insect needs to have some sort of clock to correct for the change in position of these objects in the sky. It is interesting in this regard to note that bees can readily be trained to come to a bait at a particular time of day (Beling, 1929) indicating that they do indeed have such a clock.

The planarian eye has a structure which indicates that it is sensitive to the direction of the light entering it. The cradle of muscle fibers, in which it is suspended, suggest that it can be rotated through at least a limited angular excursion. These muscle fibers are of a specialized character with substantially less than the normal number of contractile filaments per cell. In the brain at the level of the eye spots, and extending slightly to the rear, is a system of larger than average nerve fibers, some obviously linking the two lobes of the brain. To the rear of the eye spots, about at the level of the junction between the segmental nervous system and the brain, are the auricular projections. Without firm evidence it nevertheless is likely that these auricular projections

---

*Moths do not fly to their destruction because they are "attracted" by a flame or electric light bulb. In attempting to fly a course maintaining a fixed angle to a finite light source, they fly a spiral trajectory which spirals into the light if the angle is acute and away from the light if the angle is obtuse. Through most of their evolutionary history the only light sources they encountered were infinite.

contain mechano-receptor hair cells and that the nerve supply from these enters the brain somewhat behind the point of entry of the optic nerves. Such externalized mechano-receptor hair cells could serve as a kind of primitive combination vestibular, acoustical and lateral line organ system. If such were the case it would, in view of what has been said, lend considerable strength to the presumption that the planarian brain fiber system in the vicinity of, and to the rear of, the optic tracts is a primordial homolog to the vestibular-ocular complex.

In the mammals, and especially in the higher primates, one finds an enormous mushrooming and folding of the cerebral cortex. But in a primitive vertebrate such as the shark there is very little development of the cerebral cortex. This state of affairs persists even in the birds and reptiles. The cerebral hemispheres of the shark brain are thus mostly comprised of the evolution-arily more ancient basal motor ganglia, i.e. the anterior motor system discussed previously, and the rhinencephalon. The rhinencephalon, which literally translated means "smell brain," is an outgrowth of the region of the brain which receives and interprets the chemoreceptor information transmitted to it by the nerves of the olfactory tract. But this system, more recently renamed the *limbic system*, does a great deal more than perceive smells. It appears to be the primary physiological substrate for emotional affect, motivating drives, attention, habituation, stimulus novelty and reinforcement mechanisms, as well as having plastic neuronal connections (Olds and Milner, 1954; Olds, 1962; Stein, 1970; Raisman, 1969; Elazar and Adey, 1967). Thus, more than any other system of the brain or spinal cord, it appears to be the primary substrate of the psychological behavior systems of the vertebrates.

In the evolution of the mammalian brain one perceives an enormous enlargement of the cerebral cortex and cerebellum with an appreciable enlargement of the tracts serving these. The rhinencephalon does not show such a proportionate enlargement. Nonetheless it appears to be at the hub of all psychological behaviors. Although the cerebral cortex may refine, lend precision and provide more detail, it is the limbic system that is really the *sine qua non* of sentient behavior.

This limbic system is neurochemically interesting as well. It contains a relatively large proportion of synapses with dense core synaptic vesicles and, accordingly, relatively higher concentrations of serotonin and noradrenaline than other portions of the brain. Some authors, with justification, consider the hypothalamus to be a part of the limbic system. Functionally it is certainly an integral part of it, for most limbic system activity is sensitively mirrored in hypothalamic and autonomic activity. The link between anticipation of emotionally affective events and autonomic responses was recognized almost from the beginning by pioneers such as W. B. Cannon and I. P. Pavlov. Cannon described the pattern of change devolving from sympathetic nervous

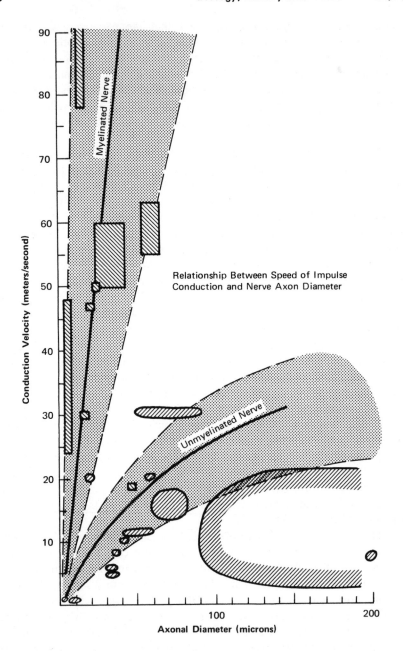

Figure 6. Relationships between speeds of nerve impulse conduction and axon diameter for myelinated and unmyelinated nerves from various animals from a variety of different phyla.

system activity as "preparatory for fight or flight." Saliva flow, a function controlled by the autonomic nervous system with goblet cells as the major effector unit was the major response focused upon in Pavlov's classical conditioning experiments (Pavlov, 1928).

Thus, one can look on these dense core synaptic vesicle systems of the autonomic nervous system as functional and neurochemical extensions of the limbic system. This limbic-hypothalamic-autonomic nervous system is probably involved in all cases where emotional affect is involved. Conversely, probably any situation entailing extensive involvement of this system is emotionally affective. Drives, motivations, appetites and psychotropic or addictive drugs probably also all involve this system in a major way.

The planarian brain receives input from chemoreceptors situated along the anterior and anterio-lateral margins of the head of the planarian which is shaped somewhat like an arrow head. In these regions receiving this chemo-receptor input there are numerous nerve endings and synapses containing dense core synaptic vesicles (Best and Noel, 1969; also cf. Fig. 7). There appear to be at least two or more size classes of these vesicles. This agrees with reports that the planarian brain contains considerable concentrations of noradrenaline and serotonin (Best, Rosenvold and Riegel, 1970; Welsh and Williams, 1969). No measurements have yet been performed to ascertain whether dopamine is also present in appreciable quantities, but it appears likely that this would be found to be the case.

On the basis of these similarities, I suggest that such regions of the planarian brain are neurochemically and anatomically similar to the rhinencephalon, the vertebrates and are actually the primordial evolutionary homolog of the vertebrate limbic system. It is these regions of the planarian brain that I think provide the physiological substrate for the psychological behavior system of the planarians. As in the vertebrates, it is the chemoreception brain with its aromatic amine neurotransmitters and dense core synaptic vesicles that primarily provides the neurological hardware for such psychological behaviors and the plasticity that goes with them.

Longitudinal Transmission System. There are large longitudinal transmission trunks of relatively long nerve fibers which connect the systems located at various levels of the vertebrate brain neuraxis with one another and with the spinal cord. These are situated toward the ventral side of the longitudinal axis of the brain system. The long fibers, especially of the longitudinal motor connection system, tend to be of large diameter.

One finds a similar situation in the planarian brain. The relatively long, large diameter, nerve fibers, forming the longitudinal transmission system, are grouped into trunks which run along the ventral portion of the brain. The more or less randomly oriented felt work of fine nerve fibers of the nauropil, in which most synaptic interconnections occur, is primarily situated in the more dorsal regions of the planarian brain neuropil.

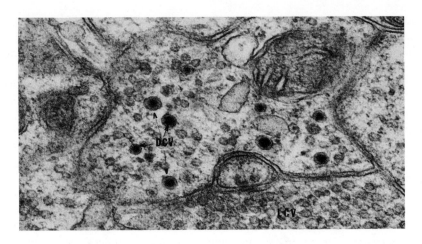

Figure 7. Electron micrograph showing dense core (DCV) and light core (LCV) synaptic vesicles in the brain of a planarian of the species *Dugesia dorotocephala*. No clearly discernible synaptic cleft is visible in this particular micrograph.

In this organization of the longitudinal transmission trunks the planarian brain appears once again to provide a primoridal homolog to the vertebrate brain and probably also to those of other bilaterally symmetric encephalized animals on the main evolutionary sequence.

These various relationships discussed above are summarized in the schematic block diagram shown in Fig. 8. I shall refer to this organizational scheme as the "Universal Brain." This diagram is intended to show those common denominator system components and relations which appear to be basic to all brains of bilaterally symmetric encephalized animals on the main evolutionary sequence. It should be interpreted in much the same kind of way that one interprets those charts used to depict the reaction pathways of intermediary metabolism. Just as in the case of intermediary metabolism some of the components and relations may be enhanced in some species and atrophied or vestigial in others. Nevertheless there appears to be a common plan or scheme underlying the neural and behavioral organization of a wide domain of animals from which that of various particular species are derivative.

BASIS OF THE HOMOLOGY BETWEEN BRAINS

Why should there exist a homological correspondence between the neurophysiological systems of a planarian and those of animals as seemingly remote as the vertebrates? A few remarks in regard to this are appropriate at this point.

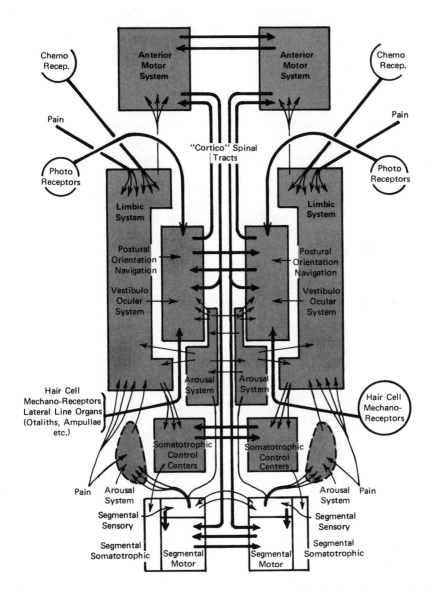

Figure 8. A schematic block diagram for the "Universal Brain" showing the anatomical systems and functional relationships that appear to be common to bilaterally symmetric encephalized animals on the main evolutionary sequence.

If the shape of an object, e.g. a particular polyhedron, is obtained by casting a material in a mold, then the shape implies little about the fine structure or molecular structure of the material comprising the object. However, if the shape of the object is the result of a molecular process such as crystallization, then the shape may imply a great deal about the molecular or fine structure of the object. Tissue dissociation and reaggregation studies (e.g. Moscona, 1957, 1962; Steinberg, 1963; Trinkaus, 1969; Trinkaus and Lentz, 1964), as well as tissue culture differentiation experiments, suggest that some kind of specific adhesion forces between the various kinds of cells are responsible for organogenesis (Trinkaus, 1969; Steinberg, 1963; Trinkaus and Lentz, 1964).

In conventional crystallization processes, the thermal motion of the elementary atoms, ions, or molecules comprising the aggregate provides a randomizing process through which various structural arrangements can be dissolved and reformed. This permits transition of the ensemble into various aggregation states. Aggregation arrangements are thermodynamically favored that have the lowest free energy compatible with the packing constraints imposed by the volumes of the elementary units. The lowest free energy arrangement is that which corresponds to perfection of the crystal. The optimal conditions for forming crystals with a high degree of perfection are usually those in which rates of dissolution and aggregation are very nearly equal for arrangements approaching perfection and in which the rates of these processes, individually, are appreciably larger than zero.

When dissociated from one another in a hospitable tissue culture medium, cells tend to engage in amoeboid movement (Trinkaus, 1969; Abercrombie, 1967). This amoeboid movement provides the randomizing process which enables transition from one aggregation arrangement to another. It thus fulfills a role analogous to that provided by thermal movements in conventional crystallization processes. Interfacial adhesion forces between cell surfaces stabilize cell aggregates by contact inhibition of amoeboid movement and by actually sticking the cells together (Steinberg, 1963; Trinkaus, 1969; Abercrombie, 1967). The various experimental results indicate two major kinds of forces favoring such cell aggregation. There is a general (and probably relatively nonspecific) force involving potassium ions that causes cells to clump together (Rappaport and Howze, 1966; Rappaport, 1969). Added to this there seem to be weaker, but more specific, forces which are dependent upon the type of cells with surfaces in contact with each other. Although it is probably these weaker forces which are really responsible for the specificity that leads to the characteristic arrangement of an organ system, the nonspecific forces play an important role by maintaining contact between the cell surface and permitting the weaker forces to be operative. There is probably a final stage in organogenesis in which some cells are subsequently laced together by intercellular filaments or by outright fusion of the cells as in the formation of muscle fibers from myoblasts or giant axons from several neuroblasts. But these processes are simply extensions and not the primary determiners of organogenesis.

It seems clear that the essential character of the significant processes which give rise to histological and gross structure of tissues, organs and the whole animal are similar to those involved in crystallization. That such processes are operative in the case of planarians, in particular, seems to be indicated by the observation of Ansevin and Buchsbaum (1961) that planarian neoblasts in tissue culture would, a portion of the time, group together and form whole planarians which, although small, were viable and motile. The gross morphology and basic symmetries of an animal thus would seem to imply a great deal about its microstructure and the kinds of cells comprising it for much the same reason that the shape of a crystal betrays the kind of compound comprising it.

The primordial flatworm ancestor of the vertebrates pioneered more for us than bilateral symmetry, distinct ventral and dorsal sides and a start toward encephalization. These are but the tip of the iceberg. There is an entire complex of microstructural implications that seem to go along with these grossly visible morphological features. This same general organizational plan of the grossly visible features seems to be mirrored in the same kind of outlining of microstructural characteristics and functional organization of the physiological systems. Thus, although the planarians total number of cells and diversity of cell types are much less than possessed by the vertebrates, all the major categories of functional specialization are present in this primitive animal. It has neurons, neurosecretory cells, neuroglia, muscle, connective tissue, goblet cells, testes, epidermal cells, neoblasts, rhabdite forming cells, gland cells, digestive cells, reticular parenchymal cells, flame cells, ovaries, chemoreceptors, optic receptors and mechanoreceptors.

The same kind of outline of the general organizational plan alluded to in the gross morphology of the planarians can be seen in the differentiation of their cell types. For example, in vertebrates one finds cardiac muscle, striate muscle and smooth muscle while the planarian has only a single kind of muscle cell resembling an archetypical common denominator of all three. Another example is provided by their neural accessory cells which appear to an archetype of the three different kinds of glial cells found in the vertebrates. A similar correspondence appears between the parenchyma of planarians and the reticuloendothelial system of the vertebrates. Vertebrates have several different kinds of stem cells while planarians have only one, the neoblast, which can differentiate into nay of the cell types of the planarians. These neoblasts normally occur in the interstitial spaces of the lattice provided by the reticular mesenchymal cells. The tissue is thus like a simplified version of the vertebrate reticuloendothelial tissue.

One might ask why this general plan outlined in the planarians should remain so stable that it should persist through the vast gulf of time and evolutionary transition required to produce the vertebrates.

From the previous discussion of the processes of organogenesis it seems evident that the genes must control and modulate organogenesis by controlling and modulating the specific and weak interfacial adhesion forces between the constituent cells. But merely changing some of the chemical groupings responsible for such selective adhesion would generally drastically alter the entire cellular ensemble. This would not permit a stepwise modification and perfection of these organ systems. Each mutation would tend to wipe out all that had been achieved by evolution up to that point. It seems unlikely that this is the way it happens.

A more likely way would be the following. The chemical groupings responsible for the interfacial adhesion forces that yield the aggregates which outline the basic organization of an animal are probably maintained unchanged and not usually altered by mutation as evolution occurs from an archetypical animal toward more highly evolved and differentiated animals. It seems far more likely that new sets of chemical surface groups are added, to produce a further partitioning of the cells within those clumps which formed the cell aggregate partitions in the archetype. One could generally expect such changes to be accompanied by an increased size and cell number in the evolving animal. Thus in the evolution from a flatworm like animal to the vertebrates one finds that the anterior motor system partitions into several distinct neural aggregates, i.e. the caudate nucleus, claustrum, putamen and globus pallidus, and this change is accompanied by increased size of the animal and an increased number of cells. It seems likely that the cell surface groups which cause the neurons of the anterior motor system to cluster there in the planarian brain are still used to sort neurons into the basal motor ganglia of the shark brain. More recently evolved surface groups then probably determine, for example, whether a particular neuron is to be incorporated into the caudate nucleus, claustrum, putamen or globus pallidus.

If this is, in fact, the way that things actually happen, then the basic layout, or ground plan or organization, of the animal will be determined by the older surface groups of the animal's cells. Once evolutionary development has proceeded, so that a number of subsequent innovations in physiological organization are predicated on such an organizational ground plan, changes in that ground plan would have a high probability of being lethal. One would thus expect that natural selection would tend to favor and select for a high degree of stability in those genes that determine the basic patterns of differentiation of cells into their major cell type categories and confer on them the surface adhesion properties that yield the basic patterns of tissue organization.

OPTIMAL DESIGN OF ORGANISMS

Considerable insight into the organization of various kinds of living creatures can be obtained by viewing them in terms of the features of engineering design that they need to have incorporated into them in order to carry on in the mode of existence they have adopted. Certain requirements will have to be fulfilled in the relation of their component parts in order for them to survive at all. These I will refer to as Minimal Design Considerations or Minimal Design Relationships. A more restrictive set of relations between their component parts must be fulfilled for optimizing their performance. This set of relationships I will refer to as Optimal Design Considerations or Optimal Design Relationships. It is intuitively obvious that the optimal design relations must satisfy the minimal design relations, but that not all of the set of conceivable relations which satisfy the minimal design considerations will necessarily satisfy those for optimal design. For those of my readers who may suspect that I am interjecting teleological mysticism in the guise of physical rationalism I would urge that they bear with me. I believe that all I have to say rests squarely on a nonmystical and entirely mechanistic rationale that I will presently outline briefly. But first let me give some examples of such minimals and optimal design relationships, as they might apply in the case of living creatures.

Trees have an aerial system of branches and twigs that support a leaf system. These leaves capture quanta of light from the sun by means of the chloroplasts they contain and use a portion of this captured energy to convert carbon dioxide, absorbed through pores in the leaf surfaces, into carbohydrates by means of the Calvin cycle. There is an unavoidable net outward flux of water vapor from the leaf surfaces. Trees also have a subterranean root system which absorbs water, trace nutrient elements, ions and substances such as nitrate ions, phosphates, etc., from the soil. The trunk of the tree provides both a structural column to support the leaf and branch system and a transport link for the exchange of materials between the root system and the leaf system. These two functions of the trunk are reflected in its anatomy. In view of these functional relationships between the various component parts of the tree there are various quantitative relationships between them that must be satisfied.

One such minimal relationship is that the maximum average rate at which water can seep into it across its bounding surface must be greater than or equal to the average rate of water transpiration from the total leaf surface of the tree. The average rate of transpiration from the leaves will, for a given shape of leaf and stomata density, be proportional to the number of leaves, n_L, the area per leaf, A_L, the "dryness" of the air, the average radiant flux $\mathcal{R}$, and the average wind velocity, v_w. The dryness can be expressed as $1 - H_R/100$ where H_R is the relative humidity. Thus, to a first approximation

Water loss through leaves $\propto n_L \; A_L \; (1-H_R/100) \; \Re \; v_w$.   (1)

One can argue that the maximum average rate at which water can seep into soil volume which the root bed subtends will be approximately proportional to the area of the enveloping surface of this bed volume, to the permeability, P, of the soil, and to the average depth, d_p, of the precipitation received. If the shape of the volume of soil containing the root bed is reasonably compact and constant, then, for practical purposes, one can characterize this root distribution by a single parameter, a_R, the average radius of the bed. The area of the enveloping surface will be proportional to $a_R{}^2$. Thus, the

$$
\begin{array}{l}
\text{maximum average rate} \\
\text{of water seepage into} \quad \propto P \; d_p \; a_R{}^2 \; . \\
\text{the root bed}
\end{array}
\tag{2}
$$

Hence, one condition which the tree must satisfy is that

$$P \; d_p \; a_R{}^2 \; > \; K \; n_L \; A_L \; (1-H_R/100) \; \Re \; v_w \; , \tag{3}$$

where K is a constant for a given kind of leaf.

Another obvious requirement that has to be satisfied concerns the relationship between the structural strength of the trunk of the tree and the loading placed on it by the leaf-branch system. There will be a pure compression loading on the trunk if there is no wind, the trunk is a vertical pillar, and the center of gravity of the leaf-branch system is aligned with the axis of the trunk. If s_c is the maximum compression stress which the wood of the tree trunk can withstand without breaking, M_b is the mass of the tree's branch-leaf system, a_T (y) is the radius of the trunk at a height y, ρ the density of the wood of the trunk, g the gravitational acceleration, and if one assumes that the trunk has a circular cross section, then the tree is going to have to satisfy the requirement that

$$a_T{}^2 \; s_c \; > \; M_b \; g/\pi + \int_y^{h_B} \rho \; a_T{}^2 \; (y') \; dy', \tag{4}$$

if it is not to have its trunk collapse under compression loading.

But there are even more demanding and restrictive requirements which the tree must satisfy. Not collapsing in a windless environment is one thing, but surviving with wind is another. The hydrodynamic drag of the wind through the leaf branch system will impose a considerable horizontal force that exerts a large bending moment of force on the tree trunk. Such forces will produce both tension and compression stresses in the trunk. If s_T is the maximum tension stress which the wood of the trunk can withstand without breaking, $f_{w\,Max}$ the maximum horizontal wind drag force on the leaf-branch system, h_g the height of the center of force of the wind drag force, then it has to be true that for all values of y from zero to h_B that

$$s_T > \frac{4\,(y_g - y)\,f_{w\,Max}}{\pi\,a_T^{\,3}} - \frac{M_b\,g}{a_T^{\,2}\,\pi} - \rho \int_y^{h_B} a_T^{\,2}\,(y')\,dy', \qquad (5)$$

$$s_c > \frac{4\,(y_g - y)\,f_{w\,Max}}{a_T^{\,3}\,\pi} + \frac{M_b\,g}{\pi\,a_T^{\,2}} + \rho \int_y^{h_B} a_T^{\,2}\,(y')\,dy'. \qquad (6)$$

One might ask why trees have trunks, since inequalities 4, 5 and 6 are more apt to be violated with trunks that are long rather than short. The answer is: to avoid being shaded by other quantum catchers with longer trunks, and to avoid having leaves eaten by hooved herbivores. One could go on and write down a number of other conditions that the tree must satisfy. Opatowski (1944, 1945, 1946) has considered a number of these. However, since my primary purpose at this point is not to consider trees, *per se*, but rather to illustrate the kinds of conditions that minimal or optimal design considerations impose on an organism, let us consider some comparable sorts of relations in animals.

A relatively small flat animal such as a planarian can rely upon diffusion to supply the oxygen necessary for the metabolism of its tissues and rid itself of the undesirable end products of the metabolism. But such reliance upon diffusion processes for transport of metabolites places an upper limit on the size of such an animal. One can deduce and write down the set of relationships which constrain the size of animals who carry on this essential business in such a fashion. If Δ is the thickness of the animal, D_O the diffusion coefficient of oxygen through its tissues, $[O_2]_e$ is the concentration of oxygen in the external milieu (usually water) and q is the rate of consumption of oxygen per unit volume of tissue, then the maximum thickness of the animal will be limited by the inequality

$$\Delta < \sqrt{2\,[O_2]_e\,D_O/q} \,, \qquad (7)$$

if the animal has a flat, leaf-like shape such as the liver flukes and polyclad flatworms.

In order for animals the size of most mammals to exist, a lung kidney and cardiovascular system are indispensible. As one might anticipate from the preceding discussion, there are a set of quantitative relationships which must be satisfied between these various component parts of the animal. For illustrative purposes it is worth deducing a few of these and writing them down. For example, if $E(O_2)$ and $E(CO_2)$ denote the exchange capacity of the lungs (or gills, in the case of a fish) in regard to oxygen and carbon dioxide, $[O_2]_s$ the saturation value of the oxygen content per unit volume of blood at the oxygen tension of the external milieu, $[O_2]_v$ the oxygen content per unit volume of venous blood, v the volume of the animal and q, as before, the rate of consumption of oxygen per unit volume of tissue, then

$$q \ V \ < \ f_b \ (\ [O_2]_s \ - \ [O_2]_v) \ < \ f_b \ [O_2]_s \ , \qquad (8)$$

$$q \ V \ < \ E(O_2) \ , \qquad (9)$$

$$q \ V \ R \ < \ E(CO_2) \ . \qquad (10)$$

The quantity R denotes the molar ratio of CO_2 produced to O_2 consumed.

If one conceives of the relevant parameters as forming a parameter hyperspace, one can visualize the various inequality relations of the type indicated as defining a set of hypersurfaces in that space. Points on one side of such a hypersurface will correspond to those configurations between the component parts of an organism that will result in exclusion by natural selection. Points on the other side of such a hypersurface will correspond to those configurations which are permitted by natural selection, in so far as the considerations which led to the particular hypersurface in question are concerned. The set of hypersurfaces generated by the various pertinent design considerations will form the boundaries of that region, or regions, of the parameter hyperspace which contain those points that correspond to permissible configurations. Any gene, or gene combination, that leads to configuration points lying outside of these permissible hyperspace volumes will be excluded by natural selection. The term "excluded" deserves some comment.

One mode of exclusion is the following. If a new gene combination, or a mutation, leads to a transition into a non-permissible (in the sense described) hyperspace volume, then the organism which is the phenotypic expression of that genotype will not survive and thus not pass on its genes to descendants. This will occur if the transition is into a non-permissible hyperspace volume that is bounded by a surface generated by a minimal design consideration. However, one might conceive that this elimination of genotypic transgressors into such non-permissible regions in no way reduces the probability of future transitions from permissible to non-permissible regions. I shall argue, without proof, that one should expect that natural selection will, in fact, operate to reduce the probability of such transitions. There are two mechanisms by which it could reasonably be expected to do this. One is by selecting for genes which have a lower mutation probability. The other is by selecting against those genes or gene combinations that yield an anticedent condition to such a transition. This latter mechanism could be expected to create a kind of semi-permissible buffer zone, inside of the limiting hypersurface generated by the minimal design consideration, that presents a dangerous, but not necessarily lethal, region for genotypic occupants. The existence of such buffer zones will generally tend to constrict the permissible region more narrowly about that demarcated by optimal design considerations.

Some prominent features of the vertebrates can be traced to their large size. If one were to rank order all known animal species with respect to size, there is little doubt that the vast majority of the largest size rankings would be occupied by vertebrates. A few leviathans of the invertebrates such as the larger varieties of squid, octopus, lobster and crab and some of the cold water anemones would be found in these larger size rankings, but they would certainly be an exceptional minority. Virtually all of these larger invertebrates are, moreover, marine forms. Large size among terrestrial animals is practically a vertebrate monopoly. It is interesting to examine some of the implications of this large size of the vertebrates in terms of the kinds of design considerations which were discussed above.

One outstanding characteristic of the vertebrates is that their soft tissues are suspended on a strong, load bearing, frame—their endoskeleton. Without such a rigid supporting frame, or the buoyant effects of sea water, animals the size of most vertebrates would collapse under their own weight. For much the same kind of engineering considerations that necessitate the use of structural steel supporting frames for buildings the size of modern sky scrapers, natural selection was forced to a similar answer in the endoskeleton of the vertebrates. But, unlike a sky scraper, the vertebrates are a fast moving bunch of creatures capable of rapid locomotion. In order to achieve such rapid locomotion it is necessary for any animal doing so to exert forces of appreciable magnitude on its environment. In the absence of a rigid framework of the type provided by an exo- or endo-skeleton, it is not possible for the animal to use any propulsion scheme which entails transverse or compressional loading of its propelling appendates. Thus, the evolutionary innovation of skeletal frameworks by the arthropods and vertebrates permitted utilization of an entire repertoire of propulsion schemes that would not be feasible for other kinds of animals.

But size and the potentiality of locomotor speed pose other serious design problems of an engineering character. Large size entails a reduction in the surface to volume ratio. Such a reduction reduces the efficacy of pure diffusion as the mechanism for exchange of metabolites with the environment. This reduction could be offset by a slower metabolic rate per unit volume, but this would not permit the exploitation of the potentiality for locomotor speed. To exploit the possibility of attaining high locomotor speed means committing a relatively large proportion of the tissue mass to muscle, and muscle with a relatively large rate of conversion of metabolic energy into mechanical work. This large rate of conversion of metabolic energy into mechanical work necessarily entails a high rate of metabolism with concomitantly high rates of exchange of metabolites with the environment. Thus, to sustain such high metabolic rates in the face of the reduced surface to volume ratio that goes along with size, it is necessary to have a transport system of the kind provided by the blood-cardiovascular-kidney-lung system.

However, size, without the ability to respond rapidly or coordinate the responses of the component parts, is probably more of a liability than an asset in an animal. Coordinated responses of the whole animal require transmission of information, usually by nerve impulses in the case of the more rapid responses. For a given velocity of propagation of a nerve impulse along a nerve fiber, the transmission time will increase as the animal becomes larger and the lengths of the nerve fibers necessitated by this increased size become greater. Generally, any complexly orchestrated response between a number of the animal's component parts, entailing feedback and correction, will involve large numbers of "messages" for its execution. Thus, if a large animal were to use the same propagation velocity for its nerve impulses as a smaller animal, it would require a much longer time to execute a response of the same degree of complexity; this would, of course, nullify the advantages of size. It would also prevent the large animal from exploiting its capabilities for high locomotor speeds since these entail a more rapid rate of closure between the animal and obstacles in its path, as well as more disastrous results from collision. Without the capability of making rapid and complex corrective responses, high locomotor speeds for a large animal would be suicidally maladaptive. Nor is a high locomotor speed of much value unless it can be attained rapidly when it is needed for pursuit of prey or escape from sudden danger.

So the large animal, as a design condition, has to achieve higher velocities of nerve impulse propagation for its longer fiber systems. One manner of doing this is to use larger diameter axons for those transmission systems necessitating such velocities. According to the theoretical considerations of Hodgkin (1954) the velocity of propagation in unmyelinated nerve should vary as the square root of the fiber diameter for constant specific membrane and axoplasmic properties. Thus, for unmyelinated nerve, the velocity of propagation of the nerve impulse should be proportional to the cross sectional area of the axonal fiber. It is found experimentally that, for unmyelinated nerve, the velocity of impulse propagation does increase as the fiber diameter increases. Some such values, excerpted from the much more inclusive list of Prosser and Brown (1961), have been used to construct Figure 6.

Besides increasing the axonal diameter, there is another important "trick" for increasing the velocity of nerve impulse propagation. Vertebrates sheath most of their longer axons of larger diameter with myelin, except for the periodic interruptions at the nodes of Ranvier. Since the whitish lipid materials of myelin insulate the regions of axonal surface, they envelop almost completely the only electrically active or excitable nerve membrane available in myelinated nerve is the portion at the nodes of Ranvier. Instead of the continuous progressive depolarization of the entire surface of the axon such as is entailed in unmyelinated nerve impulse propagation, the impulse jumps from one node of Ranvier to the next in myelinated nerve. This process, usually referred to as "saltatory transmission," yields a substantial enhancement of the speed of impulse propagation as well as altering the way in which this speed is dependent upon the axon diameter. In myelinated nerve the impulse velocity is directly proportional to the axon diameter rather than the square root of the diameter (Gasser, 1941).

As the animal increases in size, the greatest increase in transmission distances will generally occur in those fiber systems and tracts that, even in the smaller animals, are comprised of relatively long axons of relatively large diameter. It will therefore be in these fiber systems and tracts that the need will be most imperative to increase the impulse propagation velocities in order to reduce the response time to a viable magnitude, and thus, in these, that the selection pressures will be greatest for those structural modifications that yield the necessary increases in propagation velocities. Increased fiber diameter and myelinization are, as noted above, among the most important such structural modifications for achieving such increased velocities.

BIOLOGICAL CLOCKS, PREDICTION AND LEARNING

I have said that learning and/or habituation are essential properties of psychological behavior systems. Yet the only value which these properties have to an organism, in so far as its survival is concerned, resides in the fact that they permit it to predict something about the future on the basis of past events and make a response that is appropriate to that prediction. In an erratic universe, i.e. one in which the future is unpredictable from the past, nothing of value can be learned from the past that would enhance the probability of correct responses in the future. Nor would knowledge, without the capability of acting on it, have any survival value.

Some salient properties of the biosphere of our planet, Earth, derive from the fact that it 1) rotates about its axis with a period of very nearly 24 hours, 2) it orbits about the sun with a period of very nearly a year, 3) its own axis of rotation is inclined at an angle of approximately 23 degrees to the plane of its orbit about the sun, 4) it has a sizeable satellite body, the Moon, which rotates about it, 5) its orbit about the sun is not markedly eccentric. These give rise to a daily photoperiod of very nearly 24 hours, but in which the durations of the light period and the dark period shift through the seasons of the year. They give rise to an annual variation of temperature and precipitation, and to the ebb and flow of the tides. Simply possessing some kind of clock which is phased with the cycles of daily, synodial or seasonal change that are relevant to the organism provides an important element of predictability and valuable lead time in the physiological responses required to meet the changes. Such "biological clocks" have now been identified in a wide variety of species although the molecular processes which underly them are still a mystery.

Most investigators are in agreement that the following kind of mechanism is involved, at least in the daily rhythms. The organism, plant or animal, appears to have an endogenous oscillator with a natural period close to, but generally not identical with, the pertinent periodicity of the environment. In the case of the daily period this endogenous period would generally lie in the neighborhood of the value of 24 hours (for example, 23 hours, 25 hours); hence the term "circadian" is applied to such periodicities. When the organism is exposed to a periodically fluctuating environmental stimulus of the proper sort, the endogenous oscillator of the organism becomes "entrained by," i.e. synchronized with, the environmental rhythm. In the case of most of the daily rhythmicities which have been studied, the most common environmental stimulus rhythmicity for producing such entrainment appears to be the daily photoperiod (Bunning, 1964; Cloudsley-Thompson, 1961).

It would appear that the endogenous oscillator periodicity is probably genetically determined and concrete evidence that this is the case has been reported from the laboratory of S. Benzer (Konopka, 1970). Several mutants of Drosophila melanogaster have been isolated which have endogenous oscillator periods differing substantially from the range of values which could be considered to be circadian, as well as another mutant that is arhythmic. The arhythmic mutant appears to be recessive to all those allelic variants which are periodic and may be a deletion mutation. Some of these flies are chimeras in which cytogenetic markers can be used to identify the genotypic strain from which the various somatic cells are derived. The heads of some of the chimeral flies are comprised of cells which contain the genes for one endogenous oscillator periodicity while the body is comprised of cells which contain the genes for a different endogenous oscillator periodicity. It is the periodicity of the head cells that determines the endogenous periodicity of the animal as a whole.

Although the occurrence of such "biological clocks" is by no means restricted to animals, or animals with brains, it appears, nevertheless, that in the case of those phyla which do have brains, that the mechanisms of such biorhythm pacemakers and their entrainment by rhythmic external stimuli involve mediation through the brain.

If a plant seedling is germinated in the dark, maintained in the dark, but administered a brief period to exposure to light, its leaves will subsequently, for a number of days, exhibit a circadian rhythm of movement (up for maximum exposure during the "day" and dropped into a lowered posture during the "night"), e.g. cf. Bunning (1964). If now one exposes the plant seedling to another brief period of illumination during a time in which the leaves are lowering into their night position, this lowering will be reversed. In the absence of any further presentations of light the perturbation in the wave-like graph of these circadian movements will damp and disappear within a couple of cycles. However, if the illumination is again presented a "day" later, and again two "days" later, the perturbation enlarges at the expense of the original "wave crest." The result of this process is that the phase of the circadian movements is shifted to coincide with the new pattern of light and dark that has been imposed on the plant. But what is interesting in the context of the present paper is the resemblance of this phasing process to a kind of learning process, or to the process which is sometimes called "the shaping of behavior."

The relationship of such biological clock mechanisms to those involved in conventional learning processes has not been investigated or considered to any extent, yet there is some reason for thinking that this relationship may be more than trivial. Once more, it is informative to directly observe the behavior of psychologists.

One finds that it is a common practice among experimentalists who are studying learning in animals to administer the experimental training sessions at the same time each day. I do not recall ever having heard the rationale for this practice discussed in a clear and explicit manner, aside from some intuitive notions of "having everything the same," but I suspect their intuition on this matter is sound. In this regard it is interesting to recall the previously mentioned observation that bees could be readily trained to come to a food bait at a given time of day, indicating that time of day was perceived by them as some kind of relevant cue to be taken into consideration (Beling, 1929).

BEHAVIORAL MODE SWITCHING

The trumpet shaped protozoan, Stentor, normally lives with the narrow end attached to some relatively solid object, e.g. a water plant, and its flared end extended outward into the watery medium. If an experimental stimulus such as, for example, a jet of water is directed from a micropipet against its flared end, the Stentor responds in such a way as to avoid the apparently aversive stimulus by bending out of the path of the jet. Repetition of the stimulus presentation after a minute or two will produce a similar response from the subject. Continued repetition of the stimulus will, however, yield a modification of the response pattern. The Stentor contracts backward, and more vigorously with continued presentations of the stimulus. Further presentations result in the Stentor detaching itself and moving to an entirely different location some distance away. In responding in this way it would appear as though the Stentor is no longer responding merely to the individual stimulus presentations but to the entire sequence of presentations and to its previous failures to terminate the aversive stimulus presentations by responses such as bending or contracting out of the jet stream. A similar kind of response pattern can be seen in the sea anemone, a coelenterate. In both cases it would appear as though the animal, in consequence of its sequence of failures to better its situation by its previous manner of responding to the individual stimulus presentations, has switched to an entirely different kind of response. I believe that these are special instances of a general kind of phenomenon that I shall call "behavioral mode switching."

Another instance of such behavioral mode switching can be observed in the planarian *Dugesia dorotocephala* by means of a simple experiment. A flat dishpan was filled to a depth of about half an inch with aged tap water and approximately 500 planarians. When slowly tilted the water would recede to one end of the pan. This receding water line would leave some of the planarians stranded. The first time this was done virtually all of the planarians would turn directly toward the receding water line and reenter the water, provided the tilting was performed slowly enough. On the condition that several minutes were allowed to intervene before repeating the maneuver, one would find that practically all of the animals yielded this kind of response upon the second repetition also. However, upon the third or fourth repetition one would observe some appreciable proportion of the planarians now altering their response so as to head away from the receding water line. This switching from an apparently adaptive response to one that, at first glance, appears maladaptive, deserves comment. Under normal circumstances, in nature, a planarian would not be stranded by receding water levels twice within a few minutes unless it had made the mistake of entering a small bubble cut off from the main body of water. Being stranded, escaping by the most direct course to the receding water line, and then being stranded again within a few minutes, is diagnosed as a "failure" of the direct course response to alleviate the aversive situation, so the planarian engages in behavioral mode switching. In this altered mode it now responds to being stranded by heading away from the receding waterline.

Such behavioral mode switching provides both an experimental difficulty and an interesting point of linkage between the psychological behavior systems and the stereotyped behavior systems. First let me indicate the manner in which it poses experimental difficulties, especially if one is "hung up" on the conventional wisdom mentioned at the outset of this article. In the interest of statistical considerations the experimentalist likes to evoke a number of responses from individual animals in a "defined" situation. The "defined" situation usually entails putting the animal under some kind of constraint in an apparatus that presents a generally aversive and highly artificial environment for the animal. Such aversiveness of the experimental situation and the usual handling that goes with it are usually ignored. So is the possible aversiveness of the social isolation used as a standard aspect of experiments directed toward conditioning of invertebrates. After a few repetitions of the nice simple response that the experimental situation was designed to exploit and shape, the supposedly simple animal switches behavioral modes. The experimental protocol sheet which allowed spaces for checking "yes" or "no" or entry of response latency now needs a space for "none of the above." If, as is frequently the case, the experimenter has chosen the lower animal because he wanted a "simple stereotyped response" with no hanky panky about things like emotionality, the animal and response paradigm will be discarded as "unreliable."

Yet there is also another aspect of the matter that not only concerns techniques of experimental study, but the real nature of the link between the stereotyped response systems and the psychological behavior systems, too. The behaviors of higher animals, as well as those of the lower invertebrates, when studied in a certain way, can be perceived as comprised of a sequence of stereotyped blocks of behavior. This does not, however, mean that the sequence itself is stereotyped. For example, the present article is comprised of sequences of words. Each of these words is a stereotyped succession of letters. But my sentences and paragraphs can take on the full gamut of things that need to be said. One might regard the transitions from one word to the next as the point in the structure of the sentence that the psychological mechanisms enter, for it is through the specification of the word "word transitions" that the conceptual or psychological content of the sentence is manifest. But such a term "word transitions" is essentially a switching from one stereotyped sequence to the next, i.e. Mode Switching. The psychological system and its manifest behavioral content are, in the behavior of an animal, a system which controls mode switching of the stereotyped blocks of behavior.

RANDOM BEHAVIOR AND GAMES OF STRATEGY

Although I would not wish to denigrate the use of statistical design and the now routine application of statistical theory to behavioral studies, I think that certain of its terminology and notions, accepted too uncritically, have led to some widespread misconceptions of a fairly fundamental character. One of these concerns the interpretation of random elements in behavior. I should like to comment on these briefly.

If one walks into any reasonable mature and large department of psychology in a randomly chosen large university in the United States one will find that it, like Gaul, is divided into three parts: Clinical, Industrial and Experimental. Since, in our pursuit of "truth," we are not primarily interested in either "helping people," or making the industrial establishment run more smoothly, let us pass by the first two and turn down the corridor containing the laboratories and offices of those devoted to the pursuit of the truths of experimental psychology. Knock on any door and enter. Ask the inhabitant what his major objective consists of in the broad sense. One will usually receive the answer "The prediction of behavior," or, sometimes, "The control of behavior." In the case of the latter answer one can surmise that the inhabitant is a Skinnerian. But, since the former answer is by far the most common, let us probe the meaning of it in greater depth by pressing our newly found informant for further details.

One will probably find that he makes extensive use of one or more forms of the analysis of variance or analysis of covariance and that he thinks of experiments and experimental results in terms of them, so much so that it will be only with the greatest effort that one can coerce him into a discussion about behavior that is divorced from the concepts and terminology of them. In view of this pervasive influence of the general notions of the analysis of variance, or some off-shoot of it, into the concepts and goals of Experimental Psychology, it is worth examining them. One discovers that a piece of datum, score, or some individual measurement of behavioral performance is considered to be the additive resultant of several deterministic components of variation stemming from factors such as age, strain, sex, etc., of the subject plus "error variance." The name of the game is to ascertain which of the factors yield associated deterministic component estimates that are improbably different from zero, i.e. "significant," in the light of the error variance found in the experiment. The proportion of the variance which can be accounted for by the experimenter becomes, in his mind's eye, a measure of his omniscience. Error variance, in this scheme of things, is what separates men from the gods. It is not considered an intrinsic part of the behavior, but a measure of the imperfection of the experimenter. This conception is probably not true in many instances. While it is not my intent to grant absolution to sloppy experimenters I nevertheless shall develop the notion that such "error variance" may, in some instances, reflect the true nature of the behavior rather than a failure of the experimenter to take all the significant factors into account, or any failure of technique. There is reason to believe that some behavior is "designed" to be random. It is designed to be random to render it unpredictable.

When I drive the road to my house in the foothills of the Colorado Rockies, I frequently see rabbits suddenly illuminated in the beam of the lights of my car. Since I do not like to hit animals with my car I try to guess which way they are going to run in order to avoid hitting them. Predicting what they will do turns out to be impossible. Sometimes they will remain stationary. Sometimes they will run to one side and then immediately double back and run directly across the car's path. Sometimes they will run directly toward the car. At other times they will run along in front of the car. Such behavior appears to be random. Formal experiments on this kind of behavior in rabbits indicate that this is indeed the case. Why do they do this? Why do they not merely do the "right" thing? Perhaps, one may suggest, rabbits do this sort of thing because they are "stupid" animals thrown into a "panic" by the sudden appearance of the automobile headlights and that they would do the same thing if confronted with the sudden appearance of a predator. To which I would answer: exactly, provided one appreciates that this is the wisest policy for a stupid animal to follow.

In the Theory of Games of Von Neumann and Morgenstern (1947) there is an interesting problem involving two opponents in which one of the opponents is outclassed by the strength, resources, firepower, available information, speed, and/or intelligence of the other. What, in such a situation, is the optimal strategy for the weaker of the two opponents? Careful analysis of the situation shows that any deterministic strategy which the weaker opponent may adopt is certain to result in defeat. The only strategic policy which the weaker opponent can adopt that gives it any nonzero probability of avoiding defeat is to fall back on a random strategy. The adoption of such a random strategy renders the behavior of the weaker opponent unpredictable. By discarding the use of its "strategic computer" the weaker opponent renders the superior "strategic computer" of its opponent worthless. I do not wish to suggest that rabbits are familiar with the Theory of Games, or even that some Great Rabbit Designer in the Sky is familiar with the Theory of Games and deliberately incorporated such features into the cybernetic circuit design of the rabbit's brain!

In another section of this paper I suggested that the mechanism of evolution, mutation and selection provided a kind of natural Monte Carlo method for the attainment of optimal design configurations. But optimal design is not restricted to those features or configurations which are anatomically evident. It applies as much to behavior as to skeletal structure or the stream lining of cetacean mammals. To prevent misunderstanding I should be more explicit.

The selection pressures which push a species toward optimal behavioral design are only concerned with the behavioral contingencies and not the internal circuitry details which produce those behavioral contingencies. But the overt manifestation of the various behaviors of a particular species of animal depends upon the physiological substratum provided by the cybernetic configurations of the neural circuitry of its brain, which in turn is regulated by molecular processes under genetic control. Through such a causal chain, selection pressures will tend to discriminate against those genes, or gene combinations, that lead to maladaptive behavioral programming contingencies and favor those genes, or gene combinations, that yield adaptive behavioral programming contingencies. The prey-predator "game" situations that various species must engage in constitute an exceedingly important factor in their existence. Those behavioral programming contingencies which are pertinent to these prey-predator games constitute the strategy—in the Von Neumann and Morgenstern sense—which the animal "player" of one species deploys against its "opponent" of another. Thus, genes, or gene combinations, which result in significant departures from the optimal strategy for a species against its major "opponents" will be subjected to negative selection. Those genes, or gene combinations which tend to make the animal's behavior approximate the optimal strategy more closely, will be selectively favored.

Let us return to rabbits. In Colorado and its general environs, coyotes, owls, and lynxes are probably the major predators with which rabbits have to deal. Although some rabbits are killed by automobiles, this probably is not a major cause of death. Coyotes are exceedingly intelligent animals, quite capable of learning, and exploiting to their advantage virtually any deterministic strategy that a rabbit might adopt. Lynxes are fast and formidable. Owls are silent and their flying speed exceeds the running speed of rabbits. No slight increment of speed, cunning, or size is going to tip the balance for the rabbit by a significant amount in its survival "games" with its predators. It is simply "outgunned" in regard to any of its major tricks that it has at its disposal to deploy in a deterministic manner. The gene pool of the species has thus "learned" in a pragmatic way, i.e. through selection, that the best ultimate strategy when hard pressed is to "go random." This randomizing mechanism is probably activated through the same nueral mechanisms involved in the mediation of "fear."

A POSTSCRIPT

The concepts and syntheses presented above were formulated independently although I have made extensive use of the electron micrographs of my coworkers such as Kay Steen, Jack Noel and Dr. Michio Morita. Since then I became aware of the important paper by Dethier (1964), in which he compares insect behaviors and brains with those of the vertebrates. It is clear that much of the major thrust of Dethier's paper is in the same direction as the present one and it buttresses my confidence in the essential correctness of the ideas set forth in the present paper to find that they coincide with those of an investigator for whom I have so much respect. Nevertheless, I am of the opinion that the present paper goes beyond that of Dethier. Whether this represents a bolder perception or more reckless incaution on my part will be a matter for the readers and future investigators to decide.

REFERENCES

Abercrombie, M., "Contact inhibition: the phenomenon and its biological implications." *Nat. Cancer Inst. Monograph* 26, 249-277 (1967).

Ansevin, K. K., and R. Buchsbaum, "Observations on planarian cells culturated in solid and liquid medial." *J. Exp. Zool.* 146, 153-161 (1961).

Arakilian, P., "Cyclic oscillations in the extinction behavior of rats." *Jour. Gen. Psychology 21*, 137-162 (1939).

Beling, I., "Über das Zeitgedächnis der Bienen." *Z. vergl. Physiol.* 9, 259-338 (1929).

Best, J. B., and I. Rubinstein, "Environmental familarity and feeding in a planarian." *Science 135*, 916-918 (1962b).

Best, J. B., and I. Rubinstein, "Maze learning and associated behavior in planaria." *Jour. Comp. Physiol. Psychol.* 55, 560-566 (1962a).

Best, J. B., "Protopsychology." *Scientific American* (February 1963).

Best, J. B., M. Morita and J. Noel, "Fine structure and function of planarian goblet cells." *J. Ultrastructure Res.* 24, 385-397 (1968).

Best, J. B., A. B. Goodman and A. Pigon, *Science 164*, 565 (1969).

Best, J. B., and J. Noel, "Complex Synaptic configurations in planarian brain." *Science 164*, 1070-1071 (1969).

Best, J. B., V. Riegel and R. Rosenvold, Unpublished Data (1970).

Bunning, *The Physiological Clock*. Academic Press Inc., New York (1964).

Cannon, W., *The Wisdom of the Body*. W. W. Norton, New York (1932).

Clark, R. B., *Zool. Jahrb. Allg. Zool. Physiol. Tiere 68*, 395-424 (1959).

Cloudsley-Thompson, J. L., *Rhythmic Activity in Animal Physiology and Behavior*. Academic Press, New York, London (1961).

Dethier, V. G., "Microscopic Brains." *Science 10*, 1138-1145 (1964).

Edman, P., R. Faenge and E. Ostlund, Second International Symposium on Neurosecretion (W. Bargman et al., eds.). Springer-Verlag, Boston (1958).

Elazar, A., and W. R. Adey, "Electroencephalographic correlates of learning in subcortical and cortical structures." *Electroencephalography and Clinical Neurophysiology 23*, 306-319 (1967).

Florey, E., *An Introduction to General and Comparative Animal Physiology*. W. B. Saunders Co., Philadelphia and London, 1966.

Fraenkel, G. S., and D. L. Gunn, *The Orientation of Animals: Kineses, Taxes and Compass Reactions*. Dover Publications, Inc., 1961.

Galbraith, J. K., *The Affluent Society* (Chap. II). The New American Library, New York and Toronto, 1958.

Gasser, H. S., *Ohio Jour. Science 41*, 145-159 (1941).

Hall, C. S., *J. Comp. and Physiol. Psychol. 18*, 385 (1934).

Herlant, H. Meewis, *Sav. Amn. Soc. Zool. Beld. 87*, 151-183 (1956).

75394

Hodgkin, A. L., *J. Physiol. 125,* 221-224 (1954).

Hubl, H., *Arch. Entwickl. Mech. Org. 149,* 73-87 (1956).

Kenk, R., "Induction of sexuality in the asexual form of *Dugesia Tigrina.*" *Jour. Exp. Zool. 87* (1941).

Morita, M., and J. B. Best, "Electron microscopic studies of planaria." II. "Fine structure of the neurosecretory system in the planaria-Dugesia Dorotocephala." *J. Ultrastructure Res. 13,* 396 (1965).

Morita, M., and J. B. Best, "Electron microscopic studies of planaria." III. "Some observations on the fine structure of planarian nervous tissue." *J. Exp. Zool. 161,* 391-411 (1966).

Moscona, A. A., "Development in vitro of chimaeric aggregates of dissociated embryonic chick and mouse cells." *Proc. Natl. Acad. Sci. U.S. 43,* 184-194 (1957).

Moscona, A. A., "Analysis of cell recombinations in experimental synthesis of tissues in vitro." *J. Cellular Comp. Physiol. 60 Suppl. 1,* 65-80 (1962).

Munn, N. L., *Handbook of Psychological Research on the Rat.* Houghton Mifflin, Boston, p. 99 (1950).

Von Neumann, J., and O. Morgenstern, *Theory of Games and Economic Behavior.* Princeton University Press, Princeton, N. J. (1947).

Olds, J., and P. Milner, *J. Comp. Physiol. Psychol. 47,* 419 (1954).

Olds, J., *Physiol. Rev. 43,* 554 (1962).

Opatowski, I., *Bull Math. Biophysics 6,* 113-153 (1944).
    Bull Math. Biophysics 7, 1 (1945).
    Bull Math. Biophysics 8, 41 (1946).

Pavlov, I. P., *Lectures on Conditioned Reflexes.* Liveright Publishing Corp., New York (1928).

Prosser, C. L., and F. A. Brown, Jr., *Comparative Animal Physiology.* 2nd Ed. W. B. Saunders Co., Philadelphia (1961).

Raisman, Geoffrey, "Neuronal Plasticity in the septal nuclei of the adult rat." *Brain Research 14,* 25-48 (1969).

Rappaport, C., and G. B. Howze, *Proc. Soc. Expt'l Biol. Med. 121,* 1010-1016 (1966).

Rappaport, C., Seminar delivered at Colorado State University (1969).

Steinberg, M. S., "Reconstruction of tissues by dissociated cells." *Science 141,* 401-408 (1963).

Trinkaus, J. P., and J. P. Lentz, "Direct observation of type specific segregation in mixed cell aggregates." *Develop. Biol. 9,* 115-136 (1964).

Trinkaus, J. P., *Cells into Organs.* Prentice Hall (1969).

Welsh, J. H., and L. D. Williams, "Monoamine containing neurons in planarian." *Journal of Comparative Neurology 138,* 103-116 (1969).

CHAPTER V

The Evolutionary Significance of Biological Templates

George Ledyard Stebbins

University of California

In his exciting autobiography, *The Double Helix*, James Watson cites an old but vague hypothesis about the nature of heredity as follows (p. 127): "The argument went that gene duplication required the formation of a complementary (negative) image where shape was related to the original (positive) surface like a lock to a key. The complementary negative image would then function as the mold (template) for the synthesis of a new positive image." The remainder of his book tells the story of how he demonstrated that the gene is, in fact, a pair of templates which duplicates its pattern by making a new negative complementary to the existing positive simultaneously with a new positive complementary to the existing negative. The template principle is, therefore, the basis for the transmission of information in biological heredity.

In the eighteen years since Watson and Crick proposed their model of the chemical structure of the gene, DNA has become a household word, and the template principle has come to be recognized as basic to our understanding of many other biological phenomena in addition to hereditary transmission. In the present contribution, I should like to explore the implications which recognition of this principle have in respect to our understanding of evolution.

Properties Basic to the Template Principle

The existence and functioning of biological templates is made possible by three properties of the macromolecules of which living organisms chiefly consist, both nucleic acids and proteins. In the first place, each macromolecule consists of a large number of subunits, which are of a relatively few different kinds: four in the case of nucleic acids and twenty in the case of proteins. These units are arranged in a linear sequence.

Secondly, the properties of the macromolecules depend not upon the kinds of subunits which they contain, or even upon the relative proportions of different subunits, but upon the *order* in which they are arranged. In ribonucleic acid (messenger RNA), for instance, the sequence of nucleotides guanine-adenine-adenine is the informational code for inserting the amino acid known as glutamic acid into a protein chain—the reverse order adenine-adenine-guanine codes for lysine—an amino acid having completely different chemical properties from those of glutamic acid. In the polypeptide chain of a protein, reversal of a sequence of amino acids such as cysteine-alanine-lysine-isoleucine-valine, so that it became valine-isoleucine-lysine-alanine-cysteine, would give the molecule completely different properties. The chains of both nucleic acids and proteins have at either end distinctive chemical groups, so that their "left" ends can always be distinguished from their "right" ends.

In the third place, specific relations between macromolecules are established by a phenomenon known as *colinearity*. This results from the fact that the subunits of the macromolecules have arrangements of their atoms which fit the arrangements of certain other subunits in very specific ways. Thus the nucleotide base adenine matches or fits with thymine but not with cytosine; while guanine matches cytosine but not thymine. The "fit" is based upon hydrogen bonding between hydrogen atoms and those of either oxygen or nitrogen when the atoms concerned are placed in certain specific positions relative to each other. This colinear "fit" establishes the complementary relationships between the two strands of the DNA double helix. If, for instance, one strand has the order ATAGCTCAG, the order of nucleotides in the other strand must be TATCGAGTC. In this and other examples A-adenine, T-thymine, G-guanine, C-cytosine, and U-uridine (in RNA).

Template Function in Replication of DNA, Transcription of RNA, and Protein Synthesis

The biochemical mechanisms concerned with the replication of DNA, the transcription of the information contained in DNA to the messenger RNA, as well as the synthesis of proteins through translation of the genetic code have now been described so many times in scientific papers and books, in popular works, and in textbooks that these descriptions hardly need to be repeated here. Readers who are unfamiliar with them should consult such works as Watson's *The Molecular Theory of the Gene*, or in a more popular vein *The Language of Life* by George and Muriel Beadle. I shall mention only those points which are essential to the present argument.

In connection with the replication of DNA, which makes possible the distribution of an identical code of genetic information to every one of the trillions of cells in our body, two facts must be kept in mind. First, contrary to opinions which have sometimes been expressed, DNA is not a self replicating molecule. For separating the two strands of the double helix, for inserting the new subunits or nucleotides, and for binding them together to form the "backbone" of the new helix, at least two different enzymes are required. These enzymes are, of course, proteins which can be synthesized only on the basis of information coded by certain genes. Thus any one gene can be duplicated only with the aid of products derived from two other genes. In addition, these enzymes can catalyze the replication and chain binding processes only with the aid of energy which is supplied to them by that ever present energy carrying molecule, adenine triphosphate (ATP). When DNA is replicated in a test tube, as in the brilliant experiments of Arthur Kornberg, the enzymes and ATP must be added before the synthesis will take place. In living cells, they must be produced by metabolic processes which take place elsewhere in the organism. Thus it is the living system, not any one of its molecules, which is self replicating. The function of the informational template of DNA is to ensure then when this replication occurs, its pattern will be transmitted intact to the newly formed molecules.

Secondly, replication of DNA cannot begin until certain preparatory chemical reactions have taken place. Its subunits or nucleotides are themselves compound molecules. They consist of three parts: the base (A, T, G, or C) which provides the unit of information, a sugar group containing 5 atoms of carbon, and a phosphate group. The sugar group, by chemical bonding, holds the base in its proper position, while chemical or covalent bonds between sugar and phosphate groups form the form "backbone" of the molecular helix. In many living cells, the bases and sugar groups must be synthesized from smaller molecules, and then bound chemically to each other and to the phosphate group. These syntheses are catalyzed by a battery of enzymes, each one of which must be assembled on the basis of information provided by its own particular gene. Furthermore, the nucleotides cannot be assembled to form a new DNA helix until two additional phosphate groups have been added to the one which will eventually form a part of the DNA molecule. This phosphorylation requires an additional enzyme, and in some cells the signal for DNA replication to begin is a buildup in the content of the enzyme which phosphorylates thymidine. DNA replication is, therefore, different in many respects from the simple growth of an inorganic crystal.

The transcription of RNA from the DNA template is essentially similar to the replication of DNA. Specific enzymes are also needed, and the ribose nucleotides must be present, either through synthesis or breakdown of preexisting RNA, and then phosphorylated before transcription can begin. The important fact about this process is that it goes in only one direction. DNA serves as a template for RNA synthesis, but RNA can never serve as a template for the informational pattern of DNA. This irreversibility applies also to protein synthesis through translation of the code provided by the RNA template. Its evolutionary significance will be discussed after the discussion of translation.

The facts which must be emphasized about translation are its complexity and its indirect nature. In addition to the RNA, known as messenger, which provides the informational code for the sequence of amino acids in the protein chain, two other kinds of RNA are required. One of these, known as ribosomal RNA, becomes complexed with protein to form particles known as ribosomes, which serve as platforms to which the messenger RNA must be chemically bound before translation can begin. In higher animals and plants, ribosomal RNA is produced in one or a few highly localized regions on particular chromosomes. These are often associated with spherical, strongly straining organelles known as nucleoli, which are conspicuous features of cell nuclei as we see them through the microscope. The third kind of RNA, known as transfer RNA, consists of much smaller molecules than those of the other two kinds, and plays a very special role. It serves as the indirect link between the informational code of messenger RNA and the placement of the amino acid units into their proper position in the protein chain. One portion of a transfer RNA molecule bears a triplet of nucleotide bases which is complementary to a particular triplet of messenger RNA. Another part of each molecule of transfer RNA has a particular conformation which matches, through a stereochemical "fit," a portion of the molecule of one and only one kind of "activating" enzyme. Each of these latter enzymes can become attached by another part of its molecule to one and only one of the twenty different kinds of amino acids which form the units of a protein chain. By means of this double attachment, the activating enzyme catalyzes the binding of a particular molecule of transfer RNA to a particular amino acid.

Thus the genetic code, which specifies the association of certain triplets of bases in RNA with certain activated amino acids as they become incorporated into the growing protein chain, is in a sense a catalogue which classifies both the more than twenty different kinds of transfer RNA molecules needed for protein synthesis, as well as the equal number of enzymes required for binding these molecules to their respective molecules of amino acid.

The most important implication of these facts to an evolutionist is that the universal method by which the working molecules of living systems, the proteins, are synthesized requires successful interaction between scores of different enzymes and their substrates. No simpler self-replicating system exists today. Did simpler systems exist in the remote past, and if so, how did they operate? This question will be considered later.

The Molecular Concept of the Segregation of the Germ Plasm

The other highly significant fact about both transcription of RNA and translation or protein synthesis is the irreversibility of both of these processes.

Neither RNA nor protein can provide a code for the pattern of information found in DNA. This irreversibility gives us a molecular version of the concept of the segregation of the germ plasm from the somatoplasm, first developed on a cellular basis by August Weismann in the 1890's, on the basis of facts then known about animal development. Weismann used his theory as a fatal objection to Lamarck's hypothesis of evolution through the inheritance of acquired adaptive modification. To a general biologist, Weismann's theory had a serious weakness in that the germ cells are not segregated from the body cells in plants as they are in animals. Moreover, in unicellular organisms as well as many fungi and algae germ cells are not formed until immediately preceding the union of the gametes in sexual fertilization. Weismann's concept, therefore, can be applied only to multicellular animals. On the other hand, the nuclear DNA is segregated from all other parts of the cell in all organisms at all stages of development. The molecular separation of the informational "germplasm," the DNA, is universal and permanent. No matter what changes in molecular structure or organization might be induced by adaptive modifications of RNA or proteins in the cytoplasm, membranes, or walls of the cells, produced through the influence of either the cellular or external environment, the transfer of such modifications to the self replicating-pattern of information coded in DNA, and hence to the permanent hereditary thread of life, is inconceivable.

There is, however, at least one way in which the total content of genetic information existing in a cell might be modified, i.e. at least semi-permanently, through the influence of the environment. In all organisms except bacteria and blue green algae, a small amount of DNA exists quite separate from their cell nuclei, in small organelles known as mitochondria. In addition, green plants possess similar amounts of DNA in their plastids, which contain the light-absorbing chlorophyll. Chemical analyses of the DNA contained in mitochondria and plastids have shown that it contains different information from that in the nuclear DNA. Furthermore, the replication of the mitochondrial and plastid DNA is not always synchronized with the replication of the nuclear DNA. Hence one could conceive of differential replication of nuclear, mitochondrial, and plastid DNA. We can imagine that some particular, and presumably drastic environmental condition could produce such differential replication, and consequently a different balance between organellular and nuclear DNA. This altered balance could affect the protein content and hence the metabolism of the cell as a unit. If, then, a return to a more normal environment should restore a synchronized replication of mitochondrial, plastid, and nuclear DNA, the altered condition could be inherited, at least in a semi-permanent fashion.

I must admit that the argument presented above is pure speculation, and I know of no example which is best explained in this fashion. Furthermore, even if changed balances between organellular and nuclear DNA could be environmentally induced, the amount of DNA contained in mitochondria and plastids is so small that a pronounced and continued effect of such changes is hard to imagine. Nevertheless, their possible evolutionary significance cannot be ignored.

Replicative and Positional Templates

Most of the definitions of the word "template" which appear in Webster's Unabridged Dictionary describe various devices for making accurate copies of a particular design. One of them is, however, as follows: "any of various locating devices (as for placing rivets or applying airplane trim)." Man-made templates can, therefore, serve either as models for making correct copies of a design, or for locating complex parts into a larger structure, according to a predetermined design. We can, therefore, designate templates which are regularly copied as *replicative templates*, and templates which determine the precise location or position of the parts of a complex structure as *positional templates*, following the lead of Tracy Sonneborn, who first applied this term to the example in Protozoa to be described below. Do we find analogies to positional templates in living organisms?

A good deal of evidence now exists to show that several kinds of positional templates do indeed exist. Furthermore, their presence may well be universal and, in fact, essential for life. The example which has been most carefully worked out is that of a virus, bacteriophage T4. Like most viruses, T4 phage consists of a molecule of DNA which is tightly packed into a protein coat. The coat of T4, is, however, much more complex than is that of any other known virus. It consists of a polygonal "head," which contains the DNA; a columnar "tail," consisting of a central tube surrounded by several ring-like "tail pieces," and six angular "tail fibers." A T4 particle infects a bacterial cell by landing on it with its tail fibers down, and resting on the surface of the cell as if on stilts. The bottom of the tail then penetrates the cell wall and the viral DNA is injected into the bacterial cell by means of an action similar to that of a syringe.

The "morphogenetic pathway" leading to the formation of this structure has been worked out in detail, first by Edouard Kellenberger and more recently by Robert Edgar and William B. Wood. The phage DNA contains genic information for synthesizing each of the several different kinds of structural protein of which the coat consists. Forty-one different genes are known to participate in the process, and there may be more. The assembly of head, tail and tail fibers into a whole particle is, however, not guided directly by the action of the DNA. Instead, information existing presumably in the form of particular sequences of amino acid residues in the various kinds of proteins determines how the molecules will be fitted together. This hypothesis is inferred from the following facts. If particles of virus in the early stages of coat assembly are removed from bacterial cells and placed in a test tube, coats can be assembled even if the DNA has been destroyed. Nevertheless, particles assembled under these conditions, as well as those of normal phage being assembled in the host cell, must be put together in a definite sequence of order. This fact has been demonstrated by means of mutations which alter the code for one of the parts in such a way that the protein produced will not fit into the structure. When such parts are defective, the morphogenetic pathway may be almost completely destroyed, or it may proceed almost to completion, depending upon whether the defective part is inserted early or late in the sequence. The former is true of the bottom piece of the tail column, while the latter is true of the tail fibers.

These facts tell us that the sole function of the DNA of bacteriophage T4 is to contain genes which code for the amino acid sequences of structural proteins, as well as for various enzymes essential to the processes of replication and coat synthesis. When formed, the structural proteins serve as positional templates which determine not only the way in which the various parts of the coat will fit together, but also the order in which they will be assembled.

A variety of different lines of indirect evidence for positional templates exists in several different kinds of animals and plants. The most extensive evidence has been obtained in one-celled animalcules (protozoa) such as the "slipper animalcule" (*Paramecium*) and related forms. Much of the activity of these animals depends upon the structure and function of an elaborate system of organelles distributed over their outer surface or cortex. These organelles all consist of protein, and hundreds or even thousands of different kinds of structural protein molecules must go into their makeup. Although some organelles, such as the cilia, contain small bits of extranuclear DNA, many and perhaps most of the structural proteins composing them have structures which are coded by nuclear genes. This assumption is based partly upon the existence in several species of mutations of nuclear genes which affect the structure and action of the cilia. Furthermore, experiments with radioactively labelled bases indicate that RNA synthesized in the nucleus travels to the cortical region, where it presumably is active in protein synthesis.

Nevertheless, when alterations in the pattern of cortical organelles are produced, either by microsurgery or by accidents which occasionally occur while the cells are conjugating, these alterations persist unchanged over many cycles of cell division, and may even be retained after conjugation with another genetically different animal has provided the nucleus with a different complement of genes. Moreover, when animals bearing different patterns are crossed, by means of conjugation, each cell after conjugation retains the same pattern which it possessed previously. In spite of the fact that it has received genes from an animal having a different pattern, its own pattern is not altered by the action of these genes.

The complete explanation of these results is still a mystery, and the evolutionist is at a loss to explain how cortical patterns in protozoa can evolve. Nevertheless, one logical explanation of at least a part of the observed phenomena is that the existing cortical pattern of any protozoa is an elaborate series of positional templates. Some of the nuclear genes, then, code for proteins which, because of specific amino acid sequences, have stereochemical properties which fit into certain definite positions in the existing structure, and enable it to replicate its pattern. This hypothetical explanation still assumes that all of the basic information needed to make a protozoan cell is coded in the genes, and chiefly in those of the nucleus. Nevertheless, as in bacteriophage T4, the shape and organization of the final structure depends upon precise interactions between gene products which have been coded by many different genes. In protozoa, moreover, the different products which interact with each other as positional templates may have been coded at very different times in the life cycle of the organism.

Evidence for the existence of positional templates in multicellular animals and plants is fragmentary, but still substantial. In the first place, some tissues such as the lining of the intestine and the retina of the eye are made up of cells which contain patterns of cilia and related organelles having intracellular fine structures much like those of protozoa, and perhaps are built up in much the same fashion. Secondly, certain organelles, such as the chloroplasts in a leaf of barley, are made up largely of structural proteins which are organized into an elaborate fine structure, and many of these proteins are known to be coded by nuclear genes. Hence the best explanation for the method of choloroplast synthesis is that it depends upon the functioning of structural proteins as positional templates.

A third line of evidence for positional templates in higher organisms comes from the structure and mode of assembly of certain membranes or membrane-like structures, and certain kinds of cell walls. An example is collagen, a substance found in the connective tissues of all higher animals. Although genes which code specifically for the protein molecules of which collagen consists have not yet been identified, one can hardly avoid the assumption that, like all protein molecules, those of collagen are gene coded. Yet a number of different experiments have shown that when first synthesized, these molecules are not arranged into any definite pattern. Their arrangement first into fibrils, and then into precisely ordered groups of fibrils, occurs after their synthesis is completed, and appears to be a process of self assembly similar to that described for bacteriophage T4. Moreover, in this and probably other examples, the environment surrounding the molecules during the period of self assembly has a great effect upon the organization of the final product. When collagen molecules are placed in a test tube under conditions favorable for self assembly, the organization of the resulting fiber can be radically changed by altering the concentration of salt or other properties of the surrounding medium.

In my opinion, more intensive studies of the mechanics of animal and plant development will bring to light many more examples of positional templates. In particular, the development of our nerve cells and of the connections, or synapses, between them may involve the functioning of positional templates in the form of the structural proteins which form part of the membranes of these cells. Is it possible that such phenomena as psychological maturation, imprinting, memory, and other elements of our behavior depend in part upon the nature and functioning of positional templates? This possibility certainly deserves attention from neurophysiologists and other students of behavior.

Enzyme Proteins as Templates

Biologists may at this point be asking the question: "What about enzymes and related molecules which perform all of the important activities of the cell? Can the template concept be applied to the molecules of which they consist?"

Protein molecules certainly never function as replicative templates. They do not contain information for copying their own pattern, or for copying any related pattern. The positional template concept, however, can be applied to enzymes in a modified form. Many enzymes and related molecules, such as hemoglobin, consist of two, four or more primary chains, which are joined together after their formation to produce a compound molecule. Their gene-coded amino acid sequence determines in a precise fashion the way in which the molecular subunits will be associated with each other, and a good many mutations are known which produce non-functional molecules because the mutant subunits cannot become properly associated to form a working union. In other instances, the activity of an enzyme may be "turned on" or "turned off" by the attachment at one particular position of a certain kind of small molecule which, when attached, folds the protein chain or chains of the enzyme into different shapes from those necessary for activity. In these examples, those portions of the chains which are responsible for the association of the subunits with each other, or for the attachment of the activating or inhibiting small molecules, can certainly be regarded as positional templates.

There are, however, enzymes such as those that digest proteins in our digestive tract and consist of only one chain, and which are not activated or inactivated by the attachement of small molecules to them. Are these protein molecules not to be regarded as templates at all? If the template analogy is extended far enough to cover the association between the subunits of a compound enzyme molecule, but not the manner of folding of the individual subunits or the similar folding of the single chains which make up the simpler enzymes, a truly artificial distinction is made. This is because certain attractions, as well as actual chemical bonding between individual amino acid residues or between groups of them (disulphide bonds of cystine, hydrophobic attractions between sequences of residues containing methyl groups, hydrogen bonding) act in just the same way to produce the coiling (secondary structure) and folding (tertiary structure) of individual chains as to produce the association between chains (quaternary structure).

Nevertheless, if we do regard the individual chains of simple enzyme molecules as having template properties, we must recognize that living organisms contain templates of a kind which human ingenuity has never been able to devise. The facts are these. The primary amino acid sequence of the protein chain of an enzyme is such that in a certain kind of cellular environment it is folded into a very specific pattern. This pattern is such that certain amino acids of the chain, which are capable of cooperating to form or break a chemical bond of the substrate, are placed next to each other in the folded molecule, even though they may be far apart in the unfolded chain. Still other amino acids, which are capable of binding the enzyme molecule to the substrate in such a position that it may efficiently do its work, are also placed in their correct positions by the method of folding rather than by their position on the unfolded chain. The same holds for additional amino acids of the chain which associate with the energy releasing molecules of adenine triphosphate (ATP), for others which become associated with molecules of coenzymes or of metallic ions essential for the reaction, and for still others which by either attracting or repelling water place the molecule in an environment most suitable for performing its work. In short, the primary sequence of amino acid residues in a protein chain is a kind of information, based upon a specific pattern sequence, which determines either the folding of the molecule into its functional shape, or the attraction to it of other molecules, at specific positions, that are essential to its function. Since those parts of the chain which attract and become associated with other parts of the same chain are similar in both structure and action to the parts that attract other molecules or other subunits of a compound molecule, it seems to me logical to apply to all of these portions of the chain the concept of the positional template.

If we do this, we can recognize the primary structure of the protein chain of an enzyme as a *self-activating template.* To continue the analogy with man-made templates, a structure comparable to a simple enzyme would be a tool or machine consisting of between one and two hundred parts, which when manufactured would be joined together into a single linear chain. Some of the parts, however, would contain magnets or other attracting devices of a very specific nature, so that when the chain of parts was placed upon the bench where it was supposed to do its work, it would immediately assume a new, three dimensional shape. Furthermore, this shape would be one in which blades, clamps, attachments to an engine or a source of power, and other necessary parts were arranged in positions such that the machine could start working without the intervention of a human hand! Such marvelous labor-saving machines are probably beyond the power of human ingenuity to invent. They were, however, evolved by living systems when life first began. The DNA "blue prints" for many of them have been replicated in an almost unaltered form throughout the billions of years which have elapsed since that time.

The Template Concept in Relation to the Processes of Evolution

During the past forty years, most biologists studying the processes or mechanisms of evolution have agreed that four such processes are almost equally important in determining rates and directions of the evolution of most groups of organisms. These are mutation (in the broadest sense, including gross chromosomal changes), genetic recombination, natural selection, and reproductive isolation. One might reasonably ask the question: "What effect has our recognition of biological templates had upon our knowledge and understanding of each of these processes?"

In respect to mutation, the effects are profound. Ever since the early days of De Vries and Morgan, geneticists have defined mutations as alterations in the molecular structure or the supramolecular organization of the genic substance. Until, however, the molecular structure of DNA was established through the formation and general acceptance of the Watson-Crick model, no basis existed for understanding just what kinds of changes at the molecular level could constitute mutations. Now, however, we can say that point mutations consist of various kinds of alterations of the sequence of nucleotide bases in DNA. Four kinds of alterations can be imagined: substitution of one base pair for another; inversion or reversal of the order sequence of a small number of base pairs; deletion of one or more base pairs; or similar addition of base pairs. All of these changes have been produced experimentally, either by radiation or by the action of certain chemical substances which attack particularly the DNA molecule. Nearly all of the experimentally produced mutations produce genes which function more poorly than those of the unmutated "wild type." This is, however, less true of base pair substitutions than of the other kinds.

The action of natural selection is so powerful that it eliminates from the population the bearers of all but a tiny fraction of the mutations which occur. Hence we cannot assume that the relative frequency with which the different kinds of alterations of the DNA molecule occur in the experimental production of mutations with artificial mutagenic agents bears any relationship to the relative frequency with which these same kinds of alterations occur spontaneously and are established in natural populations, thus contributing to evolution. Molecular biology has, however, found a way out of this difficulty. By means of laborious and time-consuming but accurate procedures, bio-chemists can determine the exact sequence of amino acid residues in a protein chain. When this is done for corresponding or homologous molecules in different species having various degrees of relationship to each other, a basis exists for comparative molecular biochemistry which in its methods and results is in many ways similar to the comparative anatomy of classical evolutionists. Now that we know the genetic code, we can infer from the amino acid sequence of a protein the base pair sequence of the genic DNA which codes for it. Comparisons of this nature have been made for several different kinds of protein molecules. The one most studied has been the hemoglobin of blood. The complete amino acid sequence is now known for both of the two chains of which this molecule consists in a variety of species of vertebrates. These sequences have been compared by several workers, especially Thomas Jukes, who has also reconstructed the probable base sequence of the DNA which codes for each of these hemoglobins. His comparisons show that about 90 per cent of the alterations which hemoglobin chains have undergone during the evolution of vertebrates result from substitutions of single base pairs in the DNA which codes for them. The remainder have been additions or subtractions of triplets of base pairs, which have added or subtracted one amino acid residue of the chain. Most mutations, therefore, consist of alterations of single bases, which substitute one amino acid for another in the protein chain for which the gene codes.

At a higher level of molecular order, duplications of entire genes have from time to time played very important roles in evolution. Such duplications have caused two originally identical genes to exist in tandem series, one next to the other, on the DNA double helix. Once this duplication has been accomplished, one of the two genes can mutate in such a way that its original function is impaired, since the unaltered gene can code for the necessary functional protein. The altered gene can then acquire a new function by virtue of further mutations.

If we make the logical assumption that only a small proportion of the almost infinite number of nucleotide sequences which a gene could possess are able to code for functional proteins, a necessary corollary to this assumption is that the various functional sequences must differ from each other in respect to several nucleotides. Hence in order to convert one functional sequence to another via a series of single base pair substitutions, the gene must pass through a number of non-functional states. These "valleys" between two functional "peaks" can be traversed only if the organism possesses, along with the non-functional gene, an unaltered duplicate which can perform the necessary function. Evidence that this process of duplication-differentiation has played an important role in evolution comes from the presence in the same organism of proteins which have similar but not identical functions, and which are alike in respect to a large proportion of their amino acid sequences. The best known example is that of the various chains of hemoglobin, plus myoglobin which is located in our muscles. The synthesis about our knowledge of these molecules made by Vernon Ingram shows clearly that five different genes present in human cell nuclei are descended via duplication and differentiation from a single gene which existed in the primitive ancestor of fishes.

Now that we know what mutations are like at the molecular level, we are in a position to work out definitely the kinetics of the mutation process. Although, so far as I am aware, this has not been done in a quantitative fashion, some conclusions can be reached even at the present state of our knowledge. The most important of these is that mutation is defective copying. As was explained at the beginning of this article, gene duplication does not consist of the splitting in two of a molecule which has grown to a certain size. Rather it consists of building a new copy of each of the two original helices of DNA, following the pattern provided by the existing templates. One common source of mutation is the occasional mismating between base pairs: adenine mates with cytosine rather than with thymine. If this happens during one replication, the cytosine which has thus been inserted into the "wrong" place attracts guanine at the next replication. In all later replications of this particular double helix, the guanine-cytosine pair exists at the position which formerly bore adenine-thymine.

This defective copying method of mutation is quite different from that which most geneticists imagined was the case before the method of DNA replication became known. The assumption was generally made that in order for mutation to take place, a gene molecule would have to be broken, and its parts rearranged. Such breakage and rearrangement would require extra energy. This hypothesis was supported by the fact that until the late 1940's almost the only ways in which the mutation rate could be increased artificially was by applying to the organism a form of radiation or some other violent source of energy. Now, however, mutation producing chemical substances have been found which are much more efficient than radiations, and which act in a completely different way. By introducing chemically altered bases into the replicating DNA molecule, they "fool" the molecule into making an altered copy of itself.

Now there is no more energy needed for making a defective copy than for making a faithful copy of a helix of DNA. Furthermore, some calculations of the normal tendency of hydrogen atoms in the DNA molecule to oscillate or "wobble" about the nitrogen or oxygen atoms to which they are bound led to the conclusion that given the chemical structure of the DNA double helix, defective copying might be expected to occur at rates considerably higher than observed spontaneous mutation rates. This has led to the hypothesis that the chemical structure of DNA and its surrounding molecules is such that rather high rates of mutation would occur were it not for the stabilizing or "proof reading" action of the enzymes concerned with DNA replication. Evidence that such "proof reading" exists has come from the identification in both bacteria and bacteriophage of mutations which have the effect of greatly increasing the mutation rate of most of the genes present in the organism. The one identified in bacteriophage is known to be an alteration of the gene which codes for one of the enzymes active in DNA replication.

At present, therefore, the best hypothesis which exists to explain spontaneous mutations is that their occurrence reflects a chemical instability inherent in the DNA molecule itself, which brings about defective copying. The problem which faces an organism is not, therefore, how spontaneous mutations can occur, but how they can be kept sufficiently in check so that adaptive gene combinations can be retained. In at least some organisms, this repression of the tendency to mutate is done by enzymes which catalyze DNA replication.

Our understanding of the template nature of primary gene action has forced us to recognize the possible existence of two kinds of effects of mutations which could not previously have been suspected. One of them, which is still unobserved but almost certainly exists, is exerted by mutations which, although they change the base pair sequence of DNA, have no effect upon the amino acid sequence of the protein chain for which the gene codes. Nevertheless, many such mutations can alter the future mutational possibilities of the gene affected. Such "delayed action" mutations are made possible by the synonymy or "degeneracy" of the genetic code. An example is given in Figure 1. To what extent such "delayed action" mutations have figured in bringing about evolutionary divergence cannot be estimated at present. They cannot, however, be ignored. They might contribute substantially to the ability of small, isolated populations to diverge from each other in ways completely independent of natural selection.

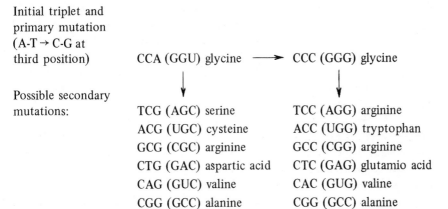

Initial triplet and
primary mutation
(A-T → C-G at
third position)   CCA (GGU) glycine ⟶ CCC (GGG) glycine

Possible secondary
mutations: TCG (AGC) serine   TCC (AGG) arginine
                        ACG (UGC) cysteine   ACC (UGG) tryptophan
                        GCG (CGC) arginine   GCC (CGG) arginine
                        CTG (GAC) aspartic acid CTC (GAG) glutamio acid
                        CAG (GUC) valine   CAC (GUG) valine
                        CGG (GCC) alanine   CGG (GCC) alanine

Figure 1. Diagram showing how a single primary mutation in the triplet codon which codes for an amino acid characterized by a high degree of degeneracy (4 different codons for the same amino acid) can be without effect except to alter the possibilities for secondary mutations. In this example, three out of the possible six secondary mutations at the first and second positions of the triplet produce divergent effects. DNA codons are placed at the left, messenger RNA in parentheses in the middle, amino acids for which they code at the right.

The second unexpected class of mutations comes from the experimental demonstration that certain genes produce primary products which have no effects other than to regulate the activity of other genes. The repressor substances first recognized by Jacques Monod and his colleagues in *Escherichia coli* are the best known examples, but the small molecule chromosomal RNA's identified by James Bonner and his associates in the chromatin of pea nuclei may also be mentioned here. The genes which code for these regulator molecules control the activity of several different enzymes. Consequently a single mutation of one of them will automatically produce as its immediate effects an alteration in several different enzyme controlled processes going on within the cell. A possible example is the lanceolate mutation in the tomato, which has been studied in detail by David Mathan. The striking effects which this mutation produces on the appearance of the plant are accompanied by a simultaneous increase in the activity of four enzymes: peroxidase, catalase, laccase, and tyrosinase. Both the morphological characteristics of lanoeolate and the increases in enzyme activity can be induced in normal tomatoes by applications of phenyl boric acid. Moreover, the antibiotic actinomycin, which inhibits the transcription of messenger RNA from the genic DNA, reduces the morphogenetic effects both of the mutant gene and of phenyl boric acid. The most logical interpretation of these facts is that "lanceolate" is a mutation of a gene which in the normal state controls by partial repression the activity of four different enzymes, and that the primary product of this mutated gene, as well as the primary product of a normal gene under the influence of phenyl boric acid, are unable to carry out the necessary regulation.

At present, we have no way of estimating the number of genes present in the normal complement of a higher animal or plant which have solely a regulatory function. Several lines of indirect evidence, however, suggest that their number may be large, perhaps in the hundreds or even thousands. I am willing to predict, therefore, that as the developmental genetics of higher organisms becomes better known, an increasing number of mutations will be found which produce abnormal individuals because they alter the rate or time at which other genes act, and so upset the harmonious interaction between genes, which is necessary for normal development.

The material just reviewed can be summarized by saying that our knowledge of genetics at the molecular level, including the template concept, has caused geneticists to recognize increasingly the overwhelming importance for normal development of harmonious interaction between many genes. As a consequence, geneticists realize that individual genes play by themselves less important roles than we ascribed to them in the past. A logical corollary to this statement is that less emphasis must now be placed upon individual gene mutations, while gene recombinations, as well as the selective pressures which establish such combinations in populations, must receive greater emphasis. Mutation, gene recombination, and natural selection are all essential for evolutionary change. Nevertheless, the rate and direction of change depend chiefly upon the factors which affect recombination and selection. Mutations are chiefly significant for maintaining a supply of genetic variability upon which recombination and selection can act.

In respect to the fourth basic process of evolution, reproductive isolation, the future contributions of molecular genetics will undoubtedly be great. Geneticists have realized for a long time that the inviability, weakness, and sterility of many interspecific hybrids are caused by disharmonious inter-actions between genes, but until normal gene action could be understood, no basis was available for exploring these interactions. Now, however, this field is being opened for investigation. One of the most dramatic ways in which this is being done is via the transplantation of nuclei from the fertilized egg of one species to the enucleated egg of a different species. These experiments, many of which have been performed by John and Betty Moore and their associates on species of frogs, have shown that the genes belonging to one species are not able to direct the normal development of the egg cytoplasm which has been synthesized in the ovary of a different species. Moreover, if a nucleus has divided several times in the cytoplasm of the wrong species, its daughter nuclei have lost their ability to direct the cytoplasm of their own species! Coupled with the recent experiments of John Gurdon, which have shown that the behavior of nuclei in developing frog eggs is influenced strongly by signals which they receive from the cytoplasm, the experiments on interspecific nuclear transplantations have shown clearly that the harmonious gene interaction which is necessary for normal development involves not only genes which are acting simultaneously at a particular stage of development, but also interaction between genes and primary or secondary products of genes which have acted at much earlier stages in the life cycle! In relation to the problem of gene interaction, and the subsidiary problem of how genic disharmony produces hybrid inviability and sterility, the time factor and the integration of successive stages of development are becoming increasingly important for our understanding of the phenomena involved.

The Template Concept and the Origin of Life

Our understanding of the molecular structure of living organisms has greatly stimulated the thinking and experiments of biochemists in respect to the origin of life on our planet. Equipped with modern knowledge, they can test the plausibility of many hypotheses by actual experiments, and on the basis of correct hypotheses may eventually be able to synthesize living systems from inorganic chemical substances. Experiments by such workers as Harold Urey, S. L. Miller, Melvin Calvin, Sidney Fox and others have demonstrated the fact that macromolecules similar to those of proteins and nucleic acids can be synthesized from simple compounds such as must have existed on the primeval earth with the aid of energy sources similar to those which must have been present. These experiments have failed, however, to answer two critically important questions. The first is: "How did organisms acquire the ability to utilize a continuous source of energy, and to convert it into a vehicle for continued growth and reproduction?" Second and even more important is the question: "What was the origin of the precise relationships between the ordered sequences of units in different macromolecules which are the basis of vital processes in all contemporary forms of life?"

Several hypothetical answers have been given to both of these questions. In respect to the first one, two divergent points of view are currently adopted. The most popular viewpoint, which can be designated the "organic soup" hypothesis, was first suggested by the late J.B.S. Haldane, and has been further developed by Norman Horowitz. It postulates the gradual accumulation of nonliving organic macromolecules in the primeval earth by the action of various more or less accidental bursts of energy, such as electric storms and volcanic eruptions, before the appearance of any living systems. When, therefore, the first self-replicating systems made their appearance, they obtained the energy needed for their metabolism by breaking down nonliving organic molecules. The use of solar radiation as a permanent source of energy for life was, according to this hypothesis, a relatively late acquisition by living organisms.

A second hypothesis has been put forward by Sam Granick, and is supported to a certain degree by J. D. Bernal as well as Melvin Calvin. According to this hypothesis, the extremely complex series of biochemical processes which now constitute the photosynthesis of green plants as well as of some bacteria, and which must certainly be regarded as having evolved long after life first appeared, was not the only method utilized by organisms for capturing radiant energy and converting it into the chemical energy necessary for life. These biochemists have postulated the existence of much simpler and less efficient mechanisms, which could have existed even in subvital molecular systems. The details of these hypotheses cannot be reviewed in a general discussion like the present one, but they seem to be reasonably plausible. They have the advantage of postulating the existence of a continuous supply of energy which could have been absorbed and utilized for building up complex systems of order even before anything resembling modern organisms had evolved.

The second question, regarding the origin of molecular order in living systems, has also been answered in two rather different ways. One group of geneticists and biochemists have assumed that DNA, which is the universal basis of genetic information in cellular organisms today, has always been the only kind of molecule to play this role. Based upon this assumption, they have postulated that life began when the first DNA molecule was constructed which could transmit functionally significant genetic information. A corollary to this postulate is that this first DNA molecule replicated without the aid of the enzymes required for the replication of all DNA molecules in modern organisms. Such replication, while chemically possible, is of necessity very slow and inefficient. Furthermore, the transcription and translation of the DNA information without the battery of enzymes which modern organisms possess for this purpose would be even more difficult. In addition, the synthesis without the aid of enzymes, of an information containing DNA molecule would require a succession of highly improbable events. Although the origin of life by this method is certainly not impossible, it involves such a long succession of highly improbable events that many biochemists and biologists have sought other methods.

Once we recognize that DNA is not the only kind of molecule which can form a replicative template, such other methods are not hard to find. The capacity for complementary hydrogen bonding exists in RNA just as it does in DNA. In tobacco mosaic virus, a double helix of RNA does, in fact, preserve and replicate its genetic information. The only advantage of DNA over RNA for this purpose is its chemical stability, which reduces the probability of the destruction of its functions via unfavorable mutations. When we assume that DNA has necessarily been the only repository of genetic information, we are tacitly accepting the assumption that the genic stability and low mutation rates possessed by modern organisms have always been essential for life. Do we need to accept this second assumption?

A number of scientists, particularly the physicist H. H. Pattee, have shown that the assumption of precise replication of complex, highly specific information is not necessary for a logical conception of the evolution of life. In modern systems, most mutations are deleterious because they interfere with a highly complex, precisely ordered system of relationships between the proteins produced by different genes. This complexity and precision is necessary for the high rates of synthesis and replication found in modern organisms. Living systems having lower levels of molecular complexity, and hence slower, less efficient methods of synthesis and replication could not survive because they could not compete with existing organisms. Nevertheless, they could have existed during the undoubtedly long period of time which elapsed between the first appearance of macromolecules and the evolution of the earliest cellular organisms similar to primitive modern bacteria. The principle that inexact copying is much less detrimental to simple systems than to complex ones can be illustrated by an analogy to man-made structures. In the construction of a log cabin or a wooden ox cart, many slight errors of judgment or measurement can be made without seriously harming the finished product, but a modern factory for making precision instruments, as well as a modern automobile, must be built according to very precise specifications, from which no deviations can be tolerated.

There is, in fact, positive evidence to suggest that RNA preceded DNA as a replicative template. No direct chemical relationships between DNA and proteins are known anywhere in the living world. Even viruses which contain only DNA must direct the synthesis of RNA and protein from host materials before they can be replicated. One could postulate that this situation arose through interpolation of RNA into a system which previously involved direct relationships between DNA and protein, but the addition of DNA to a system in which RNA-protein relationships were already well established appears to be a much more plausible hypothesis.

Even if we assume that RNA preceded DNA as the master replicative template, a basic problem still remains. This is that even a relatively short sequence of either nucleotides or amino acids can exist in an almost infinite number of combinations. Since only a small proportion of the amino acid sequences which a protein might possess can have any function at all in a living system, and since the proportion of sequences which could code for the vital function of catalyzing the replication of RNA or DNA is much smaller, the most puzzling question of all is: "How did a nucleotide sequence bearing the code for a replicative enzyme arise before the amino acid sequence necessary for replicating part or all of this information had evolved?"

We might postulate that both the nucleotide sequence bearing the necessary code and the amino acid sequence for which it was later the template arose independently in separately synthesized nucleic acid and protein molecules, and that the two different molecules "found" each other through some lucky accident. This hypothesis, however, requires a compounding of improbably events which adds up almost to impossibility.

It seems to me that the best way out of this impasse is to assume that the first relationship between nucleic acid and protein molecules was relatively unspecific in its information content. Since many enzymes, such as papain, can perform their function in a test tube on the basis of only a fraction of the amino acid units existing in the natural molecule, one can hypothesize the existence in early living or prevital systems of functional proteins which consisted of relatively few amino acid residues, and which could therefore be assembled on the basis of equally small nucleic acid templates. Moreover, such simple systems might well tolerate very high degrees of replicative inaccuracy, such as the existence at a particular position of any one of four or five different amino acids having similar properties (e.g. leucine, isoleucine, valine, alanine, phenylalanine), or the presence of a particularly active and essential amino acid (e.g. histidine or serine) at any one of four or five different positions on the chain. Systems of this kind could have become widespread in the prebiological world through slow, inefficient, and inaccurate replication because they could not be broken down by the then nonexistent degradative enzymes, or eaten by animals, which also had not yet appeared.

In my opinion, therefore, the origin of modern biological templates is best explained by an evolutionary succession not unlike that which we see in the fossil record of organisms which appeared at much later dates. The first templates, like the first organisms, were relatively small, simple in organization, slow in reproduction, and generally inefficient. Step by step, they became modified in the direction of larger size, greater complexity of organization, increase in speed of replication, and in general efficiency. The fact that the molecular organization of DNA, RNA, proteins, and the processes of replication, transcription, and translation are monotonously uniform in all cellular organisms supports the generally accepted hypothesis that the evolution of the replicative template system which now characterizes life had been completed by the time that cellular organisms first became widespread over the earth.

Biological Templates and Evolutionary Progress

One of the puzzling questions which evolutionists have faced is the following: "Granted that organic evolution has seen the successive appearance of groups of organisms having ever increasing degrees of structural or organizational complexity, how can this be explained on the basis of random mutations, random processes of genetic recombination, and random fluctuations in the external environment?" Various answers have been given to this question, most of which are quite unsatisfactory. Some biologists have suggested that greater complexity in itself can give a species an advantage in colonizing a variety of habitats, a point of view which is hard to reconcile with the ubiquitous distribution of such structurally simple organisms as bacteria and blue green algae. Others have postulated the existence of unexplained "trends" toward greater complexity, which have influenced the direction of the mutation process in a nonrandom fashion. The basis of this influence is, however, totally unexplained.

I believe that the complex system of replicative and positional templates which is apparently found in all living cells is in itself a powerful factor in guiding the course of evolution toward greater complexity. Particularly important in this connection are the interrelationships between replicative and positional templates. The latter have a distinctive feature which could not have been recognized before the facts of modern genetics became known. They determine the formation of functional structures by means of a match or fit between two complex patterns of molecular structure. These patterns exist in the primary products produced by two completely different genes. This means that a mutation of either of these genes which disrupts the pattern, unless accompanied by a corresponding mutation in the other, will produce either a poorly functioning structure or no structure at all. Because of this fact, organized structures consisting of protein molecules which fit together on the principle of positional templates will be extremely conservative and resistant to evolutionary change. The recent literature on the fine structure of cells as revealed by the electron microscope is replete with examples of this conservatism. The membrane which bounds the cell itself is a good example, since its supramolecular structure is essentially the same in all organisms from bacteria to man. Other examples are the ribosomes, organelles which support the messenger RNA; the mitochondria, which are constructed alike in all higher animals and plants; and cilia or flagella in all organisms which have them except for the bacteria.

Positional templates are, therefore, the basis for strong forces of natural selection exerted not by the external environment but by the cellular environment of the organism itself. The opinion which Lancelot L. Whyte developed as the theme of his book, *Internal Factors in Evolution*, received considerable support from this phase of molecular genetics.

The conservatism imposed by the existence of positional templates may have been the principal force which, in spite of the random nature of mutations themselves and the randomness of fluctuations in the external environment, has nevertheless driven successive groups of organisms toward levels of increasing complexity. The argument runs as follows: Mutations which interfere with the harmonious assembly of supramolecular structures or organelles will automatically be rejected by natural selection at the cell level, and will only rarely be transmitted even from one generation to the next. On the other hand, mutations which alter the number and arrangement of supramolecular structures produced by the organism during its course of development without affecting the organization of the structures themselves, are much less likely to be subject to adverse selection at the cell level, and so have a better chance of surviving. A likely hypothesis, therefore, to explain the evolution of successively greater levels of complexity is that internal selection, based upon disharmony between gene products which do not fit together as positional templates, places a floor of structural complexity below which any particular evolutionary line of organisms cannot sink. Since this floor is present, the course of evolution will be stabilized at a particular level until some combination of genes is formed which enables some of its descendants to achieve a higher level of complexity.

This argument may be extended to formulate the hypothesis that interactions between cells, tissues, organs, and in social animals between individual organisms provide similar floors which screen out degenerative mutations and make possible occasional ascents to higher levels of complexity. Discussion of this hypothesis is, however, beyond the scope of the present article.

Analogies Between Biological and Cultural Templates

An increasing number of biologists and anthropologists are beginning to realize that mankind has embarked upon a course of evolution which in many ways is quite different from the processes of organic evolution which gave rise to the various species of animals and plants. The most distinctive feature of human evolution is that whereas animals and plants became adapted by means of genetic changes in populations to environments which are basically beyond their power to control, mankind has achieved the ability to control and modify his environment to such an extent that genetic changes are no longer needed to maintain this control, except for the necessity of adaptation to new environments which he himself has created. The means by which he has achieved this control, social organization and tool using, existed already before any amount of control had been acquired. The important steps in the transition from organic to cultural evolution, therefore, were not the acquisition of the ability to make tools and to form organized societies, but the development and integration of these two abilities to an extent far surpassing that achieved by any other animal. For this development, complex learning patterns must have been necessary, and these had to be acquired and passed on by each successive generation of primitive humans or subhumans. Hence we see in the earliest evolution of man from his ape-like ancestors, the development of cultural templates superimposed upon the biological templates embodied in the DNA of his germ cells.

As the information contained in these cultural templates controlled to an ever increasing degree man's way of life, and as the information itself became increasingly complex, the adaptive value of more precise, reproducible templates increased correspondingly. The only cultural template throughout most of human evolution was the memory of individual men. Next came writing, which made possible the codification of laws and edicts, on the basis of which the first large kingdoms and empires came into being. As long as written messages could be copied only by human hands, their number was of necessity so small that literacy had to be confined to a specialized class, and democracy was well nigh impossible except in small local communities. Printing, therefore, provided the next stimulus toward the construction of new kinds of human communities on the basis of cultural templates. In very recent times, the advent of new templates, such as photography, tape recording of sound, assembly of data by computers, and spatial transmission of information by electric and electronic methods, is creating a new era in the organization of human communities. How well mankind will be able to adjust to this new era is at present an unknown problem. Whatever happens, however, the conclusion is inescapable that cultural templates will continue to dominate human life, while the biological templates embodied in DNA will be relegated to increasingly subordinate positions. In all speculations about man's future, this fact will have to be taken into account to a greater degree than it has in the past.

This paper was written while the author was a Fellow at the Center for Advanced Study in the Behavioral Sciences, Stanford, California. The author is much indebted to the Director and the personnel of the Center, who greatly facilitated the preparation of the manuscript, and to various colleagues there, for the numerous discussions which inspired much of the material presented, as well as for their critical review of part or all of the manuscript.

REFERENCES

The following books and periodicals contain most of the factual material on which the present discussion is based, as well as some discussions of the same or similar topics.

Beadle, G. and M. Beadle, *The Language of Life*. Doubleday, Garden City, N.Y. (1966).

Clowes, R., *The Structure of Life*. Penguin Press (1968).

Edgar, R. S. and W. B. Wood, "Building a bacterial virus," *Scientific American*, 217 (1):61-74 (1967).

Florkin, M. (editor), *Aspects of the Origin of Life*. Pergamon Press (1960).

Fox, S. W. (editor), *The Origins of Prebiological Systems*. Academic Press (1965).

Gross, J., "Collagen," *Scientific American*, 204(5):120-130 (1961).

Jukes, T. H., *Molecules and Evolution*. Columbia University Press (1966).

Medawar, P., "A biological retrospect," *BioScience*, 16:93-96 (1966).

Monod, J., "From enzymatic adaptations to aliosteric transition," *Science*, 154:475-483 (1966).

Moore, J. A., "Serial back-transfers of nuclei in experiments involving two species of frogs," *Developmental Biology*, 2:535-550 (1960).

Nanney, D. L., "Corticotype transmission in *Tetrahymena*," *Genetics*, 54:955-968 (1966).

Phillips, D. C., "The three-dimensional structure of an enzyme molecule," *Scientific American*, 215(5):78-90 (1966).

Schrödinger, E., *What Is Life?* Cambridge University Press (1944).

Sonneborn, T. M., "Does preformed cell structure play an essential role in cell heredity?" in *The Nature of Biological Diversity* (J. M. Allen, ed.): 165-221 (1963).

Stebbins, G. L., *Processes of Organic Evolution*. Prentice-Hall (1966).

Stebbins, G. L., *The Basis of Progressive Evolution*. University of North Carolina Press (in press).

Watson, J. D., *The Molecular Biology of the Gene*. W. A. Benjamin (1965).

Woese, C. R., *The Genetic Code: the Molecular Basis for Genetic Expression*. Harper & Row (1967).

CHAPTER VI

Evolutionary Modulation of Ribosomal RNA Synthesis in Oogenesis and Early Embryonic Development

Ronald R. Cowden
University of Denver

One of the awesome things about Biological Sciences—particularly if it is capitalized—is its incredible breadth. As matters now stand, legitimate practitioners include all levels from the molecular biologist—who may never see the critter that delivers the juice he works with—to the man concerned with whole ecosystems and communities. Philosophically, and perhaps this is an important point, it also contains some extremes in orientation: on one hand the no-holds-barred experimentalist and on the other the descriptive museum-curator mentality. If the writer were to aim one criticism about the field as it exists today, it is that all of these elements have a contribution to make in the shaping of the profession and of the young people who have announced—with admittedly differing levels of commitment and sincerity—that they wish to place themselves in our hands to be trained as professionals. It is, in fact, a matter of record that communication between the diverse elements in this fascinating field has fallen into bad repair, and that excessive concentration by a department or research group on cellular biology to the exclusion of field biology has led to some forms of tubular vision that has hampered the visualization of biological problems in their total perspective. The writer was fortunate: he attended an essentially classically oriented undergraduate school, saw the mixture of these diverse elements in Cologne and Vienna as a graduate student, and spent his first years as a faculty member in one of the "most molecular" departments in the United States.

It is with this perspective that I should like to address myself to a problem relating the molecular biology of early development to the ecological conditions under which development takes place. In keeping with the spirit of this symposium, we must agree that evolution as a concept is probably the most important generalization to emerge from the field. Well over a century's energies have been directed toward its exposition. The descriptive embryologists of the "golden age"—and by this I mean the two decades on either side of 1900—made an enormous contribution simply by grouping animals by phyla with related developmental patterns. It is among the most convincing evidence of evolutionary relationships we have today.

Having bought evolution on a grand scale, the macroevolution of the professionals, I should like to examine some of the evolutionary consequences within phyla where environmental conditions have obviously exerted selective pressure in certain directions. In the last analysis, these have affected a modulation of at least one genetic locus, the function of which is now well understood by molecular biologists. The gene or cluster of cistrons in question is the nucleolar organizer which transcribes precursors of ribosomal RNA. Thus the activation of this "gene" leads to the development of a nucleolus, a definitive visible structure. The products formed in the nucleolus eventually wind up in the cytoplasm as ribosomes, the actual sites of protein synthesis. Since the products of this gene's action is ultimately required for protein synthesis, its importance cannot be overstated. It is in comparing the function of the nucleolar organizer, both in oogenesis and early development over a broad range of organisms that some rather interesting results emerge.

It has been known for some time from experiments on as diverse organisms as hybrids of the midge, *Chironomus* (Beermann, 1960) and of anucleolate mutants of the South African clawed toad, *Xenopus laevis* (Brown and Gurdon, 1964), that the absence of nucleolar organizers is a lethal condition. The fact that the unendowed organisms survive, especially in the case of *Xenopus*, well beyond the point at which the nucleolar organizers are activated indicates that the ooplasm was sufficiently endowed with ribosomes to carry development beyond the point at which new synthesis is normally initiated.

For some reason, and the writer suspects extensive concentration on a few "favorable" species, there has been a general assumption that nucleolus formation in early embryonic development occurs in the late blastula to early gastrula stages. This is, especially if one examines the older literature, categorically untrue. Perhaps the best example is the mammal: according to Alfert (1950), Austin and Braden (1953) and more recently Mintz (1964), functional nucleoli are present by the 4-8 cell stages. While most species will go through cleavage and up to late blastula or early gastrula stages without activation of the nucleolar organizer, this is far from universal. At the opposite extreme, as will be discussed later, embryos of *Ascidia nigra* develop to swimming larvae without forming nucleoli and synthesizing new ribosomes. Thus this very fundamental and essential system is subject to modulation.

It is also time to lay a widely held fallacy to rest: that in the developing embryo, nucleolus formation takes place simultaneously in all cells at a given stage. The writer found nucleoli only in cells of the prototroch and apical tuft of the mollusc *Chiton tuberculatum* trochophores (Cowden, 1961) and only in the prospective cnidoblasts at the 5-6 hour pre-gastrula stage of the coelenterate *Pennaria tiarella* (Cowden, 1965a). Thus nucleolus initiation appears to be under control of factors in individual cells; if any generalized control mechanism is present, it is at least regional and does not affect the whole organism.

Having established these things, it becomes obvious that within each major phylum considerable evolution has taken place. Many representatives of a given group have become adapted to environmental extremes and often display considerable diversity in size and modes of reproduction. Because the writer has been interested for the past 15 years in comparative aspects of RNA and protein cytochemistry in oogenesis and early development in a wide variety of organisms, certain observations have emerged which appear at least to allow a relation of the control of ribosomal RNA synthesis to the ecology of development. To this end, I should like to present illustrations from three groups as widely diverse as coelenterates (gymnoblastic hydroids), molluscs, and ascidians. Comparisons will be drawn from species exhibiting environmental extremes.

Before going into the illustrations, perhaps it would be well to spell out the hypotheses being presented: (1) Delayed initiation of nucleolus formation—and consequently synthesis of new ribosomes—is correlated with very high levels of ooplasmic ribosomes. Conversely, low levels of ooplasmic ribosomes lead to early initiation of nucleolus formation and synthesis of new ribosomes. (2) High levels of ooplasmic ribosomal RNA and delayed nucleolus formation are associated with rapid development. The converse correlation also applies. (3) Rapid development takes place in rigorous environments while slower development occurs in protected environments. (4) With the obvious exception of the mammal, which must be treated as a special case because of placentation, slower development is in general associated with larger, more yolk laden eggs.

The first comparisons will be drawn between two gymnoblastic hydroid coelenterates: *Pennaria tiarella* (worm form) and *Eudendrium racemosum*. *Pennaria* lives only in the swiftest tidal currents and requires a high level of oxygenation. *Eudendrium* prefers more protected, less turbulent waters. In comparing the studies of Mergener (1957) on *Eudendrium* and of Cowden (1965a) on *Pennaria*, it becomes obvious that the larger *Eudendrium* embryo forms nucleoli during cleavage while this is delayed in *Pennaria* until the pre-gastrula stage. Even then nucleoli only appear in the middle layer of the pre-gastrula, the prospective cnidoblasts. The writer was fortunate to be a guest professor in Giessen (W. Germany) where Prof. Mergener was then located. We were able to compare our findings and material at length. From this it emerged that both egg size and rate of development are considerably slower in *Eudendrium* than in *Pennaria*. Because no cytochemical studies were done on *Eudendrium*, we can only guess concerning this aspect, but we do know the situation for *Pennaria*. Since egg size, rate of development, and points of nucleolus initiation in an identical pattern of early development are shown, these can be compared. To the extent that we know the parameters, the comparisons presented fit the hypotheses presented.

The second illustration is drawn from the mollusca, from as diverse groups as the primitive chiton, *Chiton tuberculatum*, to the more highly specialized pulmonate snails, *Limnaea saginalis* and *Helix pomatia*. Raven (1958) and his students, most recently Ubbels (1968), have extensively studied oogenesis and development in *Limnaea saginalis*. Contributing a necessary bit of information to this picture, Bloch and Hew (1960) demonstrated that nucleoli form in the gastrulae of *Helix aspersa*. Cowden (1961) made a cytochemical study of oogenesis and early development to the trochophore larval stage in *Chiton tuberculatum.*

Let us first examine the conditions of development. *Limnaea* lays its eggs surrounded by jelly masses on vegetation and other solid structures in ponds and streams. *Helix*, a terrestrial organism, lays its eggs in a heavy jelly envelope which later hardens on the outside. Both represent excellent examples of protected development. Further, development is direct, without an intervening larval stage. Development to a stage equivalent to the larval stage requires several days.

The situation in *Chiton* is quite different. Gametes are shed directly into the ocean at high tide. Spawning appears to depend on a lunar cycle, but when gametes can be obtained, development of *Chiton tuberculatum* proceeds to the trochophore stage at 26°C in about eleven hours. In a cytochemical study of oocyte development and early embryonic development, Cowden (1961) demonstrated that levels of cytoplasmic RNA, chiefly ribosomal RNA, in *Chiton* oocytes is extremely high. As noted earlier, embryonic development proceeds in the absence of nucleolus formation until the formation of the ectodermal ciliated structures, the prototroch and apical tuft. In these cells one finds both a small nucleolus and a conspicuous accumulation of cytoplasmic RNA. In contrast, according to Raven (1958), *Limnaea* forms nucleoli at the 24-cell stage. Both the presence of nucleoli and synthesis of cytoplasmic RNA accompany subsequent embryonic development in this organism. As has already been noted, Bloch and Hew (1960) indicated that nucleoli are present in *Helix* embryos by the gastrula stage.

Again the comparisons appear to fit the hypotheses presented. *Chiton* lives and develops rapidly in a rigorous environment while both *Limnaea* and

Helix develop under protected conditions. The eggs of the latter are larger and yolkier, and by inspection it is obvious that ooplasmic RNA levels are considerably lower (compare illustrations in Cowden, 1961; and Ubbels, 1968).

The last comparisons will be made within the Urochordata, the ascidians or tunicates, commonly called sea-squirts. As adults they bear little resemblance to chordates, but the larval forms are unquestionably chordates. They occur both as solitary forms and as colonial clusters of zooids. Within the group one encounters both oviparous (direct shedding of gametes into the ocean) and ovoviviparous (where embryonic development to the swimming larval stage takes place in the zooid) species. In ovoviviparous development, of the two species the writer has studied, *Clavelina picta* and *Ectienascidia turbinata*, development from the fertilized egg to swimming larva requires about three days. In these organisms, eggs are released individually into a brood-pouch and are released sequentially. *Ascidia nigra* is a large solitary oviparous tunicate that releases a cloud of considerably smaller gametes directly into the ocean. All three of these filter-feeding organisms attach to solid structures in swift tidal currents. It is, however, the ovoviviparous mode of development that protects embryos of *Clavelina* and *Ectienascidia*. *Ascidia nigra* requires 11-12 hours to develop from fertilization to a swimming larva.

Certain unique properties, particularly the ability to obtain both eggs and sperm from a single zooid for artificial fertilization, has led to a considerable amount of experimentation on *Ascidia nigra*. Cowden and Markert (1961) first indicated that nucleoli were not formed until larval metamorphosis is initiated. Lest there be any question about the molecular validity of this observation, unpublished autoradiographic studies of tritiated uridine incorporation, and antimetabolite studies of Smith (1967) and Markert and Cowden (1965) have confirmed the absence of ribosomal RNA synthesis in development of this organism until the larva begins to metamorphose into an adult.

In the ovoviviparous forms, a more conventional pattern of nucleolus initiation is encountered. Again, in unpublished observations by the writer, nucleoli in both *Clavelina* and *Ectienascidia* were seen to form in the late gastrula stages; synthesis of new ribosomal RNA obviously accompanies further larval histogenesis (Cowden, 1963). The same observation applies to Mancuso's (1959) studies of RNA distribution in the development of *Ciona intestinalis*.

It is also within this group that a microspectrophotometric study comparing ooplasmic RNA levels through the entire range of oocyte growth for two extreme representatives is available. Cowden (1965b) found that ooplasmic RNA concentration throughout oocyte development remained about a logarithmic order lower than in *Ascidia nigra* oocytes. The *Clavelina* egg averages about 0.24 mm. in diameter while the *Ascidia* egg has a diameter of about 0.15 mm.

Mammalian eggs offer a special example of protected conditions of development. Ooplasmic RNA levels are relatively low, and synthesis of new ribosomes is initiated between the four and eight cell stages. It accompanies all of the further stages of histogenesis. The avian embryo, which is usually a considerable cell mass at the time of laying, contains a population intensely involved in ribosome synthesis. The rate of ribosomal RNA synthesis dramatically decreases as the embryo organizes. These are examples of protected development where the larval period is completely by-passed and the emergent form consists of orders of millions of cells; in one instance support for this growth is derived from placentation, in the other from break-down of a very large accumulation of nutritive material. In both instances early initiation of ribosomal RNA synthesis is required to support protein synthetic events accompanying differentiation and histogenesis.

Returning to the principal theme of this symposium, it is obvious that as members of a particular group have come to occupy particular niches and evolve specializations, certain fundamental molecular changes are involved. The examples that have been presented to support the proposed hypotheses tend to indicate that the genetic control of ribosomal RNA production, both in the growing oocyte and early embryo, is subject to eventual modulation by environmental selective pressures.

We are still in complete ignorance concerning the events or event that activate the nucleolar organizer locus in embryonic cells. Two major possibilities must be considered: (1) that the genetic machinery is programmed to activate the nucleolar organizer at a particular stage in development regardless of other considerations; and (2) the initiation of nucleoli is under a feed-back control related to the concentration of ribosomes and protein synthesis demands of the system at any given point. The fact that an anucleolate *Xenopus* embryo can develop to the feeding stage before its condition becomes lethal would argue for the first proposition. The relationship between ooplasmic levels of RNA and the onset of nucleolus activation would suggest a feed-back control. Until the appropriate experimental information is available, these questions will remain unresolved. It is, however, significant that responses to undoubted selective pressures have arisen repeatedly in the evolution of several groups—a molecular manifestation of parallelism.

In discussing an understanding of evolution, culminating—at least in our opinion—in the evolution of man, it seems that we should make a more determined attempt to bring the diverse disciplines of biological sciences into proper communication so that its diverse capacities and philosophies could once more focus on evolution. In a sense this is a plea for the reintroduction of the comparative approach as the principal focus of biological investigations.

REFERENCES

Alfert, M., "A cytochemical study of oogenesis and cleavage in the mouse." *J. Cellul. Comp. Physiol. 36*, 381 (1950).

Austin, C. R., Nucleic acids associated with the nucleoli of living segmented rat eggs. *Exp. Cell. Res. 4*, 249 (1953).

Block, D. P., and H.C.Y. Hew, "Changes in nuclear histones during fertilization and early development in the pulmonate snail." *Helix aspersa. J. Biophys. Biochem. Cytol. 8*, 69 (1960).

Brown, D. D., and J. B. Gurdon, "Absence of ribosomal RNA synthesis in the anucleolate mutant of *Xenopus laevis.*" *Proc. Nat. Acad. Sci.* (U.S.A.) *51*, 139 (1964).

Burmann, W., "Der Nukleolus als lebenswichtiger Bestandteil des Zellkernes." *Chromosoma 11*, 262 (1960).

Cowden, R. R., "A cytochemical investigation of oogenesis and development to the swimming larval stage in the chiton." *Chiton tuberculatum L. Biol. Bull. 120*, 313 (1961).

Cowden, R. R., "RNA and protein distribution during the development of axial mesodermal structures in the ascidian." *Clavelina picta. Roux' Archiv. Entwicklungsmech. 154*, 526 (1968).

Cowden, R. R., "Cytochemical studies of embryonic development to metamorphosis in the gymnoblastic hydroid." *Pennaria tiarella. Acta Emb. Morph. Exp. 8*, 221 (1965a).

Cowden, R. R., "Cytochemical studies on cytoplasmic RNA-associated basic proteins in oocytes, somatic cells and ribosomes." *Histochemie 6*, 226 (1965b).

Cowden, R. R., and C. L. Markert, "A cytochemical study of the development of *Ascidia nigra.*" *Acta Emb. Morph. Exp. 4*, 142 (1961).

Mancuso, V., "Gli acidi nucleici nell'uovo in silvoppo di *Ciona intestinalis.*" *Acta Emb. Morph. Exp. 2*, 151 (1959).

Markert, C. O., and R. R. Cowden, "Comparative responses of regulative and determinate embryos to metabolic inhibitors." *J. Exp. Zool. 160*, 37 (1965).

Mergener, H., Die Ei—und Embryonalentwicklung von *Eudendrium rasemosum* Cavolini. *Zool. Jb.; Abt. Anat u-Ont. Tiere 76*, 63 (1957).

Mintz, B., "Synthetic processes and early development in the mammalian egg." *J. Exp. Zool. 157*, 85 (1964).

Raven, C. P., *OOgensis: The Storage of Developmental Information.* Pergamon Press, New York (1958).

Smith, K. D., "Genetic control of macromolecular synthesis during development of an ascidian: *Ascidia nigra.*" *J. Exp. Zool. 164*, 393 (1967).

Ubbels, G. A., *A cytochemical Study of Oogenesis in the Pond Snail, Limaea saginalis.* Bronder-Offset, Rotterdam (1968).

CHAPTER **VII**

Respiration as Interface Between Self and Non-Self: Historico-Biological Perspectives

Constantine J. Falliers
Children's Asthma Research Institute & Hospital, Denver

The total "surface" (the area exposed to the macroscopic physical environment) of an adult human lung is approximately 70 square meters, i.e. nearly 40 times larger than the area covered by the skin (1). With these dimensions, and in view of the enormous physicochemical activity which takes place in the bronchiolar and alveolar regions, the lung, as the instrument for gas exchange, must be considered as the major interface between the organism ("self") and the "outside world." Chemical transport mechanisms essential for the support of life, biological metabolic processes and multiple defense mechanisms combine to give us an image of respiration as the function transforming lifeless matter from the environment into an integral part of a living organism, who concomitantly returns part of its own substance to the inanimate nature. An additional important aspect of lung function (not limited to our polluted and drug-oriented era) is related to possible accidental or intentional exposures to injurious or intoxicating substances, against which the system must defend itself in order to survive.

In this presentation an attempt is made to conform to the format of the Colloquium on Biology, History and Philosophy and to elaborate selectively on the historical evolution of certain views of ventilatory function, mainly that of man. Conceptualizations of the life-giving spirit and of the conscious self are considered along this same historical perspective. A resumé of current biological data on the subject is included with a primary focus on human physiology. Emphasis is placed on concrete historical records and on specific laboratory observations. Purely philosophical speculation is mentioned very briefly, due to restrictions in the size of the text and to limitations in the author's expertise in that domain.

Distinct awareness of the relationship between breathing, individual existence and transcendental reality is vividly evident in ancient Hindu tradition. According to the Purana, when the Yoga student learns to regulate his ventilatory rhythm he begins to hear the melody of inhaling and exhaling (2). This he considers as a manifestation of the divine self settling inside him. The sound of inhalation is called *ham*, and that of exhalation *sa*. As he tries to breathe deeply in and out, the Yogi is instructed to hum the mystical syllables *Ham-Sa, Ham-Sa*, which stand for the name of the supreme divinity, symbolically represented by the holy macrocosmic gander. Miraculously, the secret of divine universal reality is then revealed to the initiate: as he continues to breathe deeply, the sequential repetition of the rhythmic sounds *ham-sa, ham-sa* discloses the ultimate truth by a reversal of the syllables to *sa ham, sa ham. Sa* in Sanskrit means "this," and *ham* means "I," so the Yogi now learns that "this (am) I." Thus, "I," the human individual of limited consciousness and existence, becomes "this," the Atman,* the divine self of infinite consciousness and existence. The world of appearances and deceiving sensations is transcended in a state of ectasy.

The student of existential experiences, who is also committed to the principles—or prejudices—of modern science, finds himself intensely fascinated by the philosophical implications of the Yogi's experiences, but also strongly tempted to explain the phenomenon in physiological terms. Undoubtedly, as long as the human mind remains unbound by the mechanistic yoke, it will pursue the effort to break through the walls of subjective existence, described with poetic detachment by Cavafy (4). Yet, the respiratory physiologist could easily tell us that the Yogi encounters no divine gander in his ectasy, but simply becomes "it," an unconscious body of flesh, by violating homeostasis, when, through voluntary hyperventilation, he reaches a state of hypocapnia and respiratory alkalosis. The neuromuscular manifestations of such a state, including tetanic carpopedal spasms, convulsions, disturbed sensorium and loss of consciousness, have been well documented clinically and experimentally.

But the voluntary control of ventilation, especially for the purpose of attaining metaphysical goals, is already a very advanced form of human behavior and of socio-cultural patterns. In a much more elementary fashion, the identification of life with breathing, and of death with apnea, or expiration (the loss of the spirit), pertains to all historical traditions and, indeed, to the beginning of homo sapiens' cognizance of himself. The breath of the Lord, Rauh Alohoim, is what, according to the biblical Genesis, makes Adam alive. And, from primitive faith to patristic philosophy, from Homer to the Bible, or from Hindu pneumatism to Hippocratic medicine, the spirit

---

*The etymological root connection between the Sanskrit *Atman* (meaning breath, soul of Life, thence the Transcendental self) and the germanic *atmen* (to breathe), Atmung (breath), etc., has been noted (3).

pervades all views of life, eventually becoming an entity—or entities— of independent existence and activity. The essential role of "winds" in living bodies, as well as in the universe, which is postulated by Hindu theories of pneumatism, evolves into the Hippocratic conceptions crystallized in the words pneuma (πνευμα)—or, for the Latins, *anima* and *spiritus* (5). The Biblical Holy Spirit shares the eternal existence of the Divine Trinity and may not be too far removed conceptually from the Homeric psyche (ψυχη), eternal soul, which abandons the body after death. Interestingly, contrasted to this psyche is *phren* (φρην), the life-spirit which is part of the body and dies with it. The word *phren*, which later became synonymous with the diaphragm, or midriff (a flat muscle essential for respiration), and was even taken to mean "mind" (cf. phrenology, the study of the mind; schizophrenia, split mind, etc.) for Homer meant the heart and parts near the heart, the breast, etc. Personality characteristics and patterns of an individual's existence and subjective perceptions were localized by Homer in the phren, perhaps both because of the sensations involved and because breathing and life seemed identical. Anatomically, however, as Sarton has aptly noted, this Homeric concept may be considered as vague an assumption as our placing of psychic attributes ("good heart," "guts") in the thorax, abdomen, etc., in modern parlance (5).

Somewhat unrelated to the recognition of the thoracic viscera as sites for subjective feeling and for vital metabolic exchanges is the identification of the lungs as distinct anatomical entities. The two cosmic forces of the Chinese, *yin* and *yang* (Pa-Kua), are thought to give rise not only to the five elements of nature, but also to the five organs of the body (5, 6). These are the spleen, the liver, the kidneys, the heart and the lungs. Although the three souls of the living are supposed to be located in the head, abdomen and feet—excluding the heart and lungs—Chinese medicine does recognize the diagnostic and prognostic significance of breath and voice: the positive element, *yang*, is reflected by a strong voice, while a weak, low-pitched voice implies a predominance of a passive element, *yin*.

Much closer to us than the mystical numerology of the Chinese is the nosologic taxonomy of the ancient Egyptians. More than 5000 years ago specific categories of disturbed respiratory function were recognized and described in sufficient detail in the Ebers and the Smith papyri (5, 6). Written early in the 16th century B.C., these hieroglyphic texts were designated by their authors as compilations of ancient knowledge of the preceding 1000 to 1500 years. In the Ebers papyrus, for example, the pulse is mentioned as a rhythmic phenomenon worthy of clinical attention 12 centuries before Hippocrates brought this sign to our attention, and, significantly, the heart is reported to be the center of the pulsation. One thousand years before Alcmaeon's pioneering biological experiments, and in contrast with many later writers, including Aristotle, the brain and not the heart is stated to be the site of sensory perception and of conscious existence. The ability of the Egyptians to outline symptoms of disturbed respiratory function, and to define diseases is truly remarkable. Cough is discussed in detail, and a list of no fewer than 21 remedies for it is included. Interestingly, asthma, a condition causing considerable concern in modern times (6, 7), is among the diseases identified with some certainty by the Egyptians. Perhaps of more than incidental interest in this respect is the fact that the first full treatise on asthma was also written in Egypt by Ben Maimon (Maimonides), physician to the Court of the Caliph of Cairo (8).

Through disease and death, of course, the significance of respiration for life becomes obvious: the living breathe, the dead do not, and ventilatory disturbances–changes in amplitude, rate and rhythm, breathlessness, cough and expectoration, chest pain, etc.–require attention and corrective measures both for the subjective discomfort they induce, and as signals of disease and danger. Strangely, however, normal respiration and cardiopulmonary function were hardly studied or interpreted in antiquity, and the modern concepts did not develop until the last two centuries.

Beyond the rudimentary biological assumptions of the ancient Asiatic and Egyptian cultures, one may justifiably stop and start with the Pythagorean Alcmaeon (6th-5th centuries B.C.) of Croton, known for his physiological experiments and dissections, who demonstrated the role of the brain as the center of sensation and of the regulation of diverse bodily functions (9, 10). It is, however, Empedocles of Akragas (5th century B.C.) famous for his extraordinary life, who is also best known for his effort to explain the function of the lungs. Describing poetically a girl playing with the clepsydra (water stealer or time-clock), he likens its mechanism to the tides of respiration. As the tide rises, the air in the body is forced out and, as it sinks, the air comes in again. Empedocles' explanation of breathing might be wrong, but, as Farrington correctly indicates (10), he did manage to demonstrate the fact that the invisible air occupied space and exerted power, thus paving the way for the atomic theory of Democritus.

The Hippocratic Corpus is rather well known and also very voluminous (6, 11) so that its detailed analysis here may be neither necessary nor possible. Nevertheless, our debt to the author or authors must be acknowledged, at least with respect to the origination of terms and rational thought processes. The medical nomenclature that still determines (or restricts?) our thinking today comes to a great extent from the Greek and Latin. Pnoe (breath) and Pneuma (a blow of air, spirit) are basic words related to pneumon (lung), as are the clinical terms pneumonia, apnea, dyspnea, etc.–not to mention the ancient science of pneumonia, That the root retains its basic meaning of air–rather than being restricted to the anatomical entity of the lung–is evident from such modern medical terms as pneumothorax, meaning air (and not lungs!) in the chest—air which has escaped from the bronchial tree and the alveoli into the pleural cavity. Although the term malaria (bad-air disease) is not Hippocratic, the "father of medicine" does pay great attention and shows appropriate concern about the influence of the environment (air, water, location, nutrition) on health and disease. Significantly, however, these extraneous influences he limits to the natural—as opposed to the super-natural—environment. For when it comes to describing the so-called "sacred disease" (epilepsy), he insists that the paroxysmal convulsive seizures pathognomonic of this condition are not caused by divine interferences any more than any other disease. He concludes that we cannot call anything sacred just because we do not understand its causes, for if we did "there would be no end to sacred things."

In this compilation of historical snapshots we can briefly pause in Alexandria in the three centuries B.C. before we leave antiquity with Galen.

In contrast with the heroic individualism of earlier Greek thinkers—Heraclitus, full of contempt for his city and its citizens; Pythagoras, fleeing the dictatorship in his native Samos and the invading Persians to establish his own mystical-political system in Croton, of southern Italy; Aristotle, pursuing his independent cogitational career despite the moves forced by changes in the political and military climate of Athens and Macedon, etc.—Alexandria, the city founded by Alexander the Great where the Nile meets the Mediterranean Sea, presents us with the impressive collective effort made possible by the generous state support of the arts and the sciences. An immense collection of erudite volumes, a succession of brilliant thinkers and systematic investigators and unique facilities for physical and biological experimentation in what was called the "Museum," provided for the learned and for the learning the ideal scientific milieu for a successive six centuries. Following the detailed and definitive anatomical observations of Herophilus, there came from Chios (via Athens?) a younger contemporary of his, Erasistratus. He was a pupil of Strato—a successor to Aristotle as Director of the Lycaeum, in Athens, and, among other, author of an incisive treatise on "pneumatics"—about whom Cicero wrote, "Strato, the physicist, was of the opinion that all divine power resides in nature, and that nature, which is a power without shape or capacity to feel, contains in itself all the causes of coming to be, of growth and of decay." Erasistratus continued his teacher's experimental approach to *physis* (nature) and is today considered as the founder of physiology. Benefitting from the anatomical discoveries of his colleague, Herophilus, he traced the subdivisions of the veins and arteries to the limits of vision and professed that they did proceed beyond what he could see. He did not, however, take the next (natural, it would now seem) step to arrive at the correct concepts of the circulation of the blood, but instead he believed that blood is contained only in the veins, while the arteries are normally filled with air. Of the numerous experiments in Alexandria, Erisistratus' demonstration that a bird in a cage, left without food, loses weight with time, can be cited an indirect evidence of biological catabolic processes in which respiration plays a major part.

Though the tradition of scholarship in the Museum of Alexandria continued, the next major developments in the study of cardio-respiratory phenomena did not take place until 450 years after Erisistratus and are connected with Galen (Galenos), the Greek born in the city of Pergamum on the Aegean coast of Asia Minor, who lived in Roman times and eventually offered his medical services to the Roman Emperor Marcus Aurelius.

Galen (131-201 A.D.) deserves our attention and respect, not only as the last brilliant light of antiquity, shining through the dark ages and on to our times, but also as a forerunner of Harvey and his great contemporaries, whom he anticipated and often superseded in his physiological views (10-13). Credit must be given to at least two forerunners of Galen: one is Empedocles—already mentioned—who supplemented his experiments with the clepsydra with a theory of respiration in which he proposed that air reaches the blood vessels through minute openings on the skin, too narrow for blood particles to get out; and the Ionian Greek philosopher, Diogenes of Apollonia (not the same as Diogenes, the Cynic philosopher), who recognized that fish absorbed air from the surrounding water, and who compared the function of the gills with that of the lungs in warm-blooded animals. Though Aristotle, in his treatise "On The Soul" appeared to misunderstand Diogenes and the function of the lungs and gills (he thought that the breath, by drawing in cool air, regulated or lowered the heat in the body, and especially the temperature of the left heart, where he assumed all heat production, as the source of all life, was taking place), Galen revived and extended the Ionian philosopher's views. Most eloquently, Galen summarizes views of many of his predecessors in a passage from his treatise "The Use of Respiration," when he says:

> What is the necessity of respiration? Is it the creation of the soul, as Asclepiades said? Certainly not its creation, but its support, as Nicarchos Praxagoras wrote? Or the cooling of the indwelling heat, as Philistion and Diocles advocated? Or its nutrition and cooling, according to Hippocrates? And if it is not so, do we breathe in order to fill our arteries with air, as Erasistratus wrote?

Galen is now certain that air (pneuma) is taken through the lungs by the adult animal (and through the placental vessels by the fetus) as it is through the gills from the water by the fish. Most importantly, this last of the great physicians of antiquity understood the functional relationship of the pulmonary blood vessels to the bronchial tubes, and he knew that the venous blood, which passes from the right heart through the pulmonary vessels into the last ventricle, becomes arterialized. He described the termination of the bronchial tubes in what we today recognize as the alveoli, and assumed the existence of invisible connections (the capillary network) between the finest branches of the pulmonary arteries and the veins. In stating that blood and air must enter the pulmonary veins simultaneously, he was emphasizing the importance of ventilation/perfusion (V̇/Q) relationships, so important in pulmonary physiology and chest medicine today. And, although Galen did not have any way of knowing that respiratory exchange involves absorption of oxygen and elimination of carbon dioxide, he did go many steps beyond his predecessors and contemporaries when he attempted to identify an element in the air which was the pneuma zotikon, the vital spirit, which was being nourished by inspiration.

With the term dynamis, Galen characterized the fundamental behavior of all living organisms (13). There were four kinds of these basic activities (dynameis), all having to do with metabolic exchanges between the living body and the environment, namely, the power to attract essential substances, to retain them, to alter them, and finally to excrete the by-product, or the nonusable portions. All these activities (anticipating modern cellular biology) were related to the function of the pneuma zotikon, or vital spirit, which is carried by the arterial blood from the left heart to the entire body. This pneuma is essential for maintaining life; it was the pneuma psychikon (the psychic, or animal spirit), which permeated the nervous system and was viewed as the carrier of nervous and mental activity.

Mankind, of course, did not stop thinking, feeling, breathing and procreating during the "Dark Ages," from the fall of Rome to the first flickers of the Renaissance. If, however, according to traditional physics, light is transmitted in a straight line, we may be justified to proceed directly from Galen to Harvey, spanning more than 1400 years, and only briefly looking also at some of his important predecessors. The great Vesalius (1514-64) may help the transition, as he revived the Galenical principles when he "wonders at the handiwork of the Almighty, by means of which the blood sweats from the right into the left ventricle through passages which escape the human vision" (11). An opponent of Vesalius, Columbo, who is mentioned as the discoverer of the pulmonary circulation (and who, of course, is not related to the explorer), did show by vivisection that the pulmonary veins contain blood and taught that, in the lungs, this becomes "spirituous" by being mixed with air. Yet, like his predecessors and contemporaries, he considered respiration as a means of refrigeration and not of combustion. The demonstration that the blood is changed from venous to arterial in the lungs was the outcome of Harvey's (1576-1657) experimental method and remains his outstanding contribution to the physiology of respiration. Beyond this major contribution there continued what Sir Clifford Allbutt called "the pathetic quest for oxygen," which, from the physician's point of view, seemed to be an attempt to answer Pepys' complaint that "it is not to this day known or concluded among physicians . . . how the action (of respiration) is managed by nature, or for what use it is" (11). Beginning with such distinguished scientists as Boyle and Hooke, and continuing with Borelli and through the end of the 18th century with Priestley, Lavoisier, Lagrange, Magnus and others, the foundations of the physiology of respiration were laid almost exclusively by physicists, chemists and mathematicians. How, within less than 100 years, the theory of the "phlogistication of dephlogisticated air" evolved into the theory of the interchange of gases in the lungs, is as spectacular an evolution of ideas as is the succession of personalities behind it. The details are part of any basic educational textbook, and their inclusion in this presentation would undoubtedly be redundant. It seems, therefore, justifiable to make a second jump which, although it spans no more than 150 years, bypasses an immensely larger body of knowledge than our previous flight through the Medieval Period. By beginning our end in the 1960's we may still find it possible to look back and trace the origins of some of our current thinking from the beginning of the "scientific revolution."

Respiratory drive *per se* has been and continues to be the subject of intensive study. Alternative pathways for neurohumoral transmission of appropriate signals, initiating and terminating a respiratory cycle of a given amplitude and frequency, have been explored. Perhaps not surprisingly, advances in the state of knowledge in both physiology and electronic data processing have made it possible, in the last decade, to extend the investigation of respiratory function beyond the breathing organism and to construct mathematical models as well as to develop programs for simulation by computer of all the multiple variables influencing or reflecting changes in respiratory patterns (14, 15).

Concepts, techniques, terms and tools have been introduced—and new ones are constantly being added—to study and test the dynamics of breathing in both health and disease. Lung volume and its subdivisions, expiratory and inspiratory flow rates, airway resistance and conductance, lung compliance and elasticity, pulmonary diffusing capacity, distribution of ventilation and lung perfusion are among the aspects of pulmonary physiology and pathology currently studied in many laboratories (1, 16). Tradition has already established spirometry as one of the fundamental procedures in the diagnostic laboratory and, despite the recent introduction of automated electronic devices with direct reading features, it does not seem that spirographs are ready to be displaced from the laboratory (17). Total body plethysmography and the forced oscillation techniques represent substantial technological advances, while they also raise again the old question of subject vs. environment. Blood gases, reflecting the adequacy or various disturbances in respiration can easily be measured in capillary or arterial samples with current micromethods, using minute electrodes and extremely small samples of blood. Feedback circuits between arterial blood gases and airway dynamics have been demonstrated (18).

Specific functions "inside" the lung, such as the regulation of contraction and relaxation of bronchial smooth muscles, the production and composition of mucus, the motility of the cilia, the passage of oxygen and carbon dioxide across the alveolar capillary membrane—and, in relation to all these, the effects of various drugs, chemicals and environmental pollutants appeared to constitute some of the most exciting and essential investigational approaches. Teleologically appealing and provocative, for instance, is the demonstration that the tone of the bronchial smooth muscle, within the range of its minute rhythmic oscillations, remains such as to maintain an optimal balance between total relaxation (which would produce an undesirable increase in "dead space," useless for respiratory exchange) and extreme contraction (which reduces dead space but increases resistance and thus makes the work of breathing unduly hard) (19). Another example of balanced "measure" (to recall the ancient Delphic maxim) is the observation that harmful effects may follow exposure of lung to both low and to high concentrations of oxygen (18). The activity of a number of cellular elements in bronchial secretions, and changes induced by diverse environmental agents also promise to be an important area of research and knowledge in the interactions between a living organism and the ecologic environment (20).

The impressive advances in the field of immunology, which took place in the last 10 years, must be considered as directly relevant to the problems of respiration, and, more specifically, to questions of self vs. non-self. From defense mechanisms against invading microorganisms to problems related to tissue transplantation, the immunologist must study how antibodies (gamma globulins) against certain foreign antigens are produced, and how, depending on whether the antigen-antibody reaction is beneficial or harmful, their production can be enhanced or abolished. Of particular interest is the fact that exocrine glands—such as the mucous glands in the bronchial mucosa—can produce substances different from what is found in the circulation. For example, autochthonous exocrine antibody, known as immunoglobulin-A, differs from the same substance in the blood or serum by being a dimer and having an attached secretory piece (21).

With so much recent emphasis on the possibility of ecological disasters, it seems impossible to conclude a presentation on respiration without considering the chemical influences on the lung, and, through the medium of respiration, on the total organism. Involuntary exposure to war gases, the use of gas chambers for extermination of entire populations, as well as the voluntary inhalation of tobacco smoke, marijuana and other substances adequately indicate how human beings may feel and act toward each other and toward themselves. Manipulations of our state of consciousness and our physiological functions are no more limited to the Yogi's *ham-sa*, to hyperventilation or to the Delphic oracle's inhalation of sacred fumes from the Earth. The health, consciousness, emotions, behavior and indeed the life of every living creature on earth can now be affected by inadvertent or planned inhalation of natural or synthetic substances, with clearly defined and sometimes overwhelming pharmacologic properties. The implications seem obvious. An individual allowing himself to be exposed to an intoxicating (in the broadest sense) aerosol or fume may be doomed to repetition of the same mistake, and eventual self-destruction, due to the induction of a self-perpetuating modification of his behavior. Society (not as a conglomerate of individuals, but as a "superior" organism itself) will find itself obligated to regulate the environment, to prevent undesirable contamination with natural, industrial or pharmacologic pollutants, and to provide the communications media or the educational system for influencing beneficially the volition of each of its members. It may be unconventional—or frightening—to consider that this willful manipulation of wills may aim not only at the preservation of the species, the prolongation of life (e.g. the anti-smoking campaign), and the enhancement of its values; current problems of overpopulation and possibly future concern about the worthlessness and misery of many lives may also induce society to devise methods to end—rather than prolong—life. Institutions that could be called "thanatoria" (i.e. places of death) may emerge as the natural extensions of "birth control" clinics. How would we react today, or twenty years from now, to the possibility of terminating the existence of millions electively, painlessly and even pleasurably through inhalation of lethal gases? We know it can be done because it has been done! Is then the loss of individual "selves" compensated by the glorification of the transcendental Atman, be that the concept of a few philosophers or the object of the veneration of an entire society? Not an inconceivable perspective, to be alleviated only by one hope: that the "Breath of the Lord," or the Holy Spirit, may continue to provide not only Life, but Wisdom.

REFERENCES

1. Comroe, J. H., *Physiology of Respiration*. Chicago: Year Book Medical Publishers, 1965.
2. Zimmer, H., *Philosophies of India*. Bollinger Series XXVI. New York: Pantheon Books, Inc., 1951.
3. Buck, C. D., *A Dictionary of Selected Synonyms in the Principal Indo-European Languages* (A Contribution to the History of Ideas). Chicago: University of Chicago Press, 1965.
4. Keeley, E., and P. Sherrard, *Six Poets of Modern Greece*. New York: A. A. Knopf, Inc., 1968.

5. Sarton, G., *A History of Science: Ancient Science Through the Golden Age of Greece.* Cambridge: Harvard University Press, 1966.
6. Major, R. H., *A History of Medicine.* Springfield: C. C. Thomas Co., 1954.
7. El Mehairy, M. M., "From History: 'Pharonic Medicine' and Allergy." Abstract No. 127, VII International Congress of Allergology. Excerpta Medica No. 211, Amsterdam, 1970.
8. Muntner, S., "Maimonides' treatise on asthma." *Dis. Chest* 54:48-52, 1968.
9. Kirk, G. S., and J. E. Raven, *The Presocratic Philosophers.* Cambridge: Cambridge University Press, 1966.
10. Farrington, B., *Greek Science.* Baltimore: Penguin Books, 1966.
11. Garrison, F. H., *An Introduction to the History of Medicine.* 4th ed. Philadelphia: W. B. Saunders Co., 1968.
12. Sarton, G., *A History of Science: Hellenistic Science and Culture in the Last Three Centuries B.C.* Cambridge: Harvard University Press, 1959.
13. Siegel, R. E., *Galen's System of Physiology and Medicine.* Basel: S. Karger, 1968.
14. Grodins, F. S., *Control Theory and Biological Systems.* New York: Columbia University Press, 1963.
15. Yamamoto, W. S., and J. R. Brobeck, *Physiological Controls and Regulations.* Philadelphia: W. B. Saunders Co., 1965.
16. Fenn, W. O., and H. Rahn, *Handbook of Physiology.* Section 3, Respiration, Vols. I and II. Washington, D.C.: American Physiology Society, 1965.
17. Falliers, C. J., J. W. Bates, and E. Berman, "Computer records of ventilatory function in asthma." *Ann. Allerg.* (in press).
18. *Proceedings of the 10th Aspen Emphysema Conference, Aspen, Colorado, June 7-10, 1967.* "Current research in chronic obstructive lung disease." U.S. Public Health Service Publication No. 1787. Washington, D.C., 1968.
19. Falliers, C. J., "Asthma and cybernetics." *J. Allerg.* 38-264-267, 1966.
20. Gell, P.G.H., and R.R.A. Coombs, *Clinical Aspects of Immunology.* Oxford: Blackwell Scientific Publications, 1968.
21. Statement by the American Thoracic Society Committee on Air Pollution: Air pollution and health. "Respiratory Diseases."

CHAPTER VIII

Measurement Theory and Biology

Martin A. Garstens

University of Maryland and the Office of Naval Research

Introduction

Although this gathering is concerned with many aspects of Biology, including the philosophical and the historical, I think it fair to say that the central and crucial problem we are trying to face is whether some general theory, outlook or perspective can be attained which will win the following of legitimate, practicing biologists in the sense of being fruitfully useful and acceptable to them in an operational way. Such an outlook has certainly been lacking in the past. It is my belief that a discussion of the measurement process and its relation with the basic problems in statistical mechanics can suggest the directions of such a perspective and theory. Biological theory must in some sense be a mixture of the macroscopic and the microscopic, where statistical mechanics of great subtlety and complexity plays an important role.

While measurement is an old activity in physics, the process of measuring atomic observables is still not clearly understood. In biology particularly, excluding strictly physical measurements, the term appears to have a special and frequently diffuse meaning. The clarification of its meaning can be a clue as to why a distinctive domain of theoretical biology has thus far not developed. The complexity of the measurement process arises because it links two separate domains. One, the observed atomic domain describable by means of quantum theory, the other classical macroscopic theory, applicable to the measuring instrument itself and to macroscopic events and variables in general. Statistical mechanics attempts to bridge these domains and measurement theory is a part of this field.

Measurement and Statistical Mechanics

The problem of measurement has two central aspects to it, both important for biological considerations. On the one hand in observing a microscopic or molecular quantity one inevitably, for quantum-mechanical reasons, perturbs the observable in such a way that the wave function describing it undergoes what is referred to as a "reduction in wave packet." It is here that statistical mechanics enters the picture. The measuring instrument is a macroscopic entity and interaction between it and the observable must ultimately cause some recognizable macroscopic change in it if it is to be a measurement. This change and the associated wave packet collapse must be explained by means of statistical mechanics. On the other hand, the foundations of statistical mechanics themselves remain unresolved. Its connection with the more established portions of physics is not at all settled. While the basic assumptions of quantum theory, applicable to the individual atoms or molecules which make up solids, have won widespread agreement and consensus, the same is not true of the statistical mechanical procedures which attempt to use quantum theory in deriving macroscopic variables and laws.

       To raise the question of the adequacy of quantum theory itself unnecessarily complicates issues which can be clarified without doing so. Therefore we will not inject this aspect of the problem into our discussion.

Statistical Mechanics

       It is important to realize that in spite of valiant efforts to do so statistical mechanics, the science of complex systems, has never been reduced to, or deduced from, the quantum theory of atomic systems. In all cases essential auxiliary hypotheses have been found necessary in order to derive known macroscopic variables and laws. The most consistent and persistent attempt to dispense with such auxiliary hypotheses is known as ergodic theory. It has not been blessed with much success. Thus Farquar ends his recent treatise on *Ergodic Theory in Statistical Mechanics*[1] with the sentence: "Although the various attempts at solution have shed much light on the problem the current verdict on ergodic theory in its relation to statistical mechanics must be given as not proved." Tolman in his *Principles of Statistical Mechanics*[2] states: "The necessity for an additional postulate arises not from any incompleteness or inexactness in the principles of classical or quantum mechanics when applied to conceptual situations for which the discipline was immediately devised, but from the real extension in theory which is needed when we are confronted by incompleteness and interactions in the specification of states for the system we wish to treat."

       Again, N. G. van Kampen in discussing irreversible processes says:[3] "there cannot be a rigorous mathematical derivation of the macroscopic equations from the microscopic ones. Some additional information or assumption is indispensable. One cannot escape from the fact by any amount of mathematical funambulism.[4] . . . Apart from some very special simple models, it is impossible to solve the microscopic equations." . . . Statistical mechanics, he continues, "should be regarded as a useful approximation method, applicable to systems with many degrees of freedom. The nature of this approximation, however, is of an unusual kind. It is not a numerical approximation to the true value of precisely defined quantities like the approximations used in celestial mechanics. On the contrary, the macroscopic quantities one wants to compute are themselves only defined in the same approximate sense, i.e. in the limiting case of many degrees of freedom. This approximate nature of the macroscopic quantities shows up in the existence of fluctuations."

Finally, in a paper on statistical mechanics, G. Ludwig[5] referring to macroscopic equations states: "All these equations have the common property that they cannot be deduced from the mechanical equations of motion, but one has to introduce more or less plausible assumptions to get them" . . . "such a deduction on the basis of quantum mechanics alone is impossible" . . . in finding these new additional postulates "we have a principle problem and not a problem of the application of well known theories" . . . amongst these new principles, he continues, "are contained the principles of living systems too. The task to find such a theory is not a hopeless one."

To get their flavor we list a few of the auxiliary assumptions which have played important roles in various formulations of statistical mechanics.

1. The postulate of equal *a priori* probability. In its classical version it says that when a macroscopic system is in thermodynamic equilibrium, its state is equally likely to be any state satisfying the macroscopic conditions of the system. In its quantum form it says that if $\psi = \sum_n C_n \Phi_n$ represents an isolated macroscopic system expressed in terms of a complete orthonormal set of stationary wave functions $\{\Phi_n\}$ and if the energy of the system lies between E and $E + \Delta$, $(\Delta \ll E)$ then the average:

$$\overline{(C_n, C_n)} = \begin{cases} 1 & \text{if} \ (E < E_n < E + \Delta) \\ 0 & \text{otherwise,} \end{cases}$$

where E_n is the eigenvalue of the Hamiltonian of the system for the nth state.

2. The postulate of random phases states with respect to the same system that the average:

$$\overline{(C_n, C_m)} = 0 \quad (\text{if } n \neq m).$$

The latter assumption enters into discussions of measurement.

3. The ergodic hypothesis states that time averages under equilibrium conditions can be replaced by space averages in phase space.

4. Master equation assumptions which concern the time development of quantities $P_n(t)$ which are the probabilities that at the instant t the system is in the state n. To justify statistical mechanics it must be shown that $P_n(t)$ approaches $\overline{(C_n, C_n)}$ when the time t is longer than say the molecular collision time. The equation is

$$\frac{dP_n(t)}{dt} = \sum_m [W_{nm} P_m(t) - W_{mn} P_n(t)]$$

with W_{nm} the transition probability per second from state n to m. Solutions are obtained when $t \to \infty$. Here it is the master equation which is assumed and must be justified. The equation is particularly interesting because it is at a level intermediate between the microscopic and the macroscopic equations. It is therefore relevant to the measurement process and also for such biological processes which are sensitive to microscopic disturbances.

5. The metric transitivity assumption characterizes a mechanical system where any energy surface can be divided into two regions such that trajectories starting in one always remain there.

6. Replacing the assumption of equal *a priori* probabilities and avoiding use of ergodic theory and representative ensembles, Jaynes and much earlier Elsasser have introduced the hypothesis of maximum entropy. In this procedure, statistical mechanics comes closer to being a technique of statistical inference than a physical theory. One obtains the same representative ensembles produced by other methods and a greater degree of flexibility in possible empirical applications, particularly in biology.

This list can be greatly extended and strongly suggests that in an expanded and generalized statistical mechanics, by addition to or modification of these auxiliary postulates, one has the foundation of a science of complex systems capable of dealing with biology. From this point of view the controversy about reducibility in biology can be resolved. The need for auxiliary hypotheses indicates non-reducibility to quantum or classical mechanical theory. But as the required base expands and new macroscopic variables are defined, biology could increasingly fall within its domain exhibiting the role of the microscopic as well as the macroscopic in living organisms.

The nature of the measurement process becomes more general from this expanded point of view since the macroscopic process triggered in a measuring instrument by microscopic events takes a different course depending on the particular statistical mechanical description which is adopted. The collapse of the wave packet can be different in biological systems than in simpler physical ones.

The Measurement Process

The measurement process in atomic physics is concerned with showing that a set of macroscopic states described by wave functions ψ_r in a measuring instrument are uniquely associated, by interaction, with corresponding states ϕ_r of an atomic system. The initial state of the system is given by

$$\Phi_0 = \Sigma_r\, C_r\, \phi_r$$

normalized by the condition:

$$\Sigma_r |C_r|^2 = 1.$$

If a_r represent the eigenvalues of the states ϕ_r (i.e. $H_A\phi_r = a_r\phi_r$ with H_A the Hamiltonian of the atomic system) then the probability P_r of observing a_r is given by:

$$P_r = |C_r|^2.$$

The transition from the superposition Φ_0 to the observed state ϕ_r which takes place as a result of the measurement is a perturbation of Φ_0 reducing it to ϕ_r, with probability P_r of being so reduced. This is the wave packet collapse caused by interaction with the macroscopic instrument. The details of what takes place demands explanation. Evidently this reduction to ϕ_r must be associated with the formation of an associated macroscopic ψ_r designating a reading of the instrument. We will leave for the appendix a detailed objective description using statistical mechanics, of how this can be thought to take place. The term objective is used since there are some subjective accounts of the process which suggest that a measurement is not accomplished until conscious recognition of the result occurs. In fact, it is maintained that the final reduction of the wave function actually occurs during such recognition. The analysis which leads to this point of view attempts to show that the reaction of the measuring instrument to an observable defined by

$$\Phi_0 = \Sigma_r\, C_r\, \phi_r$$

results in a state of the instrument given not by an individual ψ_r but, as the equations of motion of the combined system would seem to suggest, a superposition of ψ_r in the same form as above, i.e.:

$$\Psi = \sum_r g_r\, \psi_r.$$

The measurement is therefore not completed. Another measurement on Ψ is now needed, etc. Presumably the final observation by the observer would then terminate the measurement. This subjective point of view we believe to be erroneous. It is possible, as indicated in the appendix, to sketch a satisfactory objective theory where the atomic process triggers a macroscopic avalanche leading on the one hand to a final unique state ψ_r corresponding to ϕ_r and on the other hand, having its initial wave function Φ_0 describing the observable perturbed so that it collapses to some ϕ_r.

It is not the contention here that the biologic state is adequately represented by Φ_0 nor that microscopic processes play a dominant role. This is something for future analysis to reveal. There can be little doubt that the great stability in living organisms is due to its dependence on stable macroscopic processes which are not perturbed by measurements. On the other hand, particularly in the higher forms of organisms, there are present fluctuations and sensitivities in behavior and thought, suggestive of microscopic triggering of action where analyses of the type outlined above may be relevant.

Any attempts to measure these subtle and perhaps more basic microscopic variables must result in perturbations of the type discussed above, leading to reduction of the wave function representing the observable; a reduction which in a sense is a destruction, something Bohr pointed out a long time ago. How does one then measure such quantities? Can one expect some systematic understanding of them?

It seems clear that data gathered on such microscopic observables will have a distinctly statistical flavor because of the perturbations involved. But an expanded statistical mechanics can be expected to define suitable macroscopic quantities, subject to fluctuation and linked in a phenomenological theory. While the structuring of a statistical mechanics suitable to describe living organisms is a monumental task there is no reason why initial progress cannot be made.

The method of Jaynes seems particularly inviting since it appears to have great power and generality. It substitutes a statistics and a purely inductive method for representing ensembles, for the special assumptions of the type we have listed. To be inductive implies a maximum in generality since by definition any uniformity may be approached by a sampling process. In practice we still require a judicious guess as to what the uniformity might be. Like any statistical theory the method should, in biology, predict characteristics held in common by all living systems. The procedure used is to set an experimentally observed physical quantity equal to an operator representing it. The latter in turn can be shown to be equal to a matrix involving both the operator in question and the density matrix ρ of the system which is to be determined. Subject to these constraints the entropy which is a function of ρ alone is maximized and ρ is finally determined. Knowing the density matrix ρ, additional quantities can be predicted using an equation similar to the one mentioned above with new operators substituted for the old.

While the procedure is clear, a straightforward application of the method seems to be hopelessly complex involving great mathematical complication and requiring enormous numbers of parameters.

However, like all important advances in science, judicious insight and deep knowledge of biological systems should allow a selection of key variables playing a central role. There seems to be a kind of simplicity in the complexity of biological organisms which may suggest the initial variables to be chosen. Just as in the investigation of complex solids one can to some degree separate magnetic aspects from those of sound propagation to a first approximation, so similar separations may be possible here.

As the important variables are gathered, their interplay including such phenomena as fluctuations, cycles and feedback enter the picture. As insight deepens, additional variables would be required. It is conceivable and probable that a true understanding of a living system could demand an endless series of such variables.

A growth similar to that in thermodynamics can be imagined, with many more macroscopic variables and in more complex relationship. Particularly would this be true in describing the function of the brain. Here a prodigious number of macroscopic variables would seem to be necessary, easily triggered by microscopic sub-events. Fluctuation must play an important role corresponding possibly to a play of imagination, while more stable quantities could be associated with memory, habits, etc. Further it is even imaginable that there is a basic fluidity in brain structure which allows for changes in the statistical mechanism and hence in the macroscopic variables.

We have described a way of looking at biology which is in the tradition of verifiable empirical science and where the required auxiliary principles need not be reducible to quantum or classical physics. The enrichment in foundation, in the style already in practice in statistical mechanics, and the consequent delineation of new macroscopic parameters suggests an increased power in reaching an understanding of biology, hitherto unattained.

APPENDIX

The problem of measurement in modern atomic physics involves showing how a microscopic variable manifests its state in a macroscopic form. It seems clear that all knowledge of atomic states arises from "extrapolating" back from the observed macroscopic variable, which constitutes the measurement, to the atomic state itself. Quantum theory has arisen from such an extrapolation. The measurement problem then consists in using the quantum theory thus found, plus auxiliary hypotheses needed to set up an adequate irreversible statistical mechanics, to trace the process in reverse.

From the point of view of irreversible statistical mechanics the atomic event triggers an atomic avalanche culminating in a final unique equilibrium macroscopic state which corresponds with the triggering event. The process is of biological interest since many biological processes are triggered in a similar fashion. A complete mathematical solution of the process would consist in describing the details of the triggering mechanism, the resulting avalanche and the final equilibrium macroscopic state. Unfortunately it is difficult to attain this rigorous solution in a general form. Instead, by making plausible simplifying assumptions as to the course of the avalanche, an objective description of the measuring process is made possible, including an explanation of "the reduction of the wave packet." Indeed, to demonstrate the non-subjective nature of measurement one need not rigorously solve the difficult statistical mechanical problem which is involved. One need merely show, if only in a semi-quantitative form, that a one-to-one correspondence between the atomic and the macroscopic variables can be expected and that "the reduction of the wave packet" can be explained. The method we here use is based on that of Daneri, Loinger, Prosperi and Rosenfeld.[6]

Consider an atomic system in a state given by:

$$\psi_0 = \Sigma_s C_s \phi_s \tag{1}$$

with

$$\underset{r}{\Sigma}|C_s|^2 = 1. \tag{2}$$

For convenience we consider the measurement of two quantities: A at time t = 0 and B at time t. a_s, ϕ_s and b_r, χ_r can represent the eigenvalues and eigenstates of the two respectively. The probability Ps, of observing the eigenvalue a_s in the state ϕ_s at t = 0 is given by:

$$P_s = |C_s|^2. \tag{3}$$

The transition from the superposition (1) to the single state which presumably takes place as a result of the measurement *is* "the reduction of the wave packet." To see what is involved more clearly we follow the measurement of A at t = 0 with one of B at time t.

If H' represents the Hamiltonian of the atomic system, ϕ_s will have become:

$$\phi_s' = \exp\left[-\frac{iH't}{\hbar}\right]\phi_s \equiv \exp'(t)\,\phi_s \tag{4}$$

at time t. We may expand ϕ_s' as:

$$\phi_s' = \underset{r}{\Sigma}\, d_{sr}\, \chi_r. \tag{5}$$

The probability for observing b_r in an atom originally in the state ϕ_s is:

$$|C_s|^2 |d_{sr}|^2 = |C_s|^2 |\chi_r, \phi_s'|^2 = |C_s|^2 |\chi_r, \exp'(t)\phi_s|^2. \tag{6}$$

Therefore the total probability of finding an atom in the state ϕ_r is:

$$\sum_s |C_s|^2 |\chi_r, \exp'(t)\phi_s|^2. \tag{7}$$

If no measurement were made at t = 0 we would have at time t:

$$\exp'(t)\psi_0 = \sum_s C_s \phi_s' = \sum_s C_s \sum_r d_{sr}\chi_r = \sum_r (\sum_s C_s d_{sr})\chi_r.$$

The probability of finding b_r is:

$$|\sum_s C_s(\chi_r, \exp'(t)\phi_s)|^2 \tag{8}$$

which differs from (7) by the amount

$$\sum C_s C_s^* (\chi_r, \exp'(t)\phi_s) (\chi_r, \exp'(t)\phi_s')^*, \tag{9}$$

a correlation term. Its disappearance gives rise to $|C_s|^2$ in (7), the phases of C_s having been washed out. It is here that the reduction of the wave packet takes place. How does this happen?

If again at t = 0 A is measured, the total system, atoms and measuring device, at time t can be expressed in the form:

$$\sum C_s \exp'(t)\phi_s \exp''(t)\psi_s$$

where

$$\exp''(t) = \exp[-i\frac{H''t}{\hbar}]$$

and H'' is the Hamiltonian of the measuring instrument.

Let $\Omega_{\varrho j}$ represent the initial state of a body in phase space. Then $\exp''$ $(t)\Omega_{\varrho j}$ will be its state at some t later. $\Omega_{\varrho j}$, called a "cell," is a subdivision of a "shell" in phase space between the energy intervals E and (E+ΔE). Each cell contains a manifold of $S\varrho$ orthogonal states (j = 1,2,....$S\varrho$).

If now a measurement on B is carried out we ask what the probability is of finding the eigenvalues b_r while the instrument is in one of the states of the cell Ωki. As in (8) we can set

$$\sum_i |\sum_s C_s(\chi_r \Omega_{ki}, \exp'(t)\phi_s \exp''(t)\psi_s|^2$$

equal to this probability for all cells

$$\Omega_{ki} \ (i = 1,2,....S_k),$$

or if we set

$$\sum_i (\Omega_{ki}, \exp''(t)\,\psi_s)\,(\Omega_{ki}, \exp''(t)\,\psi_s')^* \ \equiv \ A_{k,ss'}; \tag{10}$$

we can express the probability in the form:

$$\sum_{s,s'} C_s C_s'^* (\chi_r, \exp'(t)\,\phi_s)\,(\chi_r, \exp'(t)\,\phi_s')^* A_{k,ss'}. \tag{11}$$

Here, by means of the term $A_{k,ss'}$ (equal to a small quantity or zero if $s \neq s'$) we have separated out the portion defining the role of the measuring instrument. Some assumption must be made as to the behavior of the latter. To do this two measures of "ergodicity" are introduced. These insure the statistical validity of the argument. If σ is the total number of states in a shell and σ_k that in a cell, then we first assume that the statistical weight of the cell is $\frac{\sigma_k}{\sigma}$. This can be represented as:

$$\sum_i |\,\Omega_{ki}, E''(t)\,\Omega_{\ell j}\,|^2 \ \approx \ \frac{\sigma_k}{\sigma}. \tag{12}$$

Secondly we assume that the statistical weight of the correlation between probability amplitudes of different states will be much smaller (or zero), i.e.: The statistical weight of

$$\left\{\sum_i (\Omega_{k\ell}, \exp''(t)\,\Omega_{\ell j})\,(\Omega_{ki}, \exp''(t)\,\Omega_{\ell'j'})^*\right\}_{\ell \neq \ell'} \ll \frac{\sigma_k}{\sigma}.$$

Expanding the macroscopic wave function ψ_s in term of Ω_{ki}^s where the Ω^s, without overlap, represent all cells which correspond to ψ_s, we get:

$$\psi_s \ = \ \sum_k \sum_{i=1}^{S_k} (\Omega_{ki}^{(s)}, \psi_s)\,\Omega_{ki}^{(s)},$$

so that finally the statistical weight:

$$A_{k,ss'} \ \approx \ \delta_{ss'} \frac{\sigma_k}{\sigma}$$

from which:

$$A_{k,ss} \ \approx \ \frac{\sigma_k}{\sigma}.$$

These expressions merely say that the correlation vanishes between different states when equilibrium is reached. Using this evaluation of $A_{k,ss'}$, in (11) we obtain:

$$\sum_s |C_s|^2 \, |\chi_r, \exp'(t)\phi_s)|^2 \tag{13}$$

in which the correlation term has been eliminated and the reason for the occurrence of the reduction of the state ψ_0 is manifested.

In summary, the above shows, what we know from experience with measurements, that each atomic event "which can be measured," causes a channeled avalanche of processes leading to a unique macroscopic outcome. The channeling and uniqueness in the above analysis is introduced by use of the notion of cells and by the defined properties of the weighting factor $A_{k,ss'}$. Similar channelings, each resulting in a unique outcome, must occur in biological systems. The discovery and tracing of these channels is the aim of theoretical biology.

NOTES

[1] I. E. Farquar, *Ergodic Theory in Statistical Mechanics*, Interscience Publishers, 1964.

[2] R. C. Tolman, *Principles of Statistical Mechanics*, Oxford, p. 64, 1938.

[3] N. G. van Kampen, *Fundamental Problems in Statistical Mechanics*, Edited by E.G.D. Cohen, pp. 173-174, 1962.

[4] Tight rope walking.

[5] G. Ludwig, "Axiomatic Quantum Statistics of Macroscopic Systems," in *Ergodic Theories*, Academic Press, pp. 57, 59, 1961.

[6] A. Daneri, A. Loinger, G. M. Prosperi, *Nuclear Physics 33*, 297, 1962, L. Rosenfeld, Nordita 1965.

CHAPTER **IX**

The Transition from Theoretical Physics into Theoretical Biology

Walter M. Elsasser
Institute for Fluid Dynamics and Applied Mathematics
University of Maryland

Introduction

The problem of our title is not usually considered as purely scientific, but as pertaining largely to philosophy and metaphysics. This means that the involvement of the scientist is partly emotional and partly rational. The need of the investigator to control or counteract his own, often unconscious tendencies is the most outstanding feature of our subject. On recognizing this, it becomes indispenable to find a purely formal criterion to which one can refer time and again to keep the intrusion of such nonrational tendencies in check.

The general idea on which I have based myself in my many years' preoccupation with the subject is this: from the viewpoint of the physicist (which I am) the science of biology may be thought of as the realm of the *utterly complex*. Physics has very successfully penetrated into the realm of the exceedingly large, at first, and later on into the realm of the exceedingly small. In either case there appeared the need for a thorough revision of our modes of thought and description, of our conceptual and mathematical apparatus. Is it then necessary to look at our entire scientific methodology again, if we go from simple systems in the laboratory to the utterly complex ones in the organism? And will such an analysis help us in turning the question of the relation of living to dead matter into an anlytical one which can be taken out of the hands of ideological parties? Both questions, I strongly believe, are to be answered in the affirmative.

The effort at penetrating this problem starts out in physics pure and simple and ends in biology. This is essential, but the implications are not usually understood: In formulating pertinent abstract problems of physics it is important that no tacit assumptions creep in which seem "self-evident" to the physicist because he is habituated to work with systems of relatively simple structure, but which are not at all obvious when one comes to deal with the utter structural complexity found in organic matter. To ferret out such implicit assumptions has been my endeavor for many years. I have published much on this, but acquaintance with my earlier work is not required for an understanding of the present paper. In order to keep the paper short, I have sketched the mathematical apparatus only in the briefest terms since it is readily available in the literature (and will not offer any serious difficulty to anybody who is somewhat acquainted with modern atomic physics). This has enabled me to present my main argument in a coherent form which shows rather clearly, I hope, where the problem now stands, and where potentialities for future development may be found.

Generalized Complementarity

We begin by assuming that the tremendous structural and dynamical complexity of organisms, so apparent everywhere in the observations, can be taken as a sufficient starting point for adapting more conventional physical analysis to the study of the organism. Now if we want to deal with complex systems on a purely physical basis (and complexity is of course a purely physical criterion) we find ourselves at once confronted with the set of ideas that Niels Bohr enunciated long ago (1933) and designated as "generalized complementarity." His terminology is unfortunate because it seems to imply some mathematical or logical generalization (i.e. modification of rules, in an explicit sense), but this is not so. The generalization consists, instead, solely in a *broader application* of the concept of quantum-mechanical complementarity, that goes beyond the standard examples of the textbooks. The latter are practically always restricted to single particles, one electron, one light quantum, etc. The broader application is now to *systems* of particles with no specific limit as to their size. While this involves no innovation in the mathematical machinery, such more general applications touch deeply upon the physical interpretation of quantum mechanics, as we shall see.

Note next how the classical concept of measurement is modified by the transition to quantum theory. The concept of measurement carries with it certain implied assumptions which are rarely if ever stated, because they are considered too "obvious." The measurement is an interaction between two bodies, the measured object and the measuring instrument. We must here assume that the measured object is not significantly influenced or altered by the process of measurement. On the other hand, the measuring instrument must be so altered if there is to be an explicit record of the measurement. Thus in classical physics there exists a profound asymmetry in the behavior of the two systems whose interaction constitutes the measurement. As we move toward smaller and smaller systems, ultimately to elementary particles, the nature of quantum theory itself forces a certain symmetry upon the process of interaction. The position of an electron is measured, say, by illumination with electromagnetic waves of sufficiently short wave length, that is, x-rays. The process may be alternatively described as a collision between the electron and an x-ray photon; in this description there is fully apparent a symmetry which violates the requirement that the measured object should not be sensibly affected by the measurement. It appears that the usual concept of a measurement cannot in a straightforward manner be transferred from macroscopic to microscopic phenomena; a more extensive analysis is needed.

This analysis has been started by Bohr who, in his later writings, has emphasized at many places that one cannot really speak of a measurement unless some imprint in the macroscopic world has been effected. The idea, so evident in its simplicity, need not be confined to measuring processes in the strict sense. Quite generally, processes in the atomic (microscopic) realm become accessible to us only by way of producing an effect in the macroscopic world which in turn makes an impression on our senses or else can once more be measured in the "classical" meaning of the term. Conversely, all quantities that are sufficiently above the level of the uncertainty relations so that they may become objects of measurement in the classical sense, can be derived from atomistic quantities by averaging processes. This suggests a formal aspect which we shall call *logical relativity*: We do not attribute a physical meaning to symbols representing atomic or molecular variables *except* when they appear related to a *macro*scopic system which contains variables that can be measured in the classical sense. This idea, which will be elaborated below, allows us to overcome the epistemological quandaries that have bedeviled the understanding of the measuring process in quantum mechanics, and leads to a systematic application of the statistical methods of quantum theory to the dynamics of organisms.

Let us return now to the question of general (in Bohr's term, generalized) complementarity. The uncertainty relations of quantum mechanics tell us that if we want to make a precise determination of the position of a particle, the spectrum of its momentum becomes very broad. This is usually interpreted by saying that pairs of conjugate variables cannot be measured with precision simultaneously; it has the additional consequence that if we want to confine the position within sufficiently narrow limits, the momentum spectrum becomes broad enough so that the expectation value of the kinetic energy of the particle will in general increase during the measurement. At this point Bohr's more general aspect of complementarity makes its appearance: Suppose we are dealing with an organism considered as a highly complicated chemical structure. If we want to make predictions for a sufficiently long time ahead it is necessary to know the structure of the system at some initial time $t = 0$, with a high degree of precision. This knowledge can be thought of as derived from a large number of simultaneous measurements carried out on many of the particles, electrons and nuclei, that constitute the system. If a sufficient accuracy in the *positions* of these particles at $t = 0$ is required, the energy communicated to them will in the average exceed the energy of the chemical bonds that keep the system in its state. Many of these bonds will be broken; the influence of the measurements upon the structure of the system is *destructive*. It will be the more so the more accurate the measurements required.

Bohr expresses this by saying that the measurements hamper the vital functions of the system; ultimately, when the influence of the measurements becomes too pervasive the organism will die, and the measurements have become worthless for biological purposes. Thus, if we are not to allow too severe an interference with the functioning of the system, we are denied a precise knowledge of its microscopic (i.e. detailed atomistic) state. In the quantum mechanics there exist for any system states of maximum definiteness: the single wave functions, commonly designated as "pure states." In this paper, the term *microstate* will be consistently used to designate a pure state (a single, nondegenerate wave function pertaining not necessarily to the energy but to some single eigenvalue of any operator whatever).

But the essence of general complementarity does not so much lie in the possibility of destroying the living system by measurements as in our inability to ascertain in practice a defined microstate. Taking a different example for the latter difficulty, by waiting long enough one can ascertain the existence of a specific microstate for which the energy is the eigenvalue; this requires careful isolation of the system; but the time for which isolation is required in order to effect the singling out of one microstate becomes so prohibitively long for systems of the size of any organism, as to be completely meaningless physically. Similar, somewhat more complicated conditions prevail if momenta of individual particles within the system are to be measured. To repeat then, the restrictions imposed by general complementarity express specifically our inability to assign a defined microstate to the system at hand if the measurements are restricted to those which are sensible in practice.

The appearance here of the term "sensible" measurements shows that we have already moved somewhat away from the traditional mode of thought of physical science. Now there would be little point in discussing this fact in a general and almost certainly vague way. It is much better to postpone the general discussion until some definite assumptions about the description of the microscopic state of systems such as organisms have been made.

To clarify what assumptions are required, we note that a great deal of information about a system can be gained by measurements that neither interfere appreciably with the dynamics of the system nor require unreasonably long times; such measurements are essentially of the "classical" type and refer to dynamical variables that are well "above" the level of quantum effects. Quantities that are measurable by classical means will be designated in what follows as *macrovariables*. We note in particular that there is no restriction on macrovariables other than that their measurement must not lead to a serious perturbation of the system. They need not be mechanical variables in the narrow meaning of the term but can equally well be chemical variables; for instance, the concentration of a chemical component, or else even the mean interatomic distance of a specific bond (measurable by x-rays), are macrovariables in the sense of this definition.

The problem is now: How can one obtain a suitable description of the system if a set of macrovariables are known to have certain numerical values? Not all microstates will be compatible with the macrovariables but only those which are within a certain range of the total energy, or else of the spatial distribution of certain chemical compounds, etc. But the restrictions imposed upon the description of the system by the knowledge of a set of macrovariables are far from determining uniquely a single microstate; instead, a very large number of microstates will in general be compatible with them. We therefore need a mathematical apparatus that allows us to formulate a description on the basis of data which are inadequate to render the description unique. Now there is a branch of mathematical science which deals with this problem: this is *the theory of inductive probabilities*. In this theory one assigns to each microstate a numerical probability of being the correct one. One so obtains a *statistical* description, or model, of the microscopic state of the system. Once such a statistical model has been successfully constructed, one can regain from it the expectation value of any macrovariable definable in the model, by means of a suitable averaging process.

The theory of inductive probabilities as adapted to quantum mechanics is the logical tool by which the implications of general complementarity can be tackled mathematically. The relationship of macrovariables and microstates is diagrammatically represented in Figure 1:

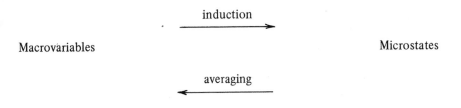

Figure 1. Illustrating Logical Relativity.

As Figure 1. indicates, the statistical representation of the microscopic structure of a system is defined only relative to a given set of macrovariables. In other words, all probabilistic statements are meaningful *only relative* to a given set of data. This is quite a general truth about inductive probabilities to which we must come back later on. It justifies for us the name of logical relativity. One needs, however, recognize a radical difference between such a scheme and that of, say, geometrical relativity. In the latter we are dealing with a set of propositions all of which are verifiable in terms of measurements. Thus, we can take two inertial systems, in motion relative to each other, and can measure in each system the mechanical properties of an object or an electromagnetic field. The results, if compared among the two systems, will empirically verify the principles of Galilean or Einsteinian relativity. But in the scheme of Figure 1. the relationship between mathematical theory and observation is different: *Only the left-hand side, the set of macrovariables, represents observational data.* Whatever stands on the right-hand side, namely, statistical propositions about microstates, represents the result of inductive inferences, and of inductive inferences only.

This has a profound influence on the physical interpretation of quantum mechanics, on the relationship between the mathematical symbolism and the observational data. If we wanted to be a bit pedantic we ought to say that the mathematical description of an atom or molecule is the description of a potential component of reality; it becomes a true component only when the atom or molecule is part of a macroscopic body, for instance if the atom is coupled to a measuring instrument which is part of the macroscopic world. This view differs obviously from the traditional one of classical physics which has been so adequate for a very long time: in it, matter, on all levels, was thought of as behaving like a system of nuts and bolts, that is, in a fully determinate manner. From the viewpoint of the inductive interpretation, matter in its microscopic aspects appears as a *potential for interaction*. The observable results of interactions are symbolized by the lower arrow in Figure 1.; only the macrovariables on the left can be said to partake in an unambiguous manner of the predicate of being physically real. To be consistent in this approach, it is desirable that one views the inductive formulation of quantum mechanics as a primary postulate, more so than any special form of the linear operator calculus which constitutes the mathematical backbone of the theory. On going in the direction of biological theory (thus avoiding problems of high-energy physics) we shall be confident that the presently available quantum-mathematics is adequate, a point to which we shall return.

Inductive Probabilities

There are many shades to the meaning of the term induction. Traditionally, since Aristotle, it has designated a form of logical inference that involves generalization. But if we look at the term in its actual usage as related to modern science, we find that it connotes two different types of inference readily distinguished by the final conclusions drawn: These may be classified as *induction for universals* and *induction for probabilities*. Although with some effort the former may be considered a limiting case of the latter, such a relationship is rather artificial, and for practical purposes the two are quite distinct. Induction for universals is characterized by the fact that the so-called universal quantifiers of logic, "all" or "every," either appear explicitly in the conclusion, or else they are implied. Newton's familiar inference that all bodies attract each other by gravity is a classical instance of this. The induction for probabilities is a process of relatively recent historical origin: The so-called problem of inverse probabilities was first studied by Thomas Bayes in 1763. An example is: Given an urn containing a large number, N, of balls; we know only that some of the balls are white, other black without knowing their proportion. Assume now that in a trial $n \ll N$ balls have been drawn of which k were white and n-k black. What is the probability, $p(K)$, that, of the N balls, K are white and N-K black? Bayes computed this probability; a few years later Laplace showed that $p(K)$ reaches its maximum value for $K = Nk/n$. Similarly (on using the terminology of statistical mechanics) we might say that the theory of inverse or inductive probabilities allows us to calculate the probability for any one of a set of possible microstates that can produce the observed numerical values of certain macrovariables.

Before we discuss this case, however, we shall very briefly review some of the history of inductive probabilities, since the subject turns out to be of more than trivial interest. Probability theory is a comparatively young discipline in the history of mathematics; it originated just a few years earlier than calculus. But while numerous applications of probability theory developed, the notion of probability itself remained ill-defined. It is only in the twentieth century that a satisfactory axiomatization of probability in terms of set theory was achieved, the first successful set of axioms having been proposed by Kolmogoroff. In this interpretation, probabilities are measures of (infinite) point sets. Now it must certainly appear that most mathematical difficulties intrinsic to the subject have been successfully overcome by making the theory of probability a branch of set theory. There is then, as it were, no place left for inductive probabilities if everything can be settled within a fully mathematical, that is, purely deductive scheme. Not very surprisingly, we find that a number of the leading probability theorists simply deny meaning to inductive probability as a separate concept.

It is perhaps more surprising that there has been a revival in the study of inductive probabilities. This began with *A Treatise on Probability* published in 1921 by John Maynard Keynes (better known as the celebrated economist he became later on). In 1939 Harold Jeffreys published his *Theory of Probability* which has run through three editions but has unfortunately not been able to match the acclaim received by Jeffreys for his great work in geophysics and astronomy. Finally, in this company we mention a paper (1946) by the theoretical physicist Richard T. Cox which he later developed into an elegant little book: *The Algebra of Probable Inference* (1961).*

What do these men want? They make the perfectly valid point that as a tool of *empirical* science, inductive inference of a statistical type is indispensable. These authors define probability as a "measure of rational belief" or expectation; probabilities are first and foremost *symbols* which receive their meaning in terms of the *operations* that are defined for them. By means of suitable axioms one arrives at a calculus, or algebra, of probabilities which turns out to be identical with the one described in every elementary textbook. One also finds that without significant loss of generality the probability symbols can be narrowed down so as to be represented by numbers between zero and one. Using the formal rules of logic (Boolean algebra) and a few additional axioms which are extremely plausible, one can thus develop a formal algebraic scheme for probabilities which turns out to be exactly the one used since the 18th century. The road to this result is also rather definite: It seems that if, on changing some axiom or other, one would try to construct a probability algebra that differs from the conventional one, then one would also be forced to deviate from "common sense" and do so to a substantial degree. This is in itself a very satisfying outcome; it shows that there are not likely to be non-Bernoullian probabilities as there are, say, non-Euclidean geometries.

But there have been long-lasting controversies about the meaning of probability throughout the history of the subject. We interpret the situation just described by saying that the difficulties are not so much mathematical as they are *epistemological*, having to do with the relationship of the mathematical apparatus to experience. It would seem that once we have reconciled ourselves to a suitable interpretation of the probabilistic world, the mathematical techniques might no longer be difficult to deal with. What is it that has created the problem here? We believe that what has been most often lacking is a suitable appreciation of *logical relativity: Probabilistic statements have a meaning only with respect to given data*, they cannot be interpreted apart from the data. If we have an urn of unknown content, we may first draw a sample of n_1 balls, then another of n_2 balls, giving an available sample of $n_1 + n_2$ balls, and so on. Eventually, we can perhaps open up the urn and find out its exact content, in which case there are no longer any probabilistic statements to be made. Each of the statements given earlier continues,

---

*Note added in 1970: A review was given by me recently (1969).

however, to have a definite meaning relative to the observations on which it was based. The old propositions are false with respect to new data but on suitable axioms they constitute a quite proper deduction (a tautology even) with respect to the original data.

The difficulty lies clearly in our implicit idea, inherited from classical physics, that there exists an objective external reality which can be described in terms of numbers and with respect to which any available numerical description is an approximation. Thus we find it quite natural that somebody measured the length of a steel rod, say, and found it to be 30.2 cm whereas another observer, having more sensitive equipment at his disposal finds 30.187 cm. On the other hand, if we infer from one sampling operation that the probability of a ball's being white is 0.60, and on a more extensive sample we find the probability to be 0.45—then we are somehow uncomfortable because the difference between the two numbers seems quite large. The so-called "collapse of the wave function" invoked in the interpretation of quantum mechanics is, from the statistical viewpoint, nothing novel at all; it is merely an expression of the fact that on novel data probability values must be redetermined relative to the data, and they might be quite substantially different from previously calculated probabilities. Analogies to this behavior can even be found in strictly deterministic classical physics. Thus it is well known that in special relativity the distinction between the electrical and magnetic components of an electromagnetic field depends on the state of motion of the coordinate system. By a mere coordinate transformation one can change the ratio of magnetic to electrical field energy almost at will.

But while in classical electromagnetism the field vectors can be claimed to have an "objective meaning" in a given coordinate system, statistical propositions (say of quantum mechanics) do not claim to be representations of any "objective" state of affairs. There does not exist a reference system (now referring no longer to geometry but to a Hilbert space) in which one could asymptotically approximate external reality. This is what the idea of general complementarity is telling us.

On the other hand, one cannot see how any of this could be changed by doctoring the *mathematical* definition of probability. The distinction between deductive and inductive probabilities is basically not a question of mathematics. It is a question of how some purely mathematical schemes can be applied to experience; it is, in other words, a distinction of epistemology. We are therefore not achieving anything by artificially restricting our mathematical techniques. By doing so we may be in danger of losing one of the eminent practical benefits of pure mathematics: the cogency of rigorous deduction. Once we give up the claim that our mathematical constructs are asymptotically a "true" description of some external reality, we are left with a great deal of latitude for the particular techniques we wish to employ. Now probabilistic statements can be experimentally verified *only* in terms of frequencies of observations. In practice, any set of observations is finite. If in a mathematical model we use an infinite point set such that the relative measures of subsets are taken as probabilities, we stand nothing to lose but we stand to gain powerful techniques of set theory that can now be employed in the treatment of probability. We believe that much of the controversy which has flared around the foundations of the theory of probability has arisen from the belief that one is dealing with mathematical difficulties where the true problems are epistemological.

Most of the criticism that has been raised against the inductive use of probability came out of a misinterpretation of the so-called "principle of insufficient reason," or principle of indifference. This principle says that when several states of a system (or substates) are indistinguishable so that the properties of the system are invariant relative to permutation of these states, then the states must be given the same statistical weight. Other things being equal, they therefore appear with the same probabilities. The classical example is the die: If the die is "good," each of its six faces has the same probability of 1/6. Clearly, however, to postulate that a die is "good" is a poor substitute for a physical investigation of the die which shows it to have the exact shape of a cube and to have its center of gravity located at the center of the cube. Or else, the die may be shown to be "good" on statistical test, by throwing it a sufficient number of times and finding each of the six faces appear in the same proportion of cases within statistical errors. The difficulty lies here clearly in mistaking a technique for an axiom of a deductive scheme: the technique has a less exalted role to play. Inductive probability has often been criticized on the grounds that it is supposed to force one to ascribe a probability of one half to an event of which nothing else is known. Thus if I have no idea whether or not it is going to rain tomorrow, I could say that the probability of rain tomorrow is one-half. This again mistakes inductive probability theory for a formal game with rules fixed in advance, whereas in reality it is a *method*; we might call it a tool: If I have a pair of pliers and if I misuse these pliers, I cannot very well invoke this failure as an argument against using pliers in general.

A method must clearly be justified pragmatically, by its results. If the method gives unsatisfactory results, the method can be changed. If one looks at the inductive use of probabilities from this viewpoint and remembers that neither are there axioms claiming the status of universals, nor is there an objective external reality to which the probabilistic statements approximate, then within these limits, inductive probability appears as a very useful tool indeed.

The Extremum Principle

At this place, let us stop for a moment to look over our path. We started out from theoretical physics and were trying to advance in the direction of theoretical biology. Very early in our march we encountered a sphinx. On sufficient prodding it explained to us that there isn't such a thing as an "objective" description of the materials of our scientific curiosity, nor is there even a definable limit to which our descriptions could be made to converge asymptotically; this is true at least when we leave "classical" physics. Having confused and disheartened us the sphinx vanished, as is the way of such creatures. All we had to fall back on, then, was Pure Method, far from Idealism, Materialism and other isms so dear to our grandfathers. At least, this being so, we might try to squeeze as much out of pure method as we can!

We have found before that no advantage is gained by restricting artificially the *mathematical* structure of probabilistic models. If we aim at a set-theoretical foundation of probability (which seems the only rigorous one) we shall use point sets in such a way that measures of their subsets represent probabilities. It is convenient to have a nomenclature: point sets whose subsets measure probabilities will be called *ensembles*. This is only a slight generalization from the manner in which the term ensemble is used in statistical mechanics and, in fact, in many cases the two meanings of the term ensemble will be coextensive. The process of finding an ensemble for specified values of a set of macrovariables will in general be far from unique; the question arises whether one can make the determination of a representative ensemble unique by suitable mathematical provisions. An answer has been found on using those properties of composite matter that are expressed in the second law of Thermodynamics: This approach suggests that we look for that ensemble for which the entropy is a maximum, compatible with the constraints expressed by the given numerical values of certain macrovariables. The result is an *extreme principle*: we maximize a general expression for the entropy of the system, with the conditions that the given macrovariables have specified values that act as constraints on this problem. Experience shows that this extremum problem with its subsidiary conditions has usually a unique solution. Hence we can say that the upper arrow in Figure 1., going from the macrovariables to a representation of the microstates, may be given an interpretation that is in general unique.

An extremely general analysis of the statistical ensemble in quantum mechanics is due to J. von Neumann in his well known book, *The Mathematical Foundations of Quantum Mechanics* (1932). Von Neumann showed that any statement of a probabilistic nature made for some quantum-mechanical ensemble (any superposition of microstates, that is, not just special ensembles like the canonical one), that any probabilistic statement can be expressed as the expectation value of a suitably defined linear operator. Of particular importance are the so-called projection operators introduced by von Neumann; these are functions which have the value unity in one region, say R, of phase space and are zero everywhere outside of R. If such a projection operator refers to the coordinates of a specific particle, then the average of the projection operator gives the probability that the particle is in R. We shall now interpret the lower arrow in Figure 1., the one labelled as averaging, to represent all the statistical conclusions that can be derived from an ensemble. The meaning we attribute to logical relativity can now be indicated thus: for any set of values of macrovariables we can construct an ensemble by the extremum principle, and once this is found we cannot only regain the values of the original macrovariables from it but also a host of other physical quantities that may be derived as ensemble averages.

Note that the statistical representation, which is the ensemble, ought not be interpreted physically as describing just one system. It is meant to describe a *class* of systems where all the members of the class are assumed to have the same Hamiltonian. Now while our mathematical scheme is mainly statistical, it has a background of traditional physics built into it by the requirement that the Hamiltonian of the system, or systems, be known. Thus we are dealing with a mixture of arguments pertaining to traditional physics and to purely statistical considerations. In the present stage of our analysis of inductive probabilities we have not been able to reduce further those components of the argument that are remnants of classical physics.

Mathematically, the construction of an ensemble proceeds as follows. Let there be a number of microstates labelled $1, 2, \ldots$ We define the entropy as usual, as

$$S = -k \Sigma P_i \ln P_i, \tag{1}$$

where the p_i are relative weights of the microstates, interpreted as the probability of a state's being present in the ensemble. The magnitude of the constant k is at present irrelevant. Formula (1) (Boltzmann, Gibbs, Shannon) needs no introduction to the theoretical physicist. Now let A be some macrovariable (in quantum mechanics, operator). Let a_i be the value which A assumes in the i-th microstate. Then

$$\overline{A} = \Sigma p_i a_i \tag{2}$$

is the expectation value of A in the ensemble. Let now a number of macrovariables have specified, measured values, say

$$\overline{A} = a, \ \overline{B} = b, \ \overline{C} = c, \ldots, \tag{3}$$

then we postulate

$$S \rightarrow \text{Max.} \tag{4}$$

with the requirements (3) as subsidiary constraints on the extremum problem (4). These conditions are introduced mathematically in the usual way, by means of Lagrangian multipliers.

Equations (1) to (3) are only approximations; in the more rigorous treatment of quantum mechanics the expression (1) is replaced by one involving the so-called statistical matrix (von Neumann, Dirac) and (2), (3) are correspondingly modified.

So far as I know, I did first propose the method of maximizing entropy long ago (1937), specifically as a tool of *inductive* inference in quantum mechanics. A number of authors later came up with formally similar suggestions (references in my 1969 paper). All these efforts made little of a stir and would have been forgotten had it not been for the independent attack on the same problem by Edwin T. Jaynes (1957). Jaynes' approach is broad, exhaustive and done with great care and attention to both the mathematical and physical aspects of the problem. Jaynes' effort has induced a number of followers to demonstrate the power of the method on other examples; there is even an undergraduate textbook on statistical thermodynamics based on Jaynes' techniques (Tribus, 1961). The following quotation from Jaynes' first paper (p. 623) illustrates his point of view:

> The maximum-entropy distribution may be asserted for the positive reason that it is uniquely determined as the one which is maximally noncommittal with regard to missing information, instead of the negative one that there was no reason to think otherwise. Thus the concept of entropy supplies the missing criterion of choice which Laplace needed to remove the apparent arbitrariness of the principle of insufficient reason, and in addition it shows precisely how this principle is to be modified in case there are any reasons for "thinking otherwise." Mathematically, the maximum-entropy distribution has the important property that no possibility is ignored; it assigns positive weight to every situation that is not absolutely excluded by the given information.

In our earlier discussion we have indicated that the division between deductive and inductive probability corresponds closely to the difference between pure and applied mathematics. To try to formalize probability without the theory of infinite sets seems the endeavor of a Don Quixote. As Jaynes remarks in a later paper (1967), it is obviously undesirable to build a mathematical theory on imprecisely defined concepts, and if this happens ... one's efforts are more than likely to be wasted. But standard set-theoretical probability theory is a perfectly rigorous branch of mathematics; one has merely used the term "probability" to designate certain measures of sets. The application of this abstract apparatus to practical situations involves assumptions; for the physicist, fortunately, these can be expressed by the simple requirement that a generalized entropy be maximized.

Jaynes has some critical comments about those "frequency theorists" of probability who dislike the methods of Laplace and Jeffreys and try to replace them by alternate "deductive" methods which sometimes might even lead to numerically erroneous results. Although I am not able to follow Jaynes here in detail for lack of specialized competence, the feeling that he might be right is very strong; it is based on the experience of a physicist who has found that knowledge of mathematical principles is not an adequate guide for solving complex applied problems. Clearly a frequency theory of probability based on set-theoretical axioms is pure mathematics.

The method of the extremum principle as here suggested for the application of the abstract, set-theoretical schemes of probability theory, has of course its roots in traditional statistical mechanics. Boltzmann and Gibbs were familiar with the fact that the entropy of a system in thermodynamical equilibrium is a maximum as compared to neighboring states. Since the calculations of statistical mechanics have in the past only concerned equilibrium states, it seems clear that the method of maximizing entropy can be as useful as one hopes only if it can be extended to nonequilibrium conditions.

Even a superficial survey shows that nonequilibrium situations in statistical mechanics are of two principal types depending on whether the conditions are stationary (in open systems) or transient. Thus the conditions (3) may specify steady flow, for instance a mean current averaged over the area of an orifice. On the other hand, one may have a transient condition prevailing after the time to which the data refer. As an example, consider two communicating vessels of which one is in the beginning filled with a gas, the other empty. We know of course what will happen and, clearly, the process of equilibration can be described in terms of a time-dependent ensemble. Proceeding beyond this, we can formulate an even more general case of nonequilibria as follows.

The case of equations (1)-(4), in which the solution is rather straight-forward in terms of conventional methods, is the one where all the measurements of microvariables are simultaneous, say at time, $t = 0$. Then the maximization of entropy leading to an ensemble will also refer to the same instant, $t = 0$. Once we have found this ensemble, the quantum-mechanical equations of motion can be integrated, and this will give us a corresponding ensemble for any later moment of time. Next, we can calculate the expectation value of any macrovariable (or any dynamical variable whatever) at any later time, by averaging over the ensemble. But this procedure is not a realistic pattern. What we should take into account is the fact that *observations are often made at different times.* The equations of motion of an ensemble enable us to extrapolate the expectation values derived from observations from one time to later times. From this viewpoint the problem embodied in (1)-(4) ought to be generalized to one where the different observations summarized in (3) can be thought of as having been made at different times. The question is whether there still exists a procedure similar to (4) that would lead to a unique representation in terms of an ensemble at times equal to or later than the last observation.

In the framework of quantum mechanics this problem appears to offer great technical difficulties. The reading of Jaynes' more recent papers (1963, 1967) and conversations with Professor Jaynes have, however, made me believe that the problem of quantum-mechanical induction, namely, the construction of an ensemble that describes statistically a system for which the measurements refer to a variety of times, may be capable of solution. Some ingenuity might be required to formulate a generalization of the extremum principle, but if this can be done at all, a mathematical solution of this general problem of induction, or at least of many special cases of it, might well be attainable.

Assume now that we can find means for solving the extremum principle when the pertinent measurements are in some way spread over the past. We then have a very general method for setting up an ensemble which describes an object characterized by certain macrovariables. In this method, the description depends on the measured values of macrovariables and on our presumed knowledge of the Hamiltonian of the system but on nothing else. These conditions express the principle of logical relativity. In ordinary language, this means that the microscopic structure of a system of atomic physics is *indeterminate*: General complementarity prevents us from arbitrarily selecting one particular microstate (which itself, of course, is only a statistical description). Instead we are forced to describe systems, except in some limiting cases, by means of bundles of microstates, that is, by ensembles; the systematic use of ensembles, in turn, has led us to a method of description, which, to use Jaynes' language, "does not ignore any possible microstate," by virtue of the extremum principle. Obviously, the microscopic indeterminacy is not unlimited; the laws of quantum mechanics admit of numerous averages that establish definite relationships among macrovariables. The indeterminacy of quantum-mechanical description is only partial, some propositions being determinate, others indeterminate. The theoretical scheme is such, however, that there are no simple and general rules which would readily separate these two cases.

The method of the extremum principle can be used to reinterpret the subject commonly designated as the "quantum theory of measurement." Although there is little disagreement in this field about the mathematics involved, there is ample controversy about the verbal expressions suitable for its interpretation. Within the framework of our previous ideas, an unambiguous interpretation can be given only to macrovariables that are well above the level of quantum effects. No significant loss of generality is incurred by thinking of such macrovariables as pointers of measuring instruments. The representation of the measuring process would then take the form of a time-dependent ensemble based on these macrovariables; in other words, what is being described now is the *measuring process* rather than any microscopic event in itself, separated from the macrovariables. The detailed description of a measuring process in terms of mathematics is undoubtedly very difficult, but the overall verbal scheme that would accompany description in terms of an ensemble is simple because claims at tying physical description into "philosophy" (such as intrusion of the "observing subject") can now be disregarded in favor of the philosophically neutral notion of logical relativity.

Relation to Biology

We are now sufficiently prepared to turn to biological theory. We soon discover that the problem of the relationship of biology to physics has remarkably little of a philosophical pre-history, far less so than the majority of scientific problems. Of course, there is the monumental work of Aristotle, a vast, organized treatment of many principles of biology, but Aristotle had almost no successors on this subject, in point of original thought. Only after the discovery of evolution has there been a gradual revival of, at least, some speculation. Since there is no decorous tradition that would have enlightened the problem at hand, it appears all but impossible to discuss its philosophy without colliding with every sort of prejudice. For this reason I will confine myself to stating simply and plainly the view I am adopting.

We assume that the laws of quantum mechanics are necessary conditions for the functioning of the organism but that the conclusions drawn from them are not sufficient to describe biological phenomena in an exhaustive manner. Of course, there are vast areas of biology where the application of the laws of physics and chemistry is perfectly adequate. We do not hold to the mechanistic view, however, that biology can be fully comprehended by the application of physics and chemistry alone. The idea that there is "more" to biology than deductive application of the principles of physics and chemistry is held, of course, by a large number of biologists and medical men. The question of what "more" means in this context can be answered *only* when one has a specific theoretical concept of what an organism is, and at present one can do no more than to grope toward such a concept.

In the period preceding quantum theory, physical science was assumed to provide an unambiguous abstract replica of the phenomena; this invariably implied that exhaustive description was equivalent to exhaustive predictability. In quantum mechanics there appears a basic indeterminacy which is sometimes glossed over by a purely verbal artifice: if one stretches the term "measurement" to mean any interference with a system, not just a type of interaction in which the effect of the measuring device upon the measured system is very small or negligible, then one can deceive oneself about the vast degree of microscopic indeterminacy inherent in the quantum-mechanical description of a complex system. If, on the other hand, one remains aware of this indeterminacy, then a natural way of relating biology to physics consists in trying to fit biological theory into the existing indeterminacy of quantum-mechanical description. (The physicist who is familiar with current objections against this view is asked to hold back his scepticism temporarily, until Sec. 5.) We might designate those features of the behavior of organisms that cannot be understood in terms of deduction from quantum mechanics as autonomous. It is better to call them *semi-autonomous* because we consistently require that they be compatible with the laws of quantum mechanics or with any deduction from these.

This theoretical approach does not require our declaring any previously established theoretical framework as *false*. Many students of the problem are baffled by this situation since they find it difficult to conceive of introducing something new into theoretical reasoning without at the same time jettisoning something old. This hesitation is not unjustified. One might ask oneself whether the introduction of such novelty must not be accompanied by a thorough analysis of the basic methodology of theory construction, and whether in the course of such an analysis ideas do not show up that make the new type of thinking more palatable. We shall discuss this point in detail later on.

First, we ask now whether we can find an approximate criterion that would distinguish between a mechanism and an organism. The following seems to go a long way in this direction. It is not meant to lead to a "definition" of an organism: The business of science is to make observations and so state propositions which connect with each other the results of these observations. In the course of these procedures we may well come across a method that specifies a class of phenomena or objects as a definite sub-class of a larger class, and this is a definition; but on dealing with such vast concepts as organisms and inanimate matter, we can only try to proceed toward some specific results rather than remain floating in the scholastic realm of "definitions."

We may describe a mechanism as a type of system characterized by a set of macrovariables in such a way that there exist definite mathematical rules, usually differential equations, which permit us to predict the values of these macrovariables for a reasonably long stretch of time ahead when they are given at an initial moment. Note, however, that this attempt to characterize a mechanism is incomplete, as may be seen by an example: The pointer of a measuring instrument is connected to a counter of fast elementary particles in such a way that the pointer is displaced whenever the counter undergoes a discharge. We shall now exclude devices of this kind from the definition of mechanism. This is wholly a matter of convention as we are free to define the term mechanism as we choose. But the example shows that, in order to have a mechanism in this more limited sense, it is not enough to be able to predict an expectation value for a macrovariable; we must add that the scatter of the expectation value be small, not just instantaneously but for a reasonably long interval of time.

It is clear that any organism contains a very large number of mechanisms; physiologists and biochemists tell us all the time how new mechanisms are being discovered. Each of these mechanisms can be studied for a finite, often considerable length of time. This does not prove, of course, that all these mechanisms can be combined in such a way that the organism as a whole, considered in the long run, can be reduced to a mechanism. To assume this would close the door to biological theory in favor of a purely mechanistic metaphysics. If we refuse the latter point of view it follows that any system of dynamical variables which fulfill some integrable differential equations describes a part-mechanism, which for some time may simulate aspects of an organism very successfully, but which is not likely to provide us by itself with any deeper insight into the nature of organic life.

Mathematicians and mathematical physicists are often in the habit of calling a theoretical scheme "constructive" if it can be derived rigorously from a definite set of axioms. As our language goes, this is a most specialized use of the term "constructive" but it is clearly the one which we physical scientists have trained ourselves to single out. Here one can find reasons why scientists in biology succeed always in picking out mechanisms for their investigation. And since life is full of mechanisms of every kind, these efforts are often extraordinarily successful. Now it is a well-known fact that any formal derivation from axioms can be duplicated by a machine (computer) and, conversely, so-called constructive propositions correspond to the theoretical description of mechanisms. We conclude from this that if any propositions of a *general* nature about organisms can be found, then these propositions can go beyond the realm of mechanism pure and simple *only* when they contain some nonconstructive components, in the meaning of constructive explained above. There is of course no guarantee that we can find meaningful propositions of this kind. Also, we might find only a small number of them. But short of having to admit that we are marooned in a natural philosophy that must forever remain purely mechanistic, it is incumbent upon us to search for such propositions.

Finite Classes

Returning now to the construction of representative ensembles, we are better able to perceive the place at which significant differences between theoretical physics and its biological counterpart might appear. First, the construction of an ensemble by the extremum principle, based on a set of known macrovariables, is a clearcut mathematical procedure which, in general, determines the ensemble uniquely. Such a scheme is mathematically "constructive" in the sense just indicated. It seems clear that there is no place where distinguishing criteria of organisms could be introduced.

But note that the interpretation, or verification, of these ensembles is essentially statistical. By verification we mean here the fact that any ensemble which was constructed from some body of data allows us to compute expectation values of numerous dynamical variables not previously measured; we can then compare these with measurements made on the systems that are represented by the ensemble. However, as Jaynes (1957) has explained in detail, the statistical character of an ensemble does not constitute a fundamental obstacle against the verification of the laws of physics. The verifiability of fundamental laws in a statistical universe clearly has an axiom at its base: the laws must be *universal*. However, to define in an altogether precise way what is meant in physical theory by this term might require the writing of a little treatise. We shall have to be content with just assuming the universality of physical laws. We do not admit, in particular, any thought of there being a difference between physics and chemistry outside and inside the organism.

Since most propositions derived from an ensemble are statistical, the meaning of an ensemble cannot be expressed by saying it is the description of a single system; the ensemble describes many similar systems, what we call a *class*. There seems no difficulty in specifying the meaning of the term "class" for this particular purpose: we shall assume that the members of the class all have the same Hamiltonian and, moreover, have a set of measured values of macrovariables in common, by assumption. In chemical terms, the Hamiltonian describes the gross chemical composition. Since we can always conceive of idealized experiments to determine what goes into and out of an organism at any stage of its life, this methodological restriction to systems with the same Hamiltonian appears innocent enough, both from the physical and from the biological viewpoint. This form of description does not seem to lend itself in the least to making a distinction between living and inanimate matter. But this is in itself a result of the greatest significance!

We do not think it admissible to change the basic principles of probabilistic induction, nor those of quantum mechanics (in the range of low energies relevant for organisms) nor the axiom of the uniformity of the laws of physics. Any such endeavor is a counsel of despair. It is reminiscent of the action of a seafarer who having lost his course would jettison his supply of fresh water in order to reduce his load. Having agreed on maintaining these principles, we may next exhibit their remarkable stringency in terms of a result first obtained by J. von Neumann (1932).

It is possible, as we already mentioned, to express any physically meaningful quantity as the expectation value (ensemble average) of a suitably defined operator. We are also at liberty to define any such operator as a dynamical variable. Assume that the ensemble has been set up to represent systems at some initial time, $t = 0$. We can then use the ensemble to determine the expectation value for any operator whatsoever at $t = 0$. Now the equations of motion can be integrated to obtain the corresponding ensemble for any later time, $t > 0$. It is true that the ensemble, being a statistical structure, spreads out more and more, it becomes more and more diffuse, as time goes on, but this does not prevent it from remaining a well-defined mathematical structure. One can then also compute the expectation value of *any* operator for all later times, $t > 0$.

From this result (whose mathematical derivation is uncomplicated) von Neumann infers that the laws of quantum mechanics are comprehensive in the sense that they exclude the existence of other universal laws. To show this, one need only remark that if these "other" laws lead to any specific numerical consequences, they do also give rise to expectation values that are functions of time. But the time-dependence of *any* expectation value in an ensemble is uniquely determined by the laws of quantum mechanics. It follows that all the expectation values derived from "other" laws must coincide with those derived from quantum mechanics; if this were not so for some variables, we would have proved that in respect to these variables there exists a contradiction between these assumed laws and quantum mechanics.

This powerful theorem shows again that there is little hope of our going from physics into biology (except for biology's purely mechanistic aspects) by a mere modification of the formalism. The formal structure of quantum physics is found to be extraordinarily compact and consistent. Again, this does not imply that the techniques of inductive representation (as distinct from quantum mechanics proper) are already in a perfectly final shape. We can expect them to be capable of improvements of a mathematical nature. What we want to indicate clearly is that within the realm of these formal constructs we can find no way that would lead us beyond physics and toward the autonomous aspects of biology. But we may at the same time perceive the direction in which we must go to tackle this problem.

An ensemble, as we said, is representative of a *class* of objects. The idea of a class used in this sense is different from that of a set of purely abstract entities. The latter can be conjured up as pure symbols, whereas the objects that the theory represents must be selected by actual physical observations. Since this involves operations beyond pure abstraction, we are in a realm where not everything is mathematically "constructive" in the sense referred to above. *We are thus led to look for the expression of the autonomous aspects of biology in the relationship between ensembles and the concrete classes which these ensembles represent.* In the remainder of this paper we shall look at the problem from this viewpoint.

One could not have predicted whether the outcome of this type of inquiry lends itself to be condensed into a single and reasonably precise statement, but this turns out to be so. I have called this statement the *principle of finite classes* and have discussed it at length in my two books on theoretical biology (1958, 1966). In order to explain this principle in a consensed form we introduce for a moment a simplified terminology: in speaking of ensembles we shall use the language of Willard Gibbs. (The reader sufficiently familiar with quantum mechanics will experience little difficulty in retranslating this into a more precise, quantum-mechanical form.) In Gibbs' statistical mechanics the ensemble appears as a probability distribution in phase space (a "cloud" of image points). This is, mathematically speaking, a continuous distribution. On the other hand, actual physical members of a class can clearly be procured only in finite numbers. In earlier times this would have appeared as a methodological postulate; at the present time we can avoid sterile arguments by invoking the finite size, both in space and time, of our universe, now so well established by astronomers. We are not, of course, interested in the absolute number of specimens of some type of organism that could be found in the universe; this would be a childish sport. To appreciate why the number of such specimens is interesting, we have to recall the tremendous structural and dynamical complexity of all known organisms. In a readily intelligible figurative language we may say that this complexity is reflected in a corresponding structural complexity of the phase space. Now each specimen taken at a given moment is in the Gibbs model represented by one point in phase space; but if the structure of the phase space is complex enough, this space appears *sparsely populated* by the points representing actual systems. This idea of sparse population will of course be of interest only if we can show that it is more than an exercise in abstraction, that it has concrete implications.

Consider for the purpose a system that has 300 measurable parameters (and for a system as complicated as an organism, even a small one, this is by no means a large number of parameters). Assume that within the range of each parameter the measuring instrument can distinguish ten intervals. The total number of possible configurations is then 10^{300}, that is, the 300-dimensional parameter space has 10^{300} distinguishable "cells." The import of such a vast number may be appreciated when we note that the astronomers are estimating that there are about 10^{80} atomic nuclei in the universe; again, the lifetime of the universe since its inception is estimated as $0.5 \cdot 10^{18}$ seconds. I have proposed the term *immense* for numbers so large that they are well beyond the number of objects or, better, events that can be thought of as occurring in the real universe. The mathematician H. J. Bremermann is using the term "transcomputable" in a similar context. Conversely, given a class containing a realistic number of specimens, the probability that any one cell in a 300-dimensional parameter space be occupied is "immensely small." Let us for instance assume an extravagantly large number of real specimens of a class, say 10^{50}; the reader will readily be able to make practical estimates to convince himself that it is justifiable to use the term extravagantly large for such a number when occurring in the real universe. The probability of a "cell's" being occupied by the image point of a real system is now 10^{-250}. The population density of this space which, to be meaningful, must be understood *relative* to the number of distinguishable "cells," is immensely small. Again, not all of the "cells" of the original sub-division might be accessible to the members of the class; thus measurements may show that of the 10^{300} "cells" orginally admitted only 10^{200} are accessible: this still leaves the relationships of immensity the same as before. For a more detailed discussion of the concept of immensity see for instance my recent book (1966).

Once more we need to emphasize that this is not meant as a mere exercise in arithmetic; the division of the representative space, or phase space, into "cells" must have a physical meaning, ultimately in terms of predictability of macrovariables. If the measurements are limited by general complementarity, one will not be able to verify the microscopic structural details without destroying the specimens. But this may be done on a limited number of samples, and one may so gain an approximate idea of the degree of structural complexity which prevails in the phase space. This again would be useless, of course, unless two systems whose image points lie in different "cells" have a finite probability of differing from each other in macroscopic dynamics after some limited interval of time has elapsed. This brings us to a point to which we have alluded before but which we are not able to discuss in more detail now, namely, the difference between the relatively homogeneous systems of physics and chemistry whose behavior can be described over long stretches of time by a limited number of macrovariables and—on the other hand— biological systems which show structure within structure within structure at all levels of geometrical extension and of energy. For biological systems, one can expect that a small difference in microscopic parameters may after some time result in major differences in the macrovariables.

We shall introduce here a terminology which exhibits this basic structural difference between the objects and classes of physics and those of biology. Two electrons are completely interchangeable, two small molecules, say, are interchangeable except for a possible minor difference in the isotopic constitution of some of their nuclei which is irrelevant for almost all purposes. Similarly, two automobiles coming off the same assembly line are interchangeable except for trivial differences. We propose to use the term *congruence classes* for classes whose members are interchangeable in this sense. Clearly, physics, chemistry, and engineering deal with objects that form congruence classes, and so does most of biochemistry and biophysics. Here, we are aiming at the introduction into biological theory of classes that are not congruence classes.

Before going into this, let us note that we have been able to avoid a dangerous trap which has caught many of those who, coming from non-biological science, have tried to indulge in speculations about biology. Machines are clearly elements of congruence classes. The particular type of mechanical or electrical connection that is called feedback can of course be designed into a machine, but this will not exempt such machines from being members of congruence classes. We only exhibit a convention universally agreed upon when we say that the realm of mechanistic explanations is coextensive with the realm of congruence classes, and this realm includes feedback.

We can readily recognize one essential property of congruence classes: if such a class is empirically given, we can study the properties of its members by *sampling*. In congruence classes a sample of sufficient size allows us to determine experimentally the properties of the class. The sample may be small as fraction of the whole but still large numerically so that all significant properties of the systems can be ascertained. It is also permissible to think of the samples as having to be destroyed in the course of the measurements. If, for instance, we are interested in NACℓ crystals we can study their properties by means of a sufficiently large set of samples. But in the case of a class of objects that is not a congruence class there are certain properties (by no means all properties, of course) whose quantitative study will be inconclusive, no matter how large the set of samples. These are the properties that depend decisively on certain features of the location of the image point in phase space, those for which one cannot successfully take an average. Traditionally, in physics, dynamical variables are divided into two kinds: the macrovariables that can be measured as functions of time, and microvariables over which one averages. For the purpose of averaging it is necessary to make certain assumptions about randomness. We may readily extend the definition of a congruence class so that it designates systems which contain microvariables that *are random* and hence can be averaged out. Thus the classes that interest us here, those that cannot be fully described as congruence classes may be alternately characterized as classes where certain averages cannot be performed or perhaps, more precisely, where essential information is lost by averaging.

In the course of my past work on these problems I have been consistently chided by many colleagues for being vague, namely holding forth propositions that are "non-constructive." In what precedes I have gone to particular lengths to show that I do not think of this as an insufficiency but as an essential part of any such analysis. Depending on the route of approach, one may think of non-constructive systems of propositions, of classes that are not congruence classes, or else of systems whose properties are intermediate between classical determinism and pure randomness. Some general aspects of this problem have been familiar to philosophers throughout history in a less quantitative form. Says Lao-Tze: "Life is that which cannot be caught in rational definitions."

Organismic Relationships

In a finite universe of discourse, and there only, one can give a formal definition of *individuality*. This results from the fact that the number of different arrangements of a set of objects soon becomes immense when the set of objects is even moderately large. Conversely, the probability for any one specific arrangement to be realized becomes immensely small. It is true that on stating this we satisfy only part of our intuitive notions about individuality. If we speak of an object as having individuality we do, in addition, usually expect that there is a certain stability with respect to time. But, no doubt, the appearance of specific but immensely rare features is one basic ingredient of individuality. The main thing to recognize here is that this notion of individuality cannot be represented within those congruence classes that are related to experience. The relation to experience is essential: it is precisely the assumption that an event or object is immensely rare which makes it impossible to collect an actual physical class of identical individuals. One can *in abstracto* form a class of any size, of identical, immensely complicated objects; but so long as this is merely abstract it is without consequence: many identical copies of an abstract structure say no more than one single copy. Here we see clearly that in the present discussion we are no longer confined to the purely formal inductive apparatus dealt with in the beginning of this paper; instead, we are now concerned with its relationship to the actual universe.

We are next able to recognize the difference between a congruence class of real objects and a class of objects in whose behavior individual traits are essential. We claim that in biology we are dealing with this last, more general concept of classes. Now we are particularly interested in the relationship of such classes to variability in the microscopic realm. Consider microscopic individual events, that is, a collection of immensely rare events in the domain of microstates. The usual way of dealing with microscopic events is to *average* them; if the system is reasonably large, the resulting averages will be macrovariables. A typical example is the pressure of a gas as defined in terms of a certain average over the motions of the molecules. The systems of physics and chemistry can as a rule be assumed homogeneous enough so that all the significant properties of the microstates can be expressed in terms of a suitable series of averages. Since each individual microscopic event is an immensely rare occurrence, and since we assume organisms to be so complicated that a set of such individual events can ultimately affect the course of the macrovariables, we can no longer take averaging for granted in the case of organisms. Structural individuality cannot be exhaustively dealt with by averaging, just as it cannot be dealt with in terms of macroscopic congruence classes.

What we have said so far has brought out the need for a "loosening up" of the traditional methods of analysis used in physical science. We have announced before that on going from physics to biology we do not propose a change of formalism; nothing that was true is just becoming false. Merely, some implicit assumptions that were not essential components of the physical theory are being dropped, and this makes room for new, and broader categories of analysis and description. Correspondingly, we now expect to replace the traditional congruence classes of physics and chemistry by broader classes. These would be the classes of biology, namely, so far as biology goes beyond its purely mechanistic aspects. Such classes will be composed of members which, in addition to having a number of mechanisms in common, also exhibit the effects of configurational individuality. Classes or processes in which the effects of configurational individuality and of the laws of quantum mechanics are inextricably mixed will be called *organismic*. Let us emphasize that the term organismic, as defined, has a very specific and essentially formal meaning. It is not the verbal correlate of a vague feeling that one is dealing with something alive. It designates the empirical existence of objects, more commonly of classes of systems whose dynamics cannot be described adequately in terms of the congruence classes of physics and chemistry alone. Instead, the presence of configurational individuality in the specimens and its effect on their dynamics is essential. *But individuality cannot be specifically segregated from the purely physical effects.* We come so to recognize wherein the radical weakness of vitalistic ideology lies: the vitalist assumes that a failure to describe organisms fully by physics and chemistry implies that a "vital principle" can be abstracted which possesses some degree of universality. But our analysis has led us only to recognize an inextricable mixture between physical determinism and individuality (in the precise sense defined), but no more.

Our results up to this point will appear somewhat negative. We have succeeded in loosening up the framework of the traditional theoretical methodology by showing that one can introduce a lot more latitude into it than is usually believed, without, in fact, changing the mathematical apparatus of the theory. On the other hand, there is implied in the foregoing the idea of *organismic classes.* The precise meaning of this term is implicit in what we have said so far: the members of these classes are objects in whose internal dynamics the effects of physical determinacy and the effects of individuality are inextricably mixed. But what the most characteristic phenomena of organismic classes will be cannot be predicted by mere manipulation of abstractions; it must be found out *by observation.*

Thus we address ourselves to the observing biologist. What does he have to say about properties of classes which he considers a specific expression of the dynamics of organisms? If we pose the question in this way, we find ourselves confronted with a rather remarkable unanimity among biologists. Those who have tried to generalize on biological observations while avoiding the traditional fixed philosophical prejudices assure us that the most distinctive type of phenomenon they observe is what they designate as biological *integration.* By this is meant the manner in which numerous physico-chemical processes in the organism appear to be mutually related in such a way as to maintain or reproduce the particular stable features of morphology and physiology that characterize a class of organisms. There are of course certain well-known mechanisms which subserve stability of this kind: various types of feedback, and the reduplication of macromolecules. From the viewpoint of a mechanistic philosophy this has been claimed to be all there would be to the stability of classes of living beings. If here we admit this stability to be organismic in the sense explained, then the purely mechanistic agents can only be part of the dynamics, and organismic stability will also involve (according to as yet unknown rules) individuality pertaining to microscopic features. But clearly, such a statement means little unless we can produce empirical evidence.

We are in the fortunate position that we can point to the beginning at least of such evidence. Here, I shall have to turn to a highly significant document entitled *Concepts of Biology* (Gerard, ed., 1958) whose importance I have discussed at greater length elsewhere (1966). This is a transcript of a conference of a dozen or so leading American biologists who discussed some of the problems indicated by the title. One of those discussions centered around the meaning of biological integration as seen mainly from the empiricist's viewpoint. Remarkably enough, the participants agreed on a notion of biological integration with features that are rather unexpected to the non-biologist. I have given literal quotations from this discussion earlier (1966, sec. 1.21) and will now summarize them as follows: in the first place, a most conspicuous feature of organic life is clearly the existence of a hierarchy of structural and dynamical orders. Next, what these biologists say is that if we study this order at two levels (for instance that of cells and the higher level of an organ composed of cells) then we find that the regularity at the higher level is superposed upon a certain "heterogeneity" of the lower level. In fact, *order above heterogeneity* seems to abridge in a simple formula what these men want to tell us about the nature of the regularities found in biological classes.

We must now of course explain what is, in this context, the meaning of "heterogeneity." If heterogeneity were synonymous with randomness, then the order at the higher level could simply be represented by averages over the microstates, and biology would be in no wise different from kinetic theory. In the spirit of the preceding analysis, however, we must assume that heterogeneity implies more than randomness; it implies that there are structures or events with *individuality*, in the sense defined, occurring at the lower level of organization, such that the effect of these events will, in the course of time, feed into the dynamics of the whole system, and will modify also the higher level. What is the result of this with respect to the classes of biology? Using a terminology coined by Paul Weiss (at the place quoted, Gerard, 1958) we shall say that organismic classes are characterized by a certain *invariance above heterogeneity*.

If we take this term to define the nature of organismic classes, we can understand its abstract connotations by looking back on our preceding discussion: invariances above heterogeneity would be an empty term in a mathematical universe where all classes are infinite. Von Neumann's theorem presents us then with an (infinite) number of quantitative propositions which can be obtained by forming expectation values over the ensemble. But in order to give to these expectation values a concrete interpretation in terms of measurements on actual systems, we would not need to have at our disposal an infinite number of real samples. But in actuality, not only is the size of any real class finite; beyond this the density of image points of actual biological systems is immensely small as related to the structural complexity of the phase space. We can now formulate a basic question in a precise fashion, as follows: Von Neumann's results have pre-empted *all universal* statements about material objects as deductively derivable from quantum mechanics. But can there not be *approximate properties of specific finite classes* which are not wholly derivable from physics and which are made possible by the fact of an immensely thin population density in their corresponding phase spaces? If we had an immense or infinite number of specimens, all such properties would be swamped out of existence by averages; as a wit has once put it: if we had an infinite set of samples of a class of organisms, the immense majority of them would have to be dead.

It will not have escaped the reader that what we did arrive at here is a different interpretation of the character of natural law: the laws of physics are always interpreted as universals in a sense well established in traditional philosophy. On the other hand, the idea of organismic laws which is emerging from our discussion is that of a set of regularities which refer only to finite numbers of objects, and moreover, in such a way that universals and contingencies (to use philosophical language again) cannot be separated. But it would help little just to state this; it has presumably been stated by many a thinker. What I have tried to show in the foregoing is that once one gains, in the inductive method, a solid basis for dealing with quantum-mechanical ensembles, then the matter of organismic regularities as *partial regularities of finite classes* becomes open to analysis by methods, solely, of mathematical statistics. By "solely" we imply that the subject matter has now been withdrawn from the field of metaphysical speculation; it is becoming an object of precise, scientific investigation. (We do not mean, of course, that one needs only mathematics; one needs equally physics and chemistry.) I know my own limitations too well to have tried this kind of mathematical task; but the problem is clearly open to attack by sufficiently experienced mathematical statisticians. Sooner or later results will be achieved that must decide the questions raised.

There is one further extremely gratifying outcome of this inquiry. That concerns our ability to recognize the importance of observational studies of individuality in biology. We need not refer by this to any specific level in the hierarchy of biological order. Experience shows individuality to be present at *all* levels of this hierarchy, from the molecular one up to the highest rungs of the ladder. There is a trickle of observational research in this field but so far only a thin trickle compared to the broad stream of investigations into biochemical and biophysical mechanisms. Some of the results are startling. I have reported (1966) on a few facets that happened to have come to my attention. This is not, of course, to suggest that the study of mechanisms has been exhausted, quite to the contrary. To quote just one example, we are in practically total darkness about the mechanisms active in developmental processes. But the study of biological individuality is certainly one readily accessible field where one may expect to meet phenomena that should permit us to penetrate farther into the nonmechanistic aspects of biology.

The author is indebted to the Office of Naval Research for support.

REFERENCES

Bohr, N. 1933. *Nature 131*, 421-423, 457-459; see also Bohr's later well-known books.

Cox, R. T. 1946. *Am. J. Phys. 14*, 1-13.

Cox, R. T. 1961. *The Algebra of Probable Inference*, 114 pp. Baltimore, Md.: The Johns Hopkins Press.

Elsasser, W. M. 1937. *Phys. Rev. 52*, 987-999.

Elsasser, W. M. 1958. *The Physical Foundation of Biology*. London and New York: Pergamon Press.

Elsasser, W. M. 1963. *Z. f. Physik 171*, 66-82.

Elsasser, W. M. 1966. *Atom and Organism, A New Approach to Theoretical Biology*, 143 pp. Princeton, N.J.: Princeton University Press.

Elsasser, W. M. 1968. *Proc. Nat'l Acad. Sci. 59*, 734-744.

Gerard, R. W., ed. 1958. "Concepts of Biology," *Behavior Science 3*, No. 2 (April); also available separately as Publ. No. 560 of Nat'l Academy of Science—Nat'l Research Council.

Jaynes, E. T. 1957. *Phys. Rev. 106*, 620-630; *108*, 171-189.

Jaynes, E. T. 1963. *Brandeis Summer Institute 1962*, 181-218. New York: W. A. Benjamin.

Jaynes, E. T. 1967. *Delaware Seminar in the Foundations of Physics*, M. Bunge, ed., pp. 77-101. Berlin and New York: Springer.

Jeffreys, H. 1939. *Theory of Probability*, 400 pp. Oxford: Clarendon Press. 2nd edition, 1948. 3rd edition, 1961—presently out of print. I have checked the last two editions against each other and find the differences between them minor.

Keynes, J. M. 1921. *A Treatise on Probability*, 466 pp. London: Macmillan. Reprinted: Harper Torchbooks, No. 557, 1962.

Tribus, M. 1961. *Thermostatics and Thermodynamics*, 649 pp. Princeton and New York: Van Nostrand.

Von Neumann, J. 1932. *Mathematische Grundlagen der Quantenmechanik*, 266 pp. Berlin: Springer. Reprinted 1943, New York: Dover. Translated 1955, Princeton: Princeton University Press.

CHAPTER X

Scientific Enterprises from a Biological Point of View

Håkan Törnebohm
University of Göteborg

I will first draw a map of scientific enterprise.

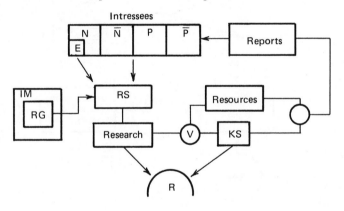

IM	=	Intellectual milieu
RG	=	Research Group
RS	=	Research strategy
V	=	either/or or both
R	=	Reality
E	=	Enterprise
N	=	Scientists in a neighborhood
$\bar{N}$	=	Other scientists
P	=	Professional uses of knowledge in society
$\bar{P}$	=	Nonprofessional enjoyers of knowledge in society
KS	=	Knowledge system

Figure 1

I wish to draw analogies between this sort of systems and individual biological systems. In a later part of my talk, systems will be considered that are composed of many systems of the kind mapped in Figure 1. They will also be compared with biological systems.

Figure 1 depicts a scientific enterprise from a fairly high altitude. Let us lower the altitude and consider the research activities in some detail.

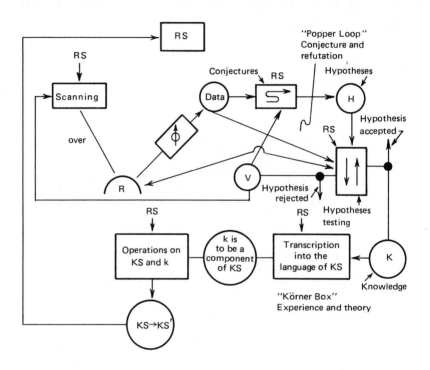

Figure 2

Scanning, data gathering, data processing and hypotheses testing correspond to the following biological operations.

An organism is looking for, preparing and testing food. The other operations correspond to intake of food and metabolic processes taking place inside the organism, finally resulting in substances being assimilated as part of living tissues and possibly enabling the organism to grow. Another analogy: Organisms are always selective with regard to nourishment. Similarly, not all data, not all corroborated hypotheses, that is, not all items of knowledge can be assimilated into a system of knowledge.

A more important similarity is this one. As an organism grows, it acquires a higher degree of organization. The foodstuff carries an amount of negative entropy which helps the organism to overcome a natural tendency to deteriorate, to become disorganized, that is, to increase its entropy, and it may even help it to build up a higher level of organization, that is, to decrease its entropy.

Similarly, data and hypotheses carry an amount of information which may be regarded as negative entropy. Indeed, the amount of information that is carried by a sentence H has been defined in this way:

$$C (H) = -\log P (H)$$

(cf. Carnap and Bar Hillel, *Outline of a Semantic Theory of Information*, 1952.)

When an item of knowledge has been transformed into a component of a knowledge system of the kind which is organized as a deductive system, it may happen: either, that it is possible to deduce it from components of KS of an earlier vintage, or, that this is not possible.

In the first case, an explanation of demonstrative type is possible and can be given, but there is a price to pay for this possibility. What appeared to be a "new" component is shown by a successful demonstrative explanation not to be new at all. The amount of information is the same afterwards as it was before. The system does not grow. This has to do with the fact that the conclusion of a deductive argument does not add information to the premises.

In the second case, on the other hand, the group of researchers RG will attempt to invent a new component, as it were, inside the "skin" of the KS so that the newcomer from outside can be deduced from other components in conjunction with the internally generated component. Such a component was Newton's hypothesis of gravitation or Planck's quantum hypothesis. In this case the system grows in amount of information, retaining its character of a deductive system, but this growth is rather precarious. The invented components are delicate ones, likely to perish. Growing areas in plants are tender and soft and easily exposed to destruction, unprotected as they are.

It is likely that cases of the second kind occur most frequently when a KS is "young" and there are few established premises from which a new component from outside might be deduced. Young ones grow. Older ones with a large number of components are less likely to grow. They are more likely, at best, to be kept in a steady state. Items of knowledge from outside serve to fortify and corroborate a system of knowledge as they add to the stock of evidence in its favor. But they do not increase the amount of information that is carried by the system.

Let us next pay attention to the box RS. RS has a steering function on the research processes which results in a more or less well organized, more or less well mapping KS. The intermediate stages of a developing KS are partly determined by RS, partly by the outcome of credibility controls which control the mapping relation from KS to R.

RS has thus a function similar to a genetic code in biological systems. The shape of a KS at a stage of its development is predetermined by RS and by the nature of its relation to R, just as a phenotype of a biological system is determined by a genotype and an environment.

Systematic studies of the structure and function of RS in scientific enterprises may thus be compared to genetic studies of biological systems.

The resource box has also biological analogies. The intellectual resources produced by one enterprise may be utilized in planning new research. The resources are thus like contributions to a gene-pool in a biological population.

Let us now extend our considerations to larger systems than single scientific enterprises, that is, to systems the components of which are such enterprises. Science in society as an historical system is the most extensive of them all.

There are two types of structures of special interest. They are neighborhoods and genealogies. We may define various types of neighborhoods. One of them is the relationship in which various enterprises stand to each other when their fields of study, their territories, their R's are the same or nearly the same.

Such a neighborhood may be compared to an ecological system. When different populations of organisms inhabit the same area, which means that they form an ecological system, there may be struggle and there may be cooperation. They may even form symbiotic systems.

Members of different schools of research hold different world-pictures and ideals of knowledge. They hold different views on what to do and on how to do it. As they are working in the same territory, such basic differences are likely to give use to produce intellectual feuds such as that between vitalists and mechanists in earlier biology.

Such feuds may end up in one school completely defeating another, as vitalism has been defeated, I believe, in microbiology. It is the same as when one animal population destroys a rival population. Or else the schools may find ways of peaceful coexistence. R is studied from different aspects (complementarity). They may find ways and means of cooperation such as that between history and theory of economic systems.

Successful enterprises are normally followed by replicas. Other research groups carry such research along similar lines as a successful one. The RS of the successors can be considered as "genetic systems" inherited from a parent. The offspring which may be quite numerous will exhibit variations. Some variants are better than others and will, in their turn, have a greater chance of producing a large number of offspring. The successful variants are selected and a population evolves along lines which lead to improved adaptations to its environment. As a population grows it will invade neighboring territories, just as an increasing animal population will migrate into new regions.

There is also an analogy to the formation of new species in the case of research enterprises. Such formations are preceded by the production of new types of RS. New problems are raised, new resources are used, either created by the RG itself or imported from other researchers. New types of plans to carry out research are made, new conceptual systems and research techniques are devised, new ideas as to what the gestalt of KS should be like are conceived by RG.

I have earlier compared RS to the genetic subsystem in a biological system. It is therefore natural to compare innovations in RS to mutations in a genetic complex. Most gene mutations are unfavorable, but a few of them give rise to modifications of the phenotypes which will increase the probability of survival of their carriers, thus generating a new species. Likewise, innovations in the RS component of scientific enterprises is a very risky business. Only a few of them are successful.

So far, no biological analogy of the boxes IM and Intressees has been offered. The intellectual environment IM will contribute to determine the world-picture and the ideals of scientific knowledge held by the RG. In other words, it will help to determine their philosophy of science. Such general IM ideas may be favorable or in conflict with the scientific enterprise. They play a part, an important part, in the intellectual processes leading up to RS. It seems natural to compare IM to a mating place for animals. Animals of certain species can mate only in very special locations which are like a romantic setting for love-making. Likewise, knowledge production can take place only in specially favorable intellectual environments.

I finally come to the box Intressees. Research initiated almost exclusively by Intressees in the left part of the box is pure research, whereas applied research can be described as knowledge production, mainly stimulated by the professional Intressees in the right part of the box. Enterprises of the second kind may be compared to biological systems of domesticated type. They are products, as it were, of genetic and environmental engineering affecting the genotypes and phenotypes of organisms.

It is time for me now to draw some general conclusions. Although some analogies may seem to be rather farfetched—they are little more than metaphors—there are others which are quite striking. It is therefore feasible to conclude that studies of scientific enterprises may benefit by adopting points of view and conceptual tools from biology.

The main features of similarity appear to be these: The systems under investigation in both fields of research are highly complex ones. Features such as control and information processing are of primary importance. This means that cybernetics and particularly the theory of information are highly relevant in both fields.

CHAPTER **XI**

Historical Observations Concerning the Relationship Between Biology and Mathematics

René Taton
École Pratique des Hautes Étude, Paris

As early as in the remotest antiquity, a certain number of the thinkers who have reflected on the problems of the philosophy of nature posed questions concerning the interactions existing between the various domains of scientific knowledge and, in particular, concerning the role that mathematics might play in the explanation of diverse types of natural phenomena, comprising those touching on the life sciences. Thus the various doctrines of ancient China or Greece bring into play concepts of a numerical or geometric order in the interpretation of certain biological phenomena. But, as a matter of fact, these apparent intrusions of mathematics into the domain of facts concerning life have hardly more than symbolic value and are related to the neo-platonic speculations which enjoyed such a great vogue during the Renaissance. Without wishing to deny the considerable importance of these conceptions for the history of ideas and without forgetting that such speculations still flourish today in certain apparently inaccurate scientific publications, I will limit myself in this exposé to the study of currents of a more specifically scientific inspiration which have attempted, more or less successfully, to apply the resources of mathematics to the study of the phenomena of life.

In my analysis I shall leave aside everything concerning the indirect application of the methods of applied mathematics by the use of models of a mechanical or physical order to which one has attempted at various times to reduce a more or less essential part of the phenomena of a biological order. Although they are domains of application favored by mathematics, the fields of mechanics and physics are actually situated outside the framework of this science. Moreover, in his recent exposé Grmek has already given us a very stimulating analysis of the mechanical interpretations of life that have appeared in the course of the two thousand years that separate the first Greek atomists from the successors of Descartes.

Beyond a few rudiments of optics and theoretical acoustics, the only fields of science in which quantitative mathematics had been able to play a truly useful role before the beginning of the 17th century were those of statics so brilliantly illustrated by Archimedes and of celestial kinematics in which the systems of Ptolomeus and Copernicus proposed concurrent descriptions. The medieval attempts at mathematization of physics are much closer to the level of intentions than to that of actual realization. Indeed, from Thomas Aquinas to Nicole Oresme, this medieval school of the intention and remission of forms had effectively attempted to mathematize the entirety of knowledge, of sensations and even of emotions by trying to reduce the intensity of qualities to a scale of measurable values. But the enormous breadth of the objective envisaged suffices to show the purely philosophic and speculative character of this attempt, approached without real care for measurements and without a clear appreciation of the fundamental differences separating such disparate notions as those of velocity, heat, whiteness, charity and grace. In the same way the geometric designs in which at the beginning of the 16th century Leonardo da Vinci and Dürer had attempted to contain the human body can have only a symbolic value. That indeed was not the case in the studies in animal mechanics pursued by Leonardo, but these works seem to have been practically unknown in their time.

So, in spite of undeniable and praiseworthy efforts, the physics of the end of the 16th century remained the physics of qualities. As for the biological phenomena that occupy us more specifically, they were appreciated only from a qualitative and eminently subjective point of view, giving place largely to metaphysical speculations and to imperative theological considerations.

While marking a decisive step in the constitution of the principal branches of the exact sciences and while making an essential contribution in the formation of the rationalist scientific spirit, the 17th century for the first time attempted to extend the quantitative and objective point of view over the entirety of knowledge.

Galileo appears doubtlessly as the most important and the most representative mover in this tendency toward mathematization, characteristic of this revolution of the human mind of the first decades of the 17th century which is at the origin of our modern science. A famous and often cited passage from his *Saggiatore* of 1623 deserves to be recalled here, less for its objective value than for its symbolic significance. He writes:

> Philosophy—that is to say, science—is written in this immense
> book which lies continuously open before our eyes—the universe—
> and which cannot be understood if one has not first learned to

understand its language and to know the characters used in writing it. This book is written in mathematical language, its characters are triangles, circles and other geometric figures, without whose intermediary it is impossible humanly to understand a single word. (Fr. transl. G. Laurent, *Les grands écrivains scientifiques*, 8ᵉ ed., Paris 1921, p. 27.)

Without doubt Galileo applied this thought first of all and explicitly to the system of the world, to mechanics and physics. Yet this intention could appear rather bold at a moment when the number of known quantitative laws in these disciplines remained extremely small and Galileo did not even know or did not admit the validity of the Keplerian laws of planetary movements. But it does seem that in his mind the living world cannot be kept outside this mathematized world. Hence, this hope of progressive mathematization pre-supposed an entrance of measurement into the domain of living things. By inventing, or at least by using, the thermoscope and the pulsometer, Galileo introduced quantitative method into physiology concurrently, it must be remembered, with Santorio who, as Grmek noted, was not moved by the same modern doctrine. If measurement is for Galileo the fundamental act of scientific research, in the field of biology numbers do not appear to him in order to be capable of interpretation in the simple framework of a mathematical analysis, but by the intermediary of a mechanical or physical model such as may be found for example in his analysis of the mechanics of animal movements. While Santorio, approaching in effect the study of animal metabolism created or propagated the basic instruments of animal physiology, at the same time rejecting mechanism, Galileo appeared as the true initiator and the ardent promoter of this doctrine. Consequently his successors will be the iatromechanicians of the 17th century Grmek has told us about.

The wish of Galileo seemed therefore somewhat deflected from his initial object and mathematics appeared only as the support of a resolutely mechanistic thought. However, the original idea, profoundly mathematical, appears very clearly in this passage by the Italian physician Biovanni Baglivi:

. . . The human body, in all that concerns animal structure, is subject to number, and therefore undergoes all the consequences that depend on it. Such was no doubt the will of God, the sovereign father of things, when, in order to make this animal machine more adept at executing the commands of the soul, He used, it seems, but the compass and the chalk of the mathemati-cian. . . . (Baglivi, *De l'accroissement de la médecine pratique*, p. 33, livre I, chap. VI, troisième obstacle.)

The work of a fervent disciple of Galileo, the posthumous *De motu animalium* of Borelli (1680-1681), presents a very elementary mathematical support and, in spite of an attempt at an explicative analysis at the level of muscular fiber does not succeed in overcoming some fundamental problems such as that of the evaluation of the work of the heart or the stomach. Jean Bernoulli, who after solid studies of medicine had been introduced by his elder brother to the infinitesimal mathematics of Leibniz and had become a fervent promoter of the new doctrine, tried in 1694 in his *De motu musculorum* to apply these powerful means to the study of fibrillary shortening. But although inspired by a renewed iatro-mathematician, this attempt could not succeed at that time and its failure marks the decline of a too theoretical conception of biological phenomena.

Indeed, the renewal of the biological application of mathematics stems from a new sector, that of the thought of probabilities and statistics whose first elements known in the time of Galileo could reveal neither its interest nor its richness. It is indeed known that the true beginnings of the calculus of probabilities date from 1654, the year when Fermat and Pascal exchanged their famous correspondence on this theme, and that three years later, in 1657, Huygens published the first essay devoted to this new calculus.

The beginnings of the study of statistics are almost contemporary with those of the theory of probabilities with which it was later to ally itself closely. It is indeed in the year 1662 that a wealthy English shopkeeper John Graunt (1620-1674) published the first statistical tables on mortality, tables which he had prepared with his friend William Petty beginning with the death registers of the city of London since the year 1592, and which soon brought him his admission into the bosom of the Royal Society. Several years later Christiaan Huygens and two other Dutch mathematicians, Jan de Witt and Jan Hudde, were interested in this problem either by approaching the interpretation of the results presented, or by publishing on their own part new tables. But it was the famous English astronomer, Edmund Halley, the friend of Newton, who gave statistic studies their real impetus. In fact, in *Philosophical Transactions* Halley (1693) published new tables founded on the interpretation of the records of births and deaths in the city of Breslau for the period 1687-1691; in this same article, aside from these data, he calculated by extrapolation tables of mortality and survival for a supposedly stationary population and in a systematic fashion approached the calculation of the annuities of life insurances, a branch of financial activity which was to experience a remarkable development in Great Britain.

In the course of the first half of the 17th century, several important advances of the calculus of probabilities, specially the one of Jacques Bernoulli, the brother of Jean, and those of Antoine de Montmort and of Abraham de Moivre developed the principles and widened application of the calculus of probabilities. These scholars began to define more closely the bonds between this new branch of mathematics and the analysis of statistics, in the way it was approached after Halley. In a parallel manner other investigations had bearing upon the first applications of statistics: on the one hand, certain problems important from the financial point of view—such as the theory of all kinds of insurance, life pensions, lotteries, gambling, etc.—on the other hand, questions of an apparently less immediate interest such as the determination of the average life-span of a man under different circumstances and certain juridical and political problems which were to experience a widespread success at the end of the 18th century and the beginning of the following one. But if the determination of the average life-span already presented an undeniable interest on the level of biology, it was the study of a more specific problem, that of smallpox inoculation, which in the course of the years 1760-1770 contributed to establishing the bases of the application of statistics to the biological and medical fields. This fact is important enough to warrant our considering the circumstances in greater detail.

Beginning with 1760 the problem of inoculation incited an enthusiastic start which was aroused by two of the principal mathematicians of the time: Daniel Bernoulli and d'Alembert. The first protagonist, Daniel Bernoulli (1700-1782), son of the aforementioned Jean Bernoulli, was the most eminent representative, in the middle of the 18th century, of the illustrious family of the Bernoulli of Basle. Following the example of his father he had undertaken serious studies of mathematics and medicine, was interested in a wide variety of subjects and could thus bring important contributions to certain mathematical subjects as well as to logic, hydrodynamics, other disciplines in physics and the calculus of probabilities. His first works, devoted to the latter discipline, had allowed him in particular to introduce important notions such as mathematical hope, to propose a solution to the famous problem of Saint Petersburg and to initiate the application of the calculus of probabilities to celestial mechanics: the study of the distribution of the inclination of planetary orbits. Professor of anatomy, botany and physics at the University of Basle, close friend of Maupertuis and Euler, Daniel Bernoulli was at that time one of the most outstanding European mathematicians. As for d'Alembert (1713-1783), the publication of his great treatises on mechanics, hydrodynamics and celestial mechanics, his very active participation in the preparation and the edition of the first volumes of the *Encyclopedia*, his activity as philosopher and polemicist earned him both a brillant reputation and strong hostilities, among them that of Euler. Certain articles already of the first volumes of the *Encyclopedia* had given d'Alembert the opportunity of presenting very critical views on certain conceptions concerning the calculus of probabilities and of contesting in particular various points of view adopted by Daniel Bernoulli. As for Maupertuis (1698-1759), who had recently died but had apparently played the role of instigator in this affair, the diversity of his work and his activity as first president of the Academy of Berlin had brought him also great prestige in a part of the scholarly world of the time, but also violently hostile reactions from certain adversaries of prestige, such as Voltaire. But it is time to approach the subject of this controversy which, for the first time, brought into full light the theoretical interest of a statistical analysis of certain practical problems and difficulties which make its application especially critical.

Smallpox was at that time at an endemic stage in Europe and constituted the cause of the largest number of deaths in the 18th century, not to mention the innumerable cases of disfiguration accompanying it. The Chinese and other Asian peoples had discovered that a person could protect oneself from this terrible disease by contracting an attenuated form, by means of pulverization of the scabs of smallpox victims in the nose. In Turkey and in Greece a more perfected method was accomplished: inoculation of the pus of smallpox victims into small scars left by pin-pricks. This procedure, introduced in London in 1717 by Lady Montagu, the wife of the English ambassador at the court of the Sultan, spread rather rapidly, first in England and then in various countries of western Europe, among them Switzerland and France. But this practice continued to encounter certain oppositions, raised especially by the fact that the weakened disease, engendered in all persons inoculated and previously nonresistant to smallpox, created a considerable risk of death.

Was it then preferable to risk death voluntarily within a short time in order to avoid all future risk of contagion by smallpox, or to await, without immediate action, a possible later contraction of the disease in its unattenuated form? Partisans and adversaries of inoculation had confronted each other for a long time, but used only qualitative arguments. The first to attempt a mathematical analysis of the problem was Daniel Bernoulli who was fully qualified for it by his training as a physician and his previous work in the field of probability.

On April 30, 1760, Daniel Bernoulli, as a foreign associate of the Royal Academy of Sciences of Paris, began to read before this assembly an important Report entitled: "Essay on a new analysis of mortality caused by smallpox, and on the advantages of inoculation for its prevention." According to his own testimony this report had been composed "upon the wish of the late M. de Maupertuis who was at that time in Basle and whom I saw," he says, "very often." This introductory note is not without interest, for on the one hand, it places the inception of this work within the few months that Maupertuis, returning to his post at the Academy of Berlin, after a leave in France, spent in Basle before his death (i.e. between October 16, 1758 and July 27, 1759), and on the other hand it seems to provide moral support of Maupertuis to the position adopted by Daniel Bernoulli.

Using the then published mortality tables, among them that of Halley, as well as the information received from various cities concerning the deaths caused by smallpox, Daniel Bernoulli attempts first to establish the risk of death from this disease in various ages and then to compare the risks run according to whether one has recourse to inoculation or not. Distinguishing on this subject individual and collective points of view, he arrives without reservations at a conclusion in favor of inoculation.

In the same year, in the open session of the Academy of Paris on November 12, 1760, d'Alembert replies to Daniel Bernoulli with a lecture "On the Application of the Calculus of Probabilities to Inoculation of Smallpox," in which he subjected his rival's essay to severe criticism, affirming in particular that one must not support a good cause with bad arguments. This lecture by d'Alembert was printed in 1761 in Vol. II of his *Short Mathematical Works* together with two complementary reports on the same subject. Daniel Bernoulli replied in the introduction of his *Essay*, when the latter was finally printed in the volume of the *History of the Royal Academy of Sciences of Paris* for 1760, published very late in 1766. D'Alembert replied again in various reports published in 1767 and 1768 in Vol. V of his *Varia of Literature and History* and in Vol. IV of his *Short Mathematical Works*. It is not within the purpose of this exposé to analyze in detail these contributions of Daniel Bernoulli and d'Alembert. I would like merely to dwell on various general observations. It must be noted first that the somewhat paradoxical position of d'Alembert earned him a rather reserved acceptance. Thus Diderot wrote that d'Alembert had read "a report that all numbskulls must take for a paper against inoculation and that all wits say that it is not in favor of."

As a matter of fact, although d'Alembert's apparently very critical position reflects in part his taste for polemics and a certain animosity toward Daniel Bernoulli, his analysis is often very well founded and brings a valuable contribution to the general methods of the use of statistics. D'Alembert rightly insists on the necessity of utilizing the most precise data possible and on the necessity of undertaking a more searching and more systematic investigation concerning the risks of contracting smallpox and concerning the correlative risks of death, according to the age levels, as well as concerning the risks deriving from inoculation. He distinguishes—and in a still clearer way than Daniel Bernoulli—the individual from the social point of view and brings a more marked interest to bear on the first case. He also distinguishes between the notion of "physical life" and "social life," considering that the individual, being a burden upon society during his childhood and his old age, has a negative social value during those ages, which becomes positive only in the intervening years. He attempts also to take into account the differences between an immediate risk and a risk at an undetermined later date.

Thus, placing into the background the polemic aspect of this controversy, the first attempt at a mathematical analysis of this medico-biological problem, confronted by Daniel Bernoulli, and its rigorous criticism at the hands of d'Alembert constitute in our view an essential stage in the elaboration of a new means of approach to certain questions of biology. In fact, the immediate conclusion as to the particular problem confronted did not appear lucidly to most of the contemporary scholars, who remained divided with respect to the efficacy of inoculation. Moreover, this polemic revealed the great complexity of the methods to be undertaken and the extreme difficult of interpretation that must follow. Considering, after all, that the rudimentary state of the collected statistical data of the time and the still insufficient level of elaboration reached by probabilistic thought prevented the attainment of a precise reply to the question posed, the then prevalent polemic nevertheless opened the way to the later development of statistics in biological and medical research.

Let us note in passing that the same problems arose, in principle at least, at the introduction of anti-smallpox vaccine by Jenner, but since the introductory remarks published in his famous work of 1798 ("An Inquiry into the Causes and Effects of the Variolae Vaccinae") seemed at the time to remove all danger of death, no argument was raised at this occasion in opposition to the generalization of this practice.

In fact, it is known now that a risk—minimal but real—does exist, and this justifies a more rigorous analysis of the problem. Nevertheless, the disproportion existing between the risks incurred in the two cases is such that the opposition to vaccination is only very exceptional.

A new approach in this direction was attempted some sixty years later, when the French physician Pierre Charles André Louis (1782-1872) in his *Anatomico-pathological Research on Tuberculosis*, published in 1825, and in a study on typhoid fever published several years later, developed the principles of the "numerical method," an attempt at a statistical analysis of the symptoms, the signs, the evolution and the consequences of diseases. Although Louis received the aid of certain mathematicians in the interpretation of his numerical series and in particular in the attempt at separating the results due to medical action from those due merely to chance, the movement set in motion by this innovator received in France only a rather reserved acceptance, as well as in the medical world at large, which did not give the experiment opportunity for satisfactory development. This was in contrast to the world of mathematicians, who were then much more preoccupied with the rise of analysis and mathematical physics, and even with the application of probabilities to the social sciences, than with the application of statistics to biological and medical problems. However, this same French probabilistic school, which remained unresponsive to the effort of Louis, indirectly instigated a still bolder undertaking.

The new impetus in this direction was indeed given by the Belgian Adolphe Quetelet, who, after having studied probability in Paris under Laplace and de Lacroix, in 1829 founded the Observatory of Bruxelles, of which he made an extremely active center of statistics. Gathering and utilizing statistics of all types available, he published in 1837 his famous *Essay on Social Physics*, in which he undertook to apply the theory of probability to the study of human qualities and of problems of social life, thus defining, thanks to the law of large numbers announced several years earlier by Poisson, his concept of the "average man" around whom oscillated the individual traits of real men. In 1846, Quetelet published a new work, *Letters to His Royal Highness the Duke of Saxe-Coburg and Gotha on the Theory of Probabilities Applied to Moral and Political Sciences*, which laid the essential foundations for further applications of the statistical method, treating specifically in the field of operations the errors of observation in connection with the deviations of natural phenomena in relationship to average elements.

Indeed, the bold attitude assumed by Quetelet was far from rallying any essential support and certain of his brilliant formulas were ironically criticized by the probabilist mathematicians of the time. That is true in particular of the famous sentence: "Now the urn that we interrogate is nature," which has been taken up and eloquently defended by certain modern authors.

While Augustin-Louis Cournot developed in France the philosophical principles of the theory of probability and its applications to the study of experimental facts and economic phenomena, John Herschel spread in Great Britain Quetelet's ideas on probabilist analysis and statistical theory, adapting them at the same time to the framework of the inductive philosophy of Stuart Mill. While the work of Cournot did not have immediate repercussions, the position of Herschel with its claims that if events seem in the long run to obey the laws of probability, it is because these laws have been established according to hypotheses conforming to the events, had a considerable and almost immediate influence. In fact, as C. Gillispie has shown, the foundations of Maxwell's application of probability to physics and very specially to the kinetic theory of gases may have been inspired in him, through the intermediary of Herschel, by the analogous principles used by Quetelet in the field of the social sciences.

In the same way, when Francis Galton gave a new impetus in 1869 to the methods of application of statistics to biological sciences, by the publication of his famous work on *Hereditary Genius,* rightly accepted as a classic in the history of biology as well as in that of statistical methods, he refers specifically to Quetelet. He quotes and reasserts very specially two examples of the application of the normal curve of Gauss to anatomical observations, presented in the *Letters* of Quetelet of 1846, the one on the girth of the chests of the soldiers in the Scotch regiments, the other on the height of French conscripts. Introducing and utilizing systematically a number of valuable notions, such as that of the coefficient of correlation between two series of facts or traits, Galton truly created the scientific method of utilizing statistical theories—a method of which his predecessors had had only an inkling.

In this exposé I shall not pursue the study of the evolution of this new branch of science, marked especially during the first half century of its development by the admirable efforts of the British school, Galton first of all, then Karl Pearson, Student, Sir Ronald Fisher, *et al.,* and, since then, by continued research undertaken in various countries, which have fostered the development of the application of statistics in the most diverse fields of biological and medical sciences: biometrics, genetics, molecular biology, heredity, animal sociology, behavioral studies, medical statistics, epidemiology, immunity, etc. In fact, we here confront fields in their full process of development whose recent evolution is well known to specialists and which call for the attention of highly qualified specialized historians. One knows in particular the spectacular development undergone by genetics since the beginning of our century and the important role that mathematics—statistics in the very first place—has played in this development, concurrently with a spectacular refinement of observation and experimentation, and with the renewal and prodigious progress of biochemical and biophysical knowledge.

Molecular biology, genetics and biometrics are certainly the present sectors of biology in which mathematics plays the most apparent and most fertile role, but the development of biocybernetics and the invasion of computers open new roads to the applications of this science, whose vast possibilities are certainly far from having been exhausted. If it may appear unreasonable to speak at this early date of biomathematics, perhaps the developments that will appear in the course of the next decades will permit this new branch of science to establish itself by gathering together elements still scattered today, thus bringing to pass, albeit in a slightly altered form, the prophecy formulated by Galileo in 1623.

CHAPTER **XII**

A Survey of the Mechanical Interpretations of Life from the Greek Atomists to the Followers of Descartes

Mirko D. Grmek
Archives Internationales d'Histoire des Sciences, Paris

The first consistent mechanical theory of nature was formulated by the Greek atomists, Leucippus and Democritus, and developed by their later followers, Epicurus and Lucretius. In their opinion, human and animal bodies are composed of atoms, and all life manifestations are only the results of complicated, but strictly determined, motions of these elementary particles. The heavy atoms of the body are moved by the soul. Yet this Epicurean "soul" is not a spiritual principle, but a collection of very light and small atoms. Mind is an ensemble of extremely small atoms. Sensations are born from movements and soul provokes all body motions in a strictly mechanical way.

Thus a completely materialistic and deterministic biological doctrine was elaborated before the end of the 4th century B.C. It contains all the basic principles of mechanical materialism, with one exception: the analogy between organism and machine. But this is quite understandable, because complicated self-moved machines had not yet been constructed. They weren't even imagined by ancient scholars.

There are, in fact, mechanical analogies in the Hippocratic corpus—for example, a comparison of the digestive organs with communicating vessels and a mechanical explanation of birth labor. Also, Plato compares vertebral movement to that of the hinge on a door (*Timaeus*, 74a). But these analogies are not followed by philosophical generalizations.

In his famous treatise *On the Motion of Animals* (707b 9-10) Aristotle compares the action of muscles and bones of the forearm with that of the classical weapon, the catapult. Aristotle's use of such a model is only partially

181

mechanistic, because this analogy serves to demonstrate the structural condition of a form of movement and not the origin and initial conditions of the movement. In ancient machines the source of energy was furnished by human or animal muscular work; thus we can understand that the analogy between organism and machine could not concern the essential problem—that is, the origin of activity. In classical machines the motion is strictly passive; in living beings it is obviously active, if not necessarily spontaneous.

Aristotle's attempts to furnish a mechanical model of some organic phenomenon are, in fact, vitalistic—that is, involving a special nonmaterial vital principle. It is vitalistic even in the very curious sentence in Aristotle's *Politics* in which a slave is designated as an "animated machine."

In Aristotle's mind all movement necessitates a prime motor (mover), which must be immobile. Thus, it can be neither a muscle nor any other part of the animal frame. The animal body is really moved by desire, a strictly vital property which can be explained only by soul. There is in Aristotelian philosophy a fundamental difference between animated and nonanimated motion. Incidentally, we can point out that in Aristotelian physics the term "motion" has a larger meaning than in modern terminology. It is not only a translation of an object from one place to another, but it embraces other kinds of changes—such as creation and destruction, alteration, and increase and decrease. Nature cannot be explained only in terms of material translation. Every motion has not only a direction, but also a purpose. The motion of celestial bodies is perfect and eternal; terrestial motions are only a search for perfection.

One consistent mechanical theory of life was conceived by Erasistratus, a great physiologist of the Alexandrian School (3rd century B.C.). In his opinion, the human body was held together and moved by three systems of channels: arteries, veins, and nerves, the last of which were also conceived of as hollow. The arteries contained pneuma, the veins blood, and the nerves a special kind of fluid called the nervous spirit. All vital functions were provoked and controlled by these three systems of organismic integration.

In Erasistratus' mind, an animal or human body was a hydraulic and pneumatic machine. The fact of the matter is that such a machine was not yet known. And it is probably not a pure hazard that the first elaborated hydraulic and pneumatic machines were constructed by two Alexandrian scholars, Ctesibios (end of the third or beginning of the second century B.C.) and Heron (about the beginning of the first century B.C.). In the work of the latter there is even an interesting testimony on the Greek theaters of automata.

Alexandrian technologists were responsible for the invention of a force pump, a water clock, and a special organ, a musical instrument in which the air, instead of being supplied by human lungs, came from a wind machine.

Action of heat on water was used as motive power in other machines. This was particularly important because it offered the possibility of an explanation of the manner in which innate animal heat could produce movement.

At the time when Rome was at its peak, a mechanistic theory of life was accepted by a special medical sect—the Methodists. Its founders were Asclepiades and Themison. The Methodists followed the teachings of the atomists and considered the human body as a network of fibers. Between the fibers were pores, in which atoms of various shapes were thought to circulate.

The anatomical achievements of the Alexandrian School, together with the Aristotelian notion of the organs, have survived in Galen's anatomo-physiology, especially in his treatise *On the Utility of Parts.* Galen offers mechanical explanations of the structure of some organs, for example, articulations of joints. His description of the hand as a perfect mechanical tool is a masterpiece. Galen's explanations are not materialistic. On the contrary, in all of his anatomical and physiological discourses his main purpose is the demonstration of a directive spiritual force, of a final and not causal determination.

During the Middle Ages nothing particularly important for our subject happened. It was at the very end of this period, i.e. in the 13th to 15th century, that new kinds of machines were constructed. It was realized that one machine can command others, but this phenomenon of a "life of relation" found no response in biological thought. The same lack of interest followed Villard de Honnecourt's construction of a mechanical "eagle." Of course, there were the wonderful achievements of Leonardo da Vinci, but his biomechanism was of the Aristotelian type, nothing more than a relatively simple mechanical imitation of some movements of articulations and extremities. Curious and far ahead of his times was Leonardo's study of the mechanics of phonation. Biomechanical analogies appeared, here and there, in the new anatomy of the 16th century, and we can guess their influence even through the title of Vesalius' *De humani corporis fabrica* (1543). With Rabelais the anatomy is overcrowded with mechanical metaphors: filters, press, cordage, pulleys, etc.

A Spanish physician, Gomez Pereira (born about 1500), tried to prove that animals were beings without sensitivity and thus very different from man with his sensitive, intelligent and immortal soul. Pereira's views were exposed in a rare and curious book entitled *Antonia Margarita; opus physicis, medicis, ac theologis non minus utile quam necessarium*, and published in Medina del Campo in 1554. Sensing how much his opinion is new and paradoxical, he chose a meaningless title. Actually, he gave to his work the name of his father and his mother! Probably it was a kind of camouflage to protect himself from the Inquisition. For the same reason he dedicated his work to the Archbishop of Toledo.

Pereira denied to the animals any kind of spontaneity. The basic statement of Pereira is that *"Bruta carere sensu."* Of course, he will not deny that a dog or a horse can "see" or "smell" his master or his enemy and that this makes him move in the direction of his master or fly from his enemy, but he affirms that this happens in an automatic way, and not as in man with the intermediary steps of psychic experience and mental propositions.

In Pereira's mind there are four modes of animal movements:

First, by direct action of present objects (*rebus praesentibus*) on sense organs and immediate responses of animals.

Second, by image of things (*phantasmatibus rerum*), that is, by action of some contents of animal memory. It is important to underline that these "images" are not imaginary in a narrow meaning of this term but real, corporeal emanations of things. These *phantasmata* are the very cause of animals' search for food and similar movements. One can guess between the first and the second of Pereira's modes the same kind of difference which Pavlov is making between unconditioned and conditioned reflexes.

Third, by others' predictions (*ab altero praedictorum*), as in the training of dogs or parakeets.

Fourth, by natural instinct (*ex instinctu naturali*), as, for example, the behavior of bees and ants. Here we are at the core of our problem, because Pereira includes in this group all internal movements of animals which we call physiological. He gives us an example, digestion.

Even if the basic supposition of Pereira is that a fundamental difference exists between animals and man, only the latter having a soul, this Spanish physician still taught that there is no basic distinction between physiological processes in both kinds of living beings: *"In hominibus ac brutis nutricatione similis motus his quarti generis conspici."* In both, the inner movements are automatic and directed by some kind of instinct. Pereira is very cautious and we cannot find in his work an explicit analogy between organism and machine. There are no allusions to any mechanical construction, but, without doubt, his book contains implicitly the idea of organic disposition as the only basis of animal external behavior and animal and human physiological processes.

But Pereira's conception of life is not mechanistic in the modern sense. His philosophy is still dominated by the notion of *qualitates occultae.* Thus, we can understand why his only explicit analogy is between animal behavior and magnetic attraction. Somebody without historical perspective could guess here a prescience of the tropism theory.

Pereira's paradox evoked no interest. He was attacked by a theologian from Salamanca, Miguel de Palacios, but he maintained his position and, in his answer, used, it seems, more clearly the notion of the beast-machine.

Descartes was accused of having been inspired by Pereira's work, but he denied it. It is true that the first edition of Pereira's work was hardly available to Descartes, but we cannot exclude some possibility of a direct or indirect knowledge of the second edition of Pereira's book (Frankfurt, 1610). Descartes, just like Pereira, called his hypothesis on animal automatism a paradox. Anyway, we shall see that Descartes was going much further, and the question of priority is in this case without significance.

One century later, when Descartes expressed his views, the philosophical background of European scientists was quite different, and the Cartesian explanation of organisms found an astonishingly receptive audience.

For a new revival of mechanical biology, it is necessary to look at the foundation of the new mechanics. No wonder that it was precisely the 17th century which gave testimony of the first mechanical theories of life in the modern sense. Under Galileo's influence a special school of medical thought was born in Italy, the so-called *iatrophysics* or *iatromathematics*. The new laws of mechanics, the introduction of quantitative experimentation, and the use of the microscope gave to the follower of the iatrophysical school the impression that all physiological phenomena could be and must be explained by simple physical properties of vital structures and movements. Two great achievements of this Galilean orientation of physiology were Harvey's discovery of blood circulation (1628) and Santorio's fundamental experiments on metabolism (1614).

We consider now that the chemical and physical interpretations of life are not contradictory, but rather complement each other. This was not the opinion of ancient scholars. For a long time the physical and chemical interpretations of life were in opposition. The mechanistic interpretation tried to explain all phenomena by the motion of body particles which were submitted to strict mechanical laws. In this view the essential properties of matter were neglected, or even denied. On the contrary, in the chemical explanation the "hidden qualities" (*qualitates occultae*) of matter were considered to be of primary importance.

We have now a strong feeling that mechanical models of life are the simplest ones. Surprisingly enough, this kind of apparent simplicity is not indicating a chronological, historical priority. Mankind began with very complicated, animistic and then pneumatic and humoral explanations which certainly were more chemical than physical. From the classical Antiquity till the Enlightenment there is in some ways a progressive simplification of biological doctrines. It reaches its peak with the faith in a complete reduction of animal and even of human organism to very rough mechanical devices.

In the 17th century the European scholars got knowledge of the construction of new types of machines in which the part responsible for order of the movement and the part generating energy are dissociated, reinforcing in this way the illusion of a kind of autonomous life. We can consider as a good example the hydraulic automata built in the royal gardens at Saint-Germain-en-Laye. A well-established tradition teaches us that it was precisely the spectacle of this mechanical microcosm in the king's gardens which, in a sudden illumination, gave to the young French philosopher René Descartes the idea of the beast-machine.

The first printed exposition of this fundamental Cartesian hypothesis is included in the very famous *Discours sur la méthode* (1637). Descartes said: "Nor will this appear at all strange to those who are acquainted with the variety of movements performed by the different automata, or moving machines, fabricated by human industry, and with the help of but few pieces compared with the great multitude of bones, muscles, nerves, arteries, veins and other parts that are found in the body of each animal. Such persons will look upon this body as a machine made by the hands of God, which is incomparably better arranged, and adequate to movements more admirable than in any machine of human invention."

In his *Traité de l'Homme* (composed probably about 1632 but published only thirty years later) Descartes proposes that the reader suppose the existence of a clay machine which in all its parts would be similar to the human body. Then he tries to explain all the functions of this machine by simple laws of mechanics, considering as motive force the heat which is centered in the cardiac cavity. Finally, Descartes leaves his reader under the impression that no difference exists between an artificial machine and a human being, except the power of reason.

At first glance Descartes' position is strange. He combines a strict mechanistic approach with a dualistic, idealistic philosophy. In his opinion there are two substances—the passive matter (*res extensa*) and the thinking soul (*res cogitans*). The soul is the essential part of the human being, and Descartes is even convinced that he knows the exact location of the soul. He believes that it is situated in the pineal gland, because this is the only important part of the brain which does not occur as a pair and is not divided into right and left. The principal function of the soul is to think, and as obviously we are not "thinking" our physiological processes, the soul is not acting on a physiological level. In Cartesian philosophy animals are without soul, and the vital reactions are not suffering from this absence.

In a letter to Henry More, dated February 5, 1649, Descartes insisted on a comparison between beasts and mechanical automata: ". . . since art is the imitator of nature, and since man is capable of fabricating various automata in which there is motion without any cogitation, it seems reasonable that nature should produce her own automata, far more perfect in their workmanship, to wit all the brutes." The most important difference between a nonliving automaton and an organism is the very special character of the central power of living beings. As Descartes explains: "Life I deny to no animal, except in so far as I lay down that life consists simply in the warmth of the heart." Thus, heat or fire is the first motor of the living machine; its primary source is the heart, or more precisely a kind of chemical process in cardiac cavities. Gilson demonstrated very well how much Descartes, by accepting the classical notion of *innate heat*, was under the influence of scholastics, which, on the other side, in his epistemology and mechanics he opposed so strongly.

If a human or animal organism is interpreted in a Cartesian way, as an automaton, one cannot escape the logical necessity to suppose divine intervention by the First Engineer. A complicated machine must be built by some superior intelligence. Then two possibilities can be envisaged. Let me express these in modern language: an animal-machine is either an automaton with cybernetic regulations and something like a program-tape inserted in it by the First Engineer; or it is a kind of car, or better a very complex factory, which cannot operate without permanent intelligent conduction and supervision. Descartes chose the first logical possibility, which was certainly very audacious on his part, for in his time nothing was known of feedback circuits and program records. We can now easily understand why he was not able to express clearly all the meanings of his beast-machine analogy: he was in search for a still nonexisting mechanical model.

Let us underline the repeating of some historical features: Like Erasistratus many centuries before, Descartes is intellectually fighting for the conception of a device which could be realized only by the engineers of the future. I guess that, at some critical stages in the development of so-called exact sciences, biology can offer a foresight to the next step because the biologists are challenged and inspired by the contemplation of the highest real existing structural systems. The beginnings of the cybernetics and the modern philosophical relationship between phenomena of life and electronics give us a new important example. And this is certainly not the end of the story.

Descartes compares the neuro-muscular system of his anthropoid machine with the church organ and its sets of pipes. This was surely not a perfect analogy. For organisms as a whole, Descartes was unable to find a better comparison than the works of a clock. And he was fully aware that this analogy is very partial too. Thus, he says that the dexterity of animals "rather shows that they have no reason at all, and that it is nature which acts in them according to the disposition of their organs, just as a clock, which is only composed of wheels and weights, is able to tell the hours and measure the time more correctly than we do with all our wisdom." A careful reading of this statement reveals that in Descartes' mind a clock can be compared with a living machine only by some of its partial aspects.

Actually, one of the common characteristics of the Cartesian beast-machine and a clock is the existence of a central and single source of energy. In one of my recent publications I tried to show how this supposition acted against the development of the fiber theory of organism in a direction which would be favorable to the foundation of the cell theory. Happily, this negative influence of Cartesian views was counteracted by the elaboration of the biological concept of irritability, created by Glisson but generally accepted only in Haller's time, i.e. one century later. Experimental evidence of the *vita propria* of parts of animals (for example the contractions of a freshly excised muscle as demonstrated by Stensen, Willis, Swammerdam, Baglivi and others) was the principal reason why, for the professional anatomo-physiologists, the Cartesian model remained an attractive but actually unacceptable hypothesis. Thus, paradoxically, the most important immediate followers of Descartes' physiological doctrine were pure philosophers and mathematicians like Mersenne or Malebranche.

In the 16th century the life sciences were dominated by descriptive methods of research, especially by description and classification of species and by anatomical investigations. Spectacular representatives of this type of biological research were Gesner and Vesalius. But in the 17th century physicians and biologists wanted to explain the processes of life and not only describe their material basis. The most important biological discovery of this period, Harvey's theory of blood circulation, was certainly more than an anatomical description. It was a new interpretation of functions, of internal dynamics. But it is profoundly linked with anatomical thinking; it is a kind of *animated anatomy*, i.e. "anatomy in motion," as Haller called it. Harvey's discovery resulted from the happy combination of two new methods, both linked with the school of Padua, where Harvey studied medicine. On the one hand was the Vesalian direct observation by a special kind of dissection, which was transmitted to Harvey by Fabricio d'Acquapendente. On the other hand was the Galilean quantitative method, a new conception of mechanics. The first method gave to Harvey a knowledge of venous valves, but it was only by application of the second that correct interpretation was possible. Harvey's discovery was not only that blood moved in closed circles, but that the motive force was a pump.

In Harvey's physiology the origin of movement was muscular contraction, a mechanical force in the strict sense. Descartes, one of the first great authorities to accept Harvey's theory of blood circulation, refused to interpret the heart as a simple pump. As we have still to see, in Descartes' mind the heart was a heat machine, something like an explosion motor. It was moved by periodical expansions of blood in the ventricles which were provoked by innate heat. For geometrical reasons (dilatation of heart when beating against the chest wall) and accepting the false medieval opinion of the synchronous "apparition" of the radial pulse and the diastole, Descartes was convinced that the expulsion of blood coincided with the diastole. Certainly if the heart is a pump, and it is one, its active phase is the systole. Harvey demonstrated easily that Descartes was on the wrong way.

The major idea of Descartes' physiology, that is, the analogy between organism and complex mechanical device, was quite tempting, and we can easily understand its success. But it must be emphasized that in concrete details, Descartes' theory was merely a series of shortcomings. Successfully applied in mathematics and physics, his deductive method was definitely misleading in biological sciences.

The human mind cannot deduce living phenomena from a few general principles. In this field of science, the Galilean and Baconian method of induction and experimental verification of hypotheses was the only suitable manner in which to progress. Thus, the Iatrophysical school in Italy and England followed this experimental way, not the Cartesian speculative philosophy. In spite of their common search for a mechanical model for life, Cartesianism and Iatrophysics have very different methodological approaches, and should not be mixed together, as is common in textbooks on the history of biology.

Galileo himself invented a good physical model to explain some biological patterns of fish. He understood perfectly the hydro-pneumatic mechanism of a fish's swimming bladder.

The most important scholar of the Iatrophysical school was a physician and mathematician of Naples, Giovanni Alfonso Borelli (1608-1679). His book *De motu animalium (On the Movement of Animals)*, published posthumously in 1680-81, was probably composed about 1660. This was the first concrete application of mathematics and mechanical principles to the study of muscular work. His mechanical interpretations of the walking of man and quadrupeds, the swimming of fish, and the flying of birds are excellent. He tried to explain by mechanical calculations the movement of internal organs, for example, the pressure of the stomach on food and the pump activity of the heart. He stated that the steady flow of blood from the arteries into the veins is the result of the elasticity of the arterial walls. Borelli failed, however, to obtain correct numerical results. This happened also to Johann Bernoulli (1667-1748), the mathematical genius who introduced infinitesimal calculus into physiology.

In spite of the scientific approach which we consider now to be essentially correct, numerical results obtained by the followers of Iatrophysics in the 17th century were very unsatisfactory. The basic method was correct, but many observational data were too roughly determined. It was only with the development of physical measurement instruments in the second half of the 19th century that the dream of the Iatrophysicists concerning the mathematical analysis of muscular movement was realized.

Borelli's teachings were different from Cartesianism in three major respects: (1) an empirical rather than speculative approach, (2) rejection of the theory of innate heat, and (3) the supposition of soul intervention in physiological processes. By a very simple and effective experiment, Borelli demonstrated the fallacy in Descartes' conception of the cardiac localization of animal heat. Measuring with a thermometer the temperature of the heart and other internal organs of a vivisected deer, Borelli discovered that the cardiac cavity is no warmer than other organs. Thus it was proved beyond all doubt that the heart is a muscular pump—a mechanical device—not a thermic machine.

In spite of his mechanistic views, Borelli refused to accept a completely determined automatism of living organisms. In his mind the animal body was not a clock, whose work could be strictly regulated mechanically. The Borellian model of an organism was an animated machine, whose activity was the result of the permanent intervention of the soul. To express it in anachronistic terms of our modern technology: his theoretical model was not a robot but a complicated car with a spiritual driver; this driver was not supposed to be identical with out intellectual principle but with a vegetative soul.

In Cartesian philosophy the creator of organisms was supposed to explain the teleological arrangement of parts, but after the first act of creation the intervention of an intelligent principle was considered superfluous. In Cartesian physiology God and soul were supposed, then immediately forgotten. In following centuries this attitude led to an uncompromising mechanical materialism, denying even divine creation. As in the time of Greek materialism, the apparently meaningful and purposeful behavior of organisms and the harmony of their structures were explained as being the result of natural selection. But this was really possible only after Darwin's patient and marvelous work. Between the time of Descartes and that of Darwin the purely mechanistic attempts to explain the phenomena of life were carried to naive extremes. Such explanations plainly oversimplified reality, and their conclusions expressed nothing more than a natural faith in theology. John Ray, writing at the end of the 17th century, expressed it in the title of his book, *The Wisdom of God Manifested in Works of Creation.* We will see that in the 18th century the analogy between organism and machine was the basis not only for materialistic thinking, but, at the same time, for most vitalistic theories, like Stahl's animism. If the world-famous Danish bishop and anatomist Niels Stensen (1638-1687) described the brain as a "marvelous machine"—of which it is essential to know the anatomical disposition—his interest is justified by his belief that the brain "is the principal organ of soul" (1669).

The general ideas of Iatrophysics were exposed in a most brilliant and provocative way by Georgius Baglivi (1668-1707), a physician from Dubrovnik in Dalmatia. He compared the jaws to pincers, the stomach to a retort, the veins and arteries to hydraulic tubes, the heart to a spring, the viscera to sieves and filters, the lung to a bellows, and muscles and bones to a system of cords and pulleys (1696). As Canguilhem pointed out, from the physical point of view these comparisons are not of the same order. Cords and hydraulic tubes are mechanisms of transmission, pulleys of transformation of movement; only the spring is a motor—that is, a device capable of originating movement. In all mechanical explanations of life the central role of the heart is obvious; Baglivi and Pacchioni tried to explain in a similar mechanical way the role of the brain. It was supposed to act by contraction of its membranous envelopes. This obsolete theory needs no further comments, but we should insist on a particular aspect of Baglivi's thought: his doctrine of the living fiber. In Baglivi's mind all physiological and pathological processes depend on fibers. They are some kind of elementary machines.

In the 18th century biology, a scientifically correct mechanical model of the living body must consider the fact that the organism has in all its parts (or, at least, in most of them) local and to some extent autonomous sources of energy. Baglivi tried a purely physical solution of this problem. He failed. Thereafter it was considered that no model of this kind can be imagined without making an appeal to chemical phenomena. In one way or another, practically all biologists of the 18th century used the old theory of Willis who explained the origins of movement in the living body by local chemical "explosions" in the muscles.

The chemical interpretation of life is older than the mechanical one. Ancient Hippocratic and Galenic humorism is in some way a chemical theory of life. But in the narrower sense the representation of life phenomena in technical terms of chemistry began only in the first half of the 16th century with the work of Paracelsus. He applied the technical knowledge of alchemy in order to understand how an organism functioned. In his mind, all living beings were composed of sulfur, mercury, and salt. But these Paracelsian constituents of the animal body were not chemical elements in the modern sense; rather they were conceived of as three functional principles. Paracelsus and his followers—Van Helmont and Sylvius, adepts of Iatrochemistry—studied vital phenomena by chemical methods, and they supposed that there was a perfect analogy between vital functions and chemical reactions taking place in retorts. Fermentation was considered to be a fundamental vital process. If Iatrophysics brought to science a better factual knowledge of animal motion, circulation of blood, the mechanics of respiration, etc. the Iatrochemical school was able to explain better the processes of digestion and the origin of animal heat.

It is a curious fact that until the 18th century the chemical conception of life was always linked to the supposition of intelligent spiritual principles. This vitalistic intelligent principle was called "archeus," "soul," or "vital force." In all the various theories, there was a basic supposition that such complicated "booking" could not take place without a chef, without the ideal pre-existence of special recipes. Iatrophysics and Iatrochemistry comprised a complementary approach to the problem of life, but followers of the two schools considered themselves as irreconcilable adversaries.

It was only in the 18th century that the chemical investigation of life began to be considered as part of a materialistic approach which was not in conflict with mechanical explanations. Hermann Boerhaave, the great Dutch scholar, was one of the men who tried an eclectic physical-chemical interpretation of physiological functions, and who understood perfectly that there is no conflict and even no fundamental difference between physics and chemistry.

In the development of mechanical theories during the 18th century, it is possible to distinguish three main tendencies. First of all, the consequent analogy between organism and machine underwent spiritualistic interpretation in Stahl's doctrine of animism. A human being was considered to be composed of an active, intelligent soul and a passive, machine-like body. The Cartesian soul was thinking, but not physiologically acting. Not so Stahl's *anima*, which was supposed to take part in all living activities. It seems paradoxical that the most vitalistic theory of living directing forces, as expressed by the school of Montpellier, was based on a mechanical interpretation of the execution of vital functions. As one example of this "vitalistic mechanism" we can consider P. J. Barthez' *Nouvelle méchanique des mouvements de l'Homme et des Animaux* (1798).

The second trend of mechanical materialism of the 18th century was represented by followers of Leibniz' philosophy. For them, matter was not something completely passive, but was interpreted as composed of animated monads, possessing the powers of perception and of motion. These new dynamic views were introduced into biology and medicine by Friedrich Hoffmann (1660-1740), who, in an extremely clever way, combined physics, chemistry, and biology. He gave his doctrine the name of "rational" system. Only the lack of greater knowledge of the facts can account for the weakness of this logically very consistent general interpretation of organisms.

The third trend was a consequent and absolute materialism, as, for example, expressed by the French philosopher and physician Julien Offray de La Mettrie (1709-1751). Descartes explained mechanically everything but intelligent action. He stated, "We may easily conceive a machine to be so constructed that it emits vocables, and even that it emits sounds correspondent to the action upon it of external objects which cause a change in its organs, . . . but not that it should emit them variously so as appositely to reply to what is said in its presence, as men of the lowest grade of intellect can do" (*Discours on the Method,* Part V). La Mettrie reacted against this spiritualistic component of Cartesianism, especially in his book *L'Homme machine* (1748). He wished to overcome (1) the Aristotelian principle of an immobile God and a God-like soul as the necessary origin of all movement, and (2) Descartes' notion of two substances. La Mettrie tried to give a mechanistic interpretation of the soul. In the introductory part of his excellent edition of La Mettrie's treatise, Vartanian emphasized that "besides the mind-body correlation, another essential feature of *L'Homme machine* is its attempt to prove that the organism as such possesses inherent powers of purposive motion." Actually, La Mettrie's concept of matter is nearer to Leibniz' dynamism than to the Cartesian notion of a passive *res extensa*. According to La Mettrie, all biological structures possess a self-moving power whose most evident expression is muscular irritability. "The human body is a machine which winds its own spring."

La Mettrie tells us that man is so complicated a machine that it is impossible to get *a priori* a correct idea about him. Descartes was wrong in reconstructing the human machine only by means of his own thinking. The study of man and animals must start with experience. "Thus it is only *a posteriori* or by trying to disentangle the soul from the organs of the body, so to speak, that one can reach the highest probability concerning man's own nature, even though one cannot discover with certainty what his nature is."

Medical practice taught La Mettrie that our feelings and behavior are dependent on drugs, disease, and other material factors. "A mere nothing, a little fiber, some trifling thing that the most subtle anatomy cannot discover, would have made two idiots of Erasmus and Fontenelle." It seems that the starting point of La Mettrie's materialistic interpretation of the soul was merely self-observation. According to Frederick the Great's eulogy on La Mettrie,

> During the campaign of Freiburg,* La Mettrie had an attack of violent fever. For a philosopher an illness is a school of physiology; he believed that he could clearly see that thought is but a consequence of the organization of the machine, and that the disturbance of the springs has considerable influence on that part of us which the metaphysicians call soul. Filled with these ideas during his convalescence, he boldly bore the torch of experience into the night of metaphysics; he tried to explain by the aid of anatomy the thin texture of understanding, and he found only mechanism where others had supposed an essence superior to matter.

Certainly the real determination of La Mettrie's thinking was more complicated. He was impressed by Newtonian physics and by Boerhaave's biochemistry and biophysics.

In the same year as the publication of La Mettrie's book, the French Academy of Science admitted to its membership Jacques Vaucanson (1709-1782), a simple mechanist whose principal claim to fame lay in his

---

*That is in Autumn, 1744.

construction of automata. He constructed a flute player; a duck that could move its wings, swim, eat, and excrete; and as asp which could hiss and dart on Cleopatra's breast. Vaucanson's machines demonstrated, or better, gave an illusion of demonstration, that man could mechanically depict living processes. Certainly all this seems now very naive. Today even a much more complicated robot than Vaucanson's flute player would not be seriously interpreted as an imitation of an organism. Thus we can hardly understand Vaucanson's philosophical background and we are surprised to see how big was the theoretical impact of his constructions upon physiologists. Recently, Doyon and Liaigre stressed the historical relationship between medical research and the construction of automata. They showed very well that the principal aim of Vaucanson's efforts was to explain the physiological phenomena by an almost perfect mechanical imitation considered indeed not as a simulation but as a real copy of nature's work. Vaucanson called his automata "moving anatomies" and actually he preceded his technical achievements by a detailed study of anatomy and physiology. His duck (constructed about 1733) was able to digest, that is, to swallow grains and, a little later, to eliminate some excrement-like matter. It was, in fact, a dishonest technical trick and not an imitation of digestion, but for us it is significant that so many people accepted Vaucanson's affirmations as a real solution of a complex physiological problem.

In 1741 Vaucanson announced to the Academy of Sciences that he has "in mind one automaton which would imitate with its movements some important animal functions, i.e., the blood circulation, the respiration, the digestion, the play of muscles, tendons and nerves, etc." The construction of a mechanical model of blood circulation was not a new idea. Soon after Harvey's discovery some attempts were made in this direction, especially in Germanic countries (for example, Salomon Reisel's plans for a "statua humana circulatoria" in 1674). And Vaucanson was not alone in France in working on "homme artificiel." A famous surgeon, Claude-Nicolas Le Cat (1700-1768) presented to the Academy of Rouen a *Dissertation sur un homme artificiel dans lequel on verrait plusieurs phénomènes de l'homme vivant* (1744). Such a mechanical automaton was supposed to have the value of a philosophical demonstration.

As I have pointed out before, all this seems rather naive. But our models of life phenomena, for example the analogy between the human brain and the computer, are actually not very different, philosophically speaking, from Vaucanson's and Le Cat's artificial man. In both cases some patterns of life are truly explained but the imitation includes only a part of the biological reality. In both cases life is compared to the machines which are not completely realized but only at the very beginning of their technological development.

Returning to La Mettrie, no doubt his materialism is, for the modern reader, shockingly simple and lacking in technical details. He confuses conditions and essence. The dependence of psychic phenomena on material conditions is not sufficient for a materialistic theory of the soul. Curiously enough, La Mettrie was still thinking in terms of Galenic humors, which "according to their nature, abundance, and different combination make each man different from another." La Mettrie is not giving the key to the great enigma, but he offers a program for the future. He influenced the materialistic philosophy of Helvetius, Holbach, and especially the psycho-physiological doctrine of Cabanis. Taken as a program of future research, La Mettrie's treatise is still valuable, and in 1928, Joseph Needham, an eminent English embryologist, published a book in which "the several arguments and observations of M. de La Mettrie are carefully considered, and although two hundred years old, are shown to be, in a sense, very justified."

Huxley, one of the most famous biologists of the 19th century and an apostle of Darwinism, defines physiology as "the physics and chemistry of life." In fact, the new physical and chemical methods of investigation introduced into biology after the Scientific Revolution in the 17th century—and applied to a greater extent in the 19th century—have become extremely useful. We are still impressed by the excellent results of physico-chemical analysis of living phenomena. Actually, the greatest progress in biology in the past 100 years has been due to this approach. The analogy between organism on the one hand and machine—or chemical factory—on the other was a powerful explanatory model.

Is this analogy only an explanatory model, and a very temporary one, or does it mean something more profound? The question is still open. Even Needham concludes with a positivistic statement, that the neo-mechanism "recognizes the supreme jurisdiction of the mechanistic theory of life, but admits at the same time to be a methodological fiction."

CHAPTER XIII

The Place of Normative Ethics Within a Biological Framework

Arne Naess
University of Oslo

The central question of my paper may be thus formulated:

Is there a place for normative ethics within the conceptual framework of biology—the term "biology" taken in a broad sense, as in G. Gaylord Simpson's view when he says that social science is a branch of biology.

Before answering, I must, I think, try to make the position *understandable* that the evaluations, norms and imperatives of normative ethics are autonomous in relation to description of how and why people act as they do in ethically relevant situations. I shall start with a reminder that not even science lacks certain normative features.[1]

Science contains prescriptions as well as descriptions

In trying to systematize *all* human knowledge, insight, and conjecture, it is clear that we do need both descriptive and prescriptive sentences or concepts of validity. There are rules of inference which can be exchanged for other rules, but to eliminate *all* rules, explicit and implicit, has never been done in a convincing way. Rule sentences form one of the classes of unavoidable prescriptive sentences in science taken as a whole.

Some rules are taken to be *valid*. The validity is not established by observation. The use of observation to establish the truth or likelihood of a hypothesis presumes some of the rules we wish to establish.

Let us take an example:

A: All whales are fish.
B: All fish have warm blood.
C: All whales have warm blood.

The inference of C from A and B, that is, the *transition* from premisses to conclusion is *valid*. The validity is independent of truth or falsity of any biological observations or theories. A reduction of validity to truth is not here possible.

Apart from *prescriptions* such as rules, systematized human knowledge contains *postulates*—descriptive sentences taken to be, and used as, true without test and perhaps in principle or at least *ad hoc* untestable. It has always been the aim of some scientist and philosophers to reduce the arbitrariness of postulates, and even to eliminate them, and let us wish them good luck. If a mass of knowledge depends in one respect on a postulate, the whole mass acquires a postulational character.[2] We don't like that state-of-affairs.

There has been among scientists a prejudice against normative ethics stemming from the unfounded and mostly implicit assumption that science or, more broadly, human knowledge and insight, is a vast ocean of plain, more or less truthful or probable information. But prescriptions, such as rules, and untestable postulates, are just as genuine parts or aspects as the informative. Scepticism toward objective validity of anything normative does not only affect normative ethics, it also touches science. The two products of human reflection, science and ethics, have prescriptive aspects in common.

Life from within. Existence of "musts," pure norms

In normative ethics we do not ask for truth, but for *a kind of* validity, just as in the general theory of inference. A rule may be invented, but not its validity. The validity is discovered, found, established. The normative ethician claims an analogous "objectivity" of his norms.

Let us try without prejudice to glance at the ethical life of man as seen from within, that is, as experience. One conspicuous feature is the existence judgments of good and evil experienced as *insights* into a realm, the realm of what is ethically right or wrong. As any other kind of insight it is experienced as *true*. One person would say to himself, "I should not steal this money from the pocket of my friend," "It *is the case*, I should not steal," "It is true, I should not steal it!" The philosopher may say: "*It is a fact that* I should not use any human being merely as a means." Sometimes we try to explain *why* one should not do a thing, sometimes not; I shall call pure norms and imperatives those the validity of which are not *completely* dependent upon the validity of descriptions and means-end norms (instrumental norms).

Suppose, as in the example of H. V. White, that shame *is to be adjusted to* by suicide. From within, the representatives of this code may correctly say "x *is* shameful, therefore he commits suicide!," not "x is *felt* to be shameful, therefore he commits suicide!", or "x is, according to the code, shameful, therefore he commits suicide!" Conceived from within, the imperative cannot be transformed without change in meaning into a statement in psychology, history, sociology or any other descriptive science. If *formulated* descriptively (x *is* shameful, therefore suicide is to be committed), one may say *ethical reality* is described, but this is perhaps only a misleading metaphor.

So much for the elucidation of the "purely" normative aspect of ethics and the understandability of the combination of objectivism and normativity.

Does biology presume the non-existence of pure norms?

The central question may now be stated as follows: does biology, as a science of life, presume that no valid norms are pure norms? That is, is the conceptual and methodological framework of human biology, as the perfectly general science of human life today, such that the existence of valid pure norms can neither be established, nor explained, if established?

Perhaps the existence of pure norms must be discounted because of the requirement within any biological research of testability. The testability of pure norms is doubtful, if not impossible. There is at least no generally recognized methodology of normative intuition or insight.

The question raised is, however, not whether a pure norm could possibly be part of biological knowledge or be stated as a biological proposition, but whether the *existence* of such norms must be denied because of the undefinability of them in terms of biology. It is, nevertheless, of some importance to assess the strength of the criticism of pure norms based upon the largely correct thesis of untestability. I shall argue that the criticism is weak: pure norms cannot be compared to particular scientific hypotheses, but rather to scientific principles—and are such principles testable?

When testability is predicated of particular hypotheses, it is done by presuming truth (or correctness) of certain theories and hypotheses made use of (but not tested) when testing the particular hypothesis. Thus, there is no *specific* test of this or that hypothesis. When testability is predicated of the larger unit of presumed propositions, it involves acceptance of basic rules and postulates within the science at issue, including its (normative) methodology. Thus, there is no definite requirement of testability, and if basic assumptions or principles are tried, the test is circular. The principle of contradiction affords the classical example: any test, if argued to be valid, seems to presuppose the validity of the principle. I do not know of any testability argument against the existence of pure norms which specify a requirement of testability that it is reasonable to expect fulfilled by any postulate or rule used in science.

Conclusion: the claims of the normative ethician have not been shown to be unsubstantiated because of lack of testability of pure norms.

Prevalent naturalism, subjectivism, utilitarianism. Reductionism

It is, I guess, a feeling among most biologists that the belief in pure norms is based on an illusion. Prevalent views are those of naturalism, utilitarianism or subjectivism, if one permits oneself to use some rather vague terms from the history of ideas. These are names both of metaethical views and ethical views. They exemplify *reductionism* in the sense that they either "explain away" the existence of pure norms—reducing them to nothing!—or offer equivalents in purely descriptive terms. Thus "x *is* shameful" is reduced to "x *is felt to be* shameful."[3]

Descriptive principles of survival and adaptation

As a fundamental principle we find among biologists some sort of assertion that all living beings necessarily try to adapt to the environment. Or that their actions are purposive, having—and necessarily having—the survival of the species as the goal (e.g. human beings). Other reactions are taken to be symptoms of dysfunctioning, sickness, mental illness.

If this kind of general assertion is taken as the major premiss, and an assertion that every specimen of *homo sapiens* is a living being as the minor premiss, the inference is valid that all men necessarily try to keep alive or keep the species flourishing. The ethical imperatives and other injunctions are, in order to conform to the inference, interpreted in a heterotelic or instrumental sense. "You shall not steal" is, for instance, taken as shorthand for "Having the survival of the species, and therefore ordered society, as a goal, you ought not to steal!"

But needs and demands are not the same for all animals, and the formidable development of the brain in human beings may have introduced something entirely or almost entirely new. For the first time in the animal kingdom, it seems, organisms are able to grasp their own situation in life. They can clearly understand that they are born and will die, that they have fellow creatures with equal needs as themselves, they are able to give a picture of the history of their own species and that of the universe. Each enlightened human being is a microcosmos. Not only needs but also demands for meaning, purpose, justice are developed along with other faculties. The imperatives thus created are clearly not translatable into instrumental norms such as "it is instrumental to the survival of the species or tribe that members do not steal from each other." Such norms are empirical hypotheses. Empirical evidence might show that stealing, preparatory for a good revolution, might be helpful in making the members effective looters during the revolution. This, in turn, might help the survival of a tribe. But survival is for enlightened human beings only one value among others.

More specifically, such norms are sometimes developed in a manner that neither the life of the individual nor that of the species is taken to be supreme. P. W. Zapffe advocates gradual extinction of *homo sapiens* through birth control. Reason: life cannot possibly satisfy the needs created in organisms with the capacity for thinking, reflection and feeling made possible by the development of the human brain.[4] Socrates did not feel individual extinction to be an ultimate evil: "You can kill me, but you cannot hurt me."

If biological principles of "adaptation," "survival," etc., are to be applied in all seriousness to human beings, there must be developed "intra-organismic" concepts, such that the voluntary death of an individual organism or a culture or nation may be compatible with the highest form of adaptation and struggle for survival—not of the species, but of a principle defined by the organism itself. (Justice, truth, honesty, mercy, etc.).

As for the utilitarian goal of "happiness" or "well-being," we remember that people have rather been letting themselves be burned at the stake than recant their convictions (Michael Servetus).[5] To say that they found they were more happy inside than outside the fire is, I think, at the best an arbitrary extension of the meaning of "happiness." They were as if confronted with a wall they could not break through: They *could* not retract. It was not a case of preference. With respect to their behavior and from the point of view of textbook biology, they could of course retract, and they certainly *wished* to retract. We have examples of people under torture *trying* to give in and convey information, but who did not succeed. They could not bring themselves to do the only "rational." Here the expression "It *is the case that* something should or should not be done" is particularily revealing. Or: "I *cannot* do this...," where it is physically and "behaviorally" quite possible to do it.

I have mentioned these cases because if biology is to cover, in principle, human life as a whole, it must explain, and not explain away, the mechanisms by which aims and goals are found and discovered which may have "biological death" as necessary consequence.

These strange mechanisms create an *inner environment* more potent than the external one and also new concepts of reality, the historical and the normative, in addition to the external. To declare that these concepts have no denotation or extension, that nothing is subsumed under them, is to take a stand within the realm of metaphysics. And whatever metaphysics a biologist is inclined to believe in, it is a prudent thing to try to keep his biological conceptual framework neutral. That is, not to *presume* the existence of pure norms and normative reality to be illusory.

What does the normative ethician require? Primarily a kind of determination by inner environment

Through the foregoing I hope to have made tolerably clear what phenomenon I am referring to as *validity of a pure norm*. That *it is so*, because something should or must be so and so.

The normative ethician requires of a biological conceptual framework, if biology in principle includes a science of human life, that the human, as a human organism, is capable of conceiving, adhering to, following or breaking pure norms. The content of these norms must be taken to be part of the (inner) environment of these peculiar organisms.

The biological framework would thus have to permit the distinction between the act of *following* the norm and the norm (itself). This does not automatically require any *Norm-an-sich* analogous to a Kantian *Ding-an-sich*, it is sufficient to conceptualize the everyday distinction between "following a rule" and "the rule itself." What is required of the biologist is here the conception of a mechanism of inner determination, something that can explain how a pure norm has an impact upon an individual and makes him try to follow it.

We look, strangely enough, for a *kind of determination*, a *lack of choice*, not only the indeterminism required for initiating alternative chains of events, as is usually required by ethicians. Socrates and others experienced *a lack of alternatives*, a fierce kind of coercion. But only after sorting potential alternatives in a preliminary way. An analogy: if I write on the blackboard two equations

$$e^{\pi \cdot i} = i^2 \qquad\qquad e^{\pi \cdot i} = -i^2,$$

mathematicians may be in doubt for a moment which one is correct, if any, but they end up with a lack of alternatives, lack of choice. They see that $i^2 = -1$ and $-i^2 = +1$, and conclude decisively that the first formula is correct (and deservedly famous), the second incorrect. They go through some potential alternatives, but end up with no choice. The *same kind* of experience is characteristic of basic ethical decisions.

Suggestions how biology could meet the requirement of the normative ethics

How *might* biology meet the requirement of the ethician?

There is a way to understand the gradual development of a self with phenomenal strength, integrity, and independence—asymptotically reaching an absolute freedom.

For such a self, self-preservation may well take forms rather different from the self-preservation of less developed organisms, but one might still talk of preservation. (Spinoza has the required conceptual framework.)

If I say to myself "Do A not B!" and I consequently do A, I require as "ethician" that I *could* have decided otherwise, that it was *my* decision. The universe *minus* "me" *was not so* that A had to be decided upon.

Let me try out a tiny formalism.[6] Suppose that the "I" (or rather the self) that makes a decision has a genesis, or history, not only a momentary structure or state, and suppose that $S_{n-1}, S_{n-2}, \ldots$ are specific numbered states at time $n-1, n-2 \ldots$ before the state S_n in which I committed myself to "Do A, not B!" The ethician will require that there is at least one S_{n-i} at time $n-i$ such that it was in part a consequence of a decision "Do C, not D!"

by me that the next state was a certain S_{n-i+1} and not $S'_{n-1+1'}$ that latter having occurred *if* I had decided to do D.

This might involve us, however, in an infinite regress. If the development from one state to another takes time, we would have to suppose never to have been born. Accepting what we have been told about our birth date and our mother and father, there must in terms of the model be a *first* "free" decision. But its sudden occurrence out of nothing would be hard to explain. We shall therefore probably have to satisfy ourselves with concepts of graded freedom.

Is then nothing gained for the ethician?

Let us just for fun think that causal weight can be measured, and that for each free (but not absolutely free) decision, 50% of the causal weight belongs to the "I decide"-factor. This means that after one previous free decision, the "I decide"-factor next time will have 50% plus 50% of the first 50%, that is, 75% freedom. At the time of the third choice, the "I decide"-factor would be responsible for 87.5% of the causal weight, then 93.75%. There is thus an asymptotic convergence toward 100%. After one thousand increasingly free decisions, a decision may without hesitation be said to be completely and fully *mine*, the self may be said to be an autonomous entity, and the inner environment built up to full capacity.

Does behavioral science of ethics presume subjectivism?

Let us assume we can accommodate such a mechanism in an organism that a near absolute freedom in the above sense, "power to decide otherwise," is reached.

Would the norm announcements or pronouncements of that organism have to be classed as "subjective" on a par with utterances such as "I like honey" or "I like you to look upon me as your king"? Is there a possibility of taking notice of a claim of ethical objectivity and validity?

My answer would be that if the organism under consideration is a scientist, e.g. an astronomer or biologist, and if we can as biologists of "knowledge" accord objective validity to that organism's utterances about the distance and composition of the sun, or about rats learning to find their way in mazes, then there is also room for attributing objective validity to the norm-pronouncements of that organism.

Behavioral science of science. Maze epistemology

The argument for this equivalidity thesis of some cognitive and some normative utterances is as follows:

Let us consider the methods of behavioral science of science, a field of inquiry well within the framework of biology. The metascientist describes the scientist in his work, his observing, deducing, experimenting, writing and speaking. The reality which the scientist tries to catch, and the agreement or disagreement of hypotheses made by him, must by the metascientists be described ultimately in terms of behavior of the scientist. The behavioral scientist of biology does not have his own absolute biology or his own private way of reaching biological reality. He is totally dependent upon what his "animal" tells him, that is, upon the biologist—in so far as he wishes to be *directly* informed about biological matters.

The situation can be dramatically illustrated in reference to the classical study of learning in rats with the help of mazes. The scientist presumes that his perception and knowledge of the maze is better than that of the rat and in a way absolute: the scientist knows how the maze *really* is and is able to describe adequately the struggle of rats to learn it. The successive approximations of learning are taken to be approximations in "adjustments" to the "environment" identifying the environment in terms of the *scientists'* conceptions. Behavioral science of science rests on the illusion that scientists are the rats of metascientists, and that the metascientist can sit back in his chair describing the struggle of scientists to adjust and find out about the environment, the solar system or the human brain. But the metascientist is cheating: he asks his rats to describe what they are *trying to find out*. The rats whisper back "we try to map out *the solar system*" or "we try to find out how *the brain* works." The metascientist has no direct access to the "world" or "reality" denied to the scientist, his biology of human knowledge is a fake.[7]

Consequently, whether an organism utters "The sun is on the average 49 million miles distant from the earth" or "You should never use another person solely as a means," the contents or function of the utterance (including the claims and tests of agreement with the so-called reality) must be accounted for by means of observation of the organism itself, that is, ultimately in terms of behavior. But this means that the level of subjectivity will, in principle, be the same, whether the object studied is the reference of "the distance to the sun is 149 million km" or "A person should never be used only as a means."

Cultural value of objectivism today

Objectivism of pure norm is a controversial topic in contemporary discussion. If one can speak of a main trend in Anglo-American ethics in the 1940's and 50's, it is a trend away from objectivism. Today the picture is more complex, however, and no such main trend is discernible. On the popular level, there is, I think, a steady increase of the tendency to look upon ethical norms as norms of utility: the good is what is useful for society, well-being, increasing pleasure and comfort and eliminating pain. It means, as far as I can see, a trend toward ignoring norms of dignity, independence, standing up against pressures of groups, personal integrity—norms which may cost immensurably in terms of pain and discomfort. Apart from philosophical reasons, there are, as I see it, general cultural reasons for keeping alive at least the understanding of what objectivism asserts and what it requires or implies if a biological conceptual framework is made a framework of life science in its complete generality.

NOTES

[1] The modest program and thesis of understandability should not be confused with the ambitious thesis that at least one pure norm is objectively valid. When we look through the list of proposed examples of valid pure norms, it is normal to find that one cannot accept them unconditionally. We find exceptions, if's and but's.

The proposed norms are, however, more or less general. For the affirmation of the existence of one pure norm, it is not necessary that any norm is discovered that holds for more than one single situation, and it is not necessary to find an interpersonally unambiguous formulation.

With reference to the controversy on pure norms, I might class the following as opposed to the non-cognitivism of Stevenson, Ayer, Hägerström, Ross, Kelsen and others: G. E. Moore (*Principia Ethica*, 1903), H. A. Prichard (*Moral Obligation*, Oxford, 1949), F. Brentano (*The Origin of the Knowledge of Right and Wrong*, transl. London 1902), M. Scheler (*The Nature of Sympathy*, transl. New Haven 1954), Nicolai Hartmann (*Ethics*, transl. London 1932), R. Reiner (*Gut and Böse*, Freiburg 1965), F.S.C. Northrop (*The Complexity of Legal and Ethical Experience*, Boston 1959), W. D. Ross (*The Foundation of Ethics*, Oxford 1939).

A sceptical epoch in these matters is manifested by G. E. Moore in his latest writing, see *The Philosophy of G. E. Moore* (ed. Schilpp, New York 1942, p. 543 *et seq.*), and by myself, in "Do we know that basic norms cannot be true or false?" (*Theoria*, 25, 1959) and in "We still do not know . . ." (*Theoria*, 28, 1962).

[2] This statement cannot hold without modifications when the "mass of knowledge" in question is increased until it covers *the whole* of knowledge. By means of certain modifications, the otherwise inevitable conclusion that knowledge as a whole is postulational, can be avoided.

[3] Thus reduced, ethical statements have the same kind of validity as *other* sociological and psychological statements. Either it is the case that I feel shame, or it is a mistake. Either a man conforms to the role of a physician in a definite kind of society or he does not. "Naturalistic objectivism" is one of the professional labels of the ethics obtained by transforming pure norms into psychological or social science statements. Ethics thus conceived does not pose

any special problems for biology. Therefore, I have not treated it in more detail.

[4]The philosopher and dramatist P. W. Zapffe compares the overdevelopment of the human brain with the overdevelopment of horns in Megaceros euryceros. Cf. *Om det tragiske*, Oslo, 1941.

[5]Michael Servetus would not admit predestination. Rather be burnt on slow fire. He might have said with Nietzsche "Woran ich zugrunde gehe, das ist für mich nicht wahr," only that the "I" here is not the narrow "I" of textbook biology. See Stefan Zweig's *Ein Gewissen gegen die Gewalt*.

[6]An alternative formalism might be tried out enlarging S. Körner's approach to "physically effective and independent choices," p. 216 *et seq.* in his *Experience and Theory*, London, 1966.

[7]For a more comprehensive treatment of "maze epistemology" see my "Science as Behaviour: Prospects and Limitations of a behavioural Metascience," in *Scientific Psychology*, ed. B. B. Wolman, Basic Books, 1965, p. 63 *et seq.*

CHAPTER **XIV**

The Evolutionary Thought of Teilhard de Chardin

Francisco J. Ayala
The Rockefeller University

It is only with considerable reluctance that I have accepted to present a summary of the evolutionary cosmogony of Teilhard de Chardin at this Colloquium. It appears hardly possible to condense in a few pages a system of ideas which attempts to provide a comprehensive account of the history of the universe and of the place of man in the scheme of things. The difficulty arises not only from the wide scope of Teilhard's undertaking but particularly from the individual characteristics of his writings. Teilhard's language is in fact largely poetic rather than philosophical or scientific; he frequently introduces neologisms and uses other terms with a different meaning than is commonly attributed to them. Teilhard is at times inconsistent—words are employed in different contexts with disparate meanings, not always specified. Finally, he freely uses poetic metaphors and analogies that are sometimes placed in contexts which represent them as proofs, and often extended beyond their original application.

To attempt, then, a systematic presentation and a fair criticism of Teilhard's ideas requires a considerable amount of exegesis and interpretation. According to some critics it is questionable whether the time and effort required by such a demanding task are warranted by the intellectual value of Teilhard's scientific or philosophical contributions. I shall limit my endeavor to attempt a brief presentation of what I consider to be the central ideas of Teilhard's cosmogony. The evaluation and systematic criticism of such a system of ideas—or rather of such a vision, since Teilhard's contribution is more visionary than systematic—I shall leave to the participants of this Colloquium during the following discussion period.

I will make use as far as possible of Teilhard's own words by selecting those passages which are most representative and precise. The quotations will be taken from the most comprehensive presentation of his ideas, *The*

Phenomenon of Man.[1] The exposition will be in the collection *Perspectives in Humanism* under the editorship of Ruth Nanda Anshen. When the galley proofs were sent to France, some 25-odd passages, many of them extremely significant, were suppressed as a condition for the publication of the volume. The tragic lesson of this incident is that a similar censorship may have been applied to other works of Teilhard.

Time: the Fourth Dimension

The greatest discovery of modern science is, for Teilhard, the discovery of time as a constitutive element of reality. We cannot define matter in terms of the three spatial dimensions any more. To understand the material world we must describe it in terms of time as well as space.

The awareness of the significance of time came about in the modern mind as a product of the Darwinian revolution, whose importance is comparable to the intellectual revolution brought about earlier by Copernicus and Galileo. Biological evolution, formulated scientifically in the middle of the XIX century, has taught modern man that organisms are continuously changing. The concept of evolution was then extended to the whole realm of matter.

It was only in the middle of the nineteenth century, again under the influence of biology, that the light dawned at last, revealing the *irreversible coherence* of all that exists. First the concatenations of life and, soon after, those of matter. . . . Time and space are organically joined so as to weave, together, the stuff of the universe. That is the point we have reached and how we perceive things today. (*The Phenomenon of Man*, p. 217. Italics as in the original.)

Evolution, understood in its broad sense, means that reality as we know it has not appeared suddenly, instantaneously, but rather that the world is the result of a gradual and progressive development lasting hundreds of millions of years, and still in progress. The theory of evolution implies that we live not in a finished world, but rather that the universe is in a process of "cosmogenesis." The universe is a world on the making. Therefore, Teilhard writes:

Is evolution a theory, a system or a hypothesis? It is much more: it is a general condition to which all theories, all hypotheses, all systems must bow and which they must satisfy henceforward if they are to be thinkable and true. Evolution is a light illuminating all facts, a curve that all lives must follow. (*The Phenomenon of Man*, p. 218.)

To understand Teilhard's thinking we must realize that for him this dynamic view of the universe conforms to and is illuminated by the Christian view of the world. Jean Daniélou, a noted Jesuit theologian, has pointed out (*Études*, February 1962) that the chief difficulty preventing Christian faith in the past from making some connection between salvation history and cosmic history has been that while salvation was viewed as a dynamic movement in time, the world of nature was understood as a static and inert mass. Theodosius Dobzhansky has noted that Christianity is a religion implicitly evolutionistic in that it believes history to be meaningful. "Its current flows from the Creation, through progressive revelation of God to Man, to Christ, and from Christ to the Kingdom of God. Saint Augustine expressed this evolutionistic philosophy most clearly." (Th. Dobzhansky, *Mankind Evolving*, New Haven and London: Yale University Press, 1962, p. 2.)

According to Teilhard the history of salvation and the history of the universe are coming into focus as two dimensions of a unique reality. The divine purpose of cosmic history is ultimately salvation history. Salvation is seen as the religious dimension of cosmic history which relates the evolution of the cosmos to a higher order of being—the supernatural realm, the domain of religion.

Universal Evolution

According to Teilhard all levels of reality—man, organisms and nonliving matter—are genetically related. The *fact* of evolution cannot be questioned, although one may ask whether such process follows a discernible *direction*.

> That there is *an* evolution of one sort or another is now, as I have said, common ground among scientists. Whether or not that evolution is *directed* is another question. (*The Phenomenon of Man*, p. 141. Italics as in the original.)

Before discussing the question of *direction* in evolutionary processes, let us examine how Teilhard sees the movement of expansion by which the various levels of reality come to be.

The history of our planet appears as a *continuous* flow of events and changing configurations. The continuity of the process makes any division into eras or stages more or less artificial. Such compartmentalizations are, however, useful since they emphasize the main events of the historical process and consequently they help us to understand it. It is, then, possible to divide the history of our planet into three main stages:

(1) The period during which the earth crust solidifies by cooling. There is not any life as yet.

(2) The appearance of life, and the gradual development and expansion of the living forms. According to present estimates, life appeared on earth at least three billion years ago, considerably earlier than was commonly estimated during Teilhard's life. Fossil remains of certain primitive organisms found in South Africa have been dated as 3.1 billion years old.[2] After its appearance life gradually became more complex and diversified. This second period of the history of our planet can in fact be characterized by this

marvelous expansion of living matter. The planet, says Teilhard, is now covered by a new envelope, that he calls the *biosphere*, i.e. the rich variety of organisms living almost everywhere on the crust of the earth.

(3) During the last two million years a new development has occurred on the earth. This phenomenon has such novel characteristics that it adds a new dimension to our planet—the dimension of the spirit. Emerging from the biosphere, Man entered the world as a rare animal somewhere in Africa. Rapidly, he has populated the earth and has given it a new appearance. The earth has been covered by a second envelope, the envelope of thought, what Teilhard calls the *noosphere*.

The three main stages in the history of the world are, then, characterized by the following three terms: matter, life, and spirit or thought. They contain the three levels of reality that we perceive around us and which constitute the total reality of the earth—the geosphere, the biosphere and the noosphere.

There is, for Teilhard, an essential difference between inorganic matter and life, and between life and thought. The three levels of reality are nevertheless intimately linked by a genetic relationship. The biosphere proceeds from the geosphere, and the noosphere develops from the biosphere.

That life originated from inorganic matter is, according to modern science, at least the most likely hypothesis. The present state of the question can easily be summarized. Science has not yet succeeded in synthesizing life in the laboratory, but a majority of scientists accept the natural transition of inorganic to organic matter as a working hypothesis. They consider the origin of life from inorganic matter as the most rational and likely explanation of the origin of life. This hypothesis is supported by some indirect evidence showing the possibility of the transition from inorganic to organic matter under the primeval conditions of the earth.[3]

The origin of life from inorganic matter must be considered, according to Teilhard, as the natural result of the process of "maturation" of matter. In the evolution of matter, higher levels of complexity are gradually reached. When a certain level of complexity is attained, altogether novel properties appear. Emergence of new properties occurs whenever a certain critical point or threshold is reached. The possibility and fact of such transitions, according to Teilhard, neither involves major philosophical difficulties nor contradicts the principle of universal causality.[4] His explanation, however, is only an analogy, hardly more than a metaphor.

> In every domain, when anything exceeds a certain measurement, it suddenly changes its aspect, condition or nature. The curve doubles back, the surface contracts to a point, the solid disintegrates, the liquid boils, the germ cell divides, intuition suddenly bursts on the piled up facts . . . Critical points have been reached, rungs of the ladder, involving a change of state—jumps of all sorts

in the course of development. (*The Phenomenon of Man*, p. 78. Italics as in the original.)

Teilhard's favorite representation of the evolutionary process is a spiral—a continuous line turning upwards in which, however, different turns or levels can be distinguished.

The transition from inorganic matter to life happened only once.

> Naturalists are becoming more and more convinced that the genesis of life on earth belongs to the category of absolutely *unique* events that, once happened, are never repeated. This is a much more credible hypothesis than would appear at first sight, if we succeed in forming a tenable idea of what is hidden in the history of our planet. (*The Phenomenon of Man*, p. 100. Italics as in the original.)

Life evolved gradually giving origin to numberless forms, the animals and plants which exist today or have become extinct in the past. Eventually, certain organisms reached a great degree of complexity. Within one of the evolutionary lives of development a new critical point, another threshold was reached. A change of state occurred and a new phenomenon appeared in the realm of life—reflective conscience, man.

Teilhard has emphasized the distinction between man and animal.

> The animal knows. *But it cannot know that it knows:* that is quite certain. If it could, it would long ago have multiplied its inventions and developed a system of internal constructions that could not have escaped our observation ... Because we are reflective we are not only different but quite other. It is not a matter of change of degree, but of a change of nature, resulting from a change of state. (*The Phenomenon of Man*, pp. 165-166. Italics as in the original.)

We may never know all the details of the evolutionary development leading to man from his nonhuman ancestors. In fact, says Teilhard, we may never understand *how* the gradual transition to reflective knowledge may occur, as we do not completely understand the development of intelligence in a child. The process is continuous. But we can see that a new turn of the spiral, a new rung of the ladder has been reached. The appearance of reflective thought was a decisive event in the history of the earth. The planet has acquired a new dimension, the dimension of the spirit, the dimension of thought.

The Parameter of Complexity—Consciousness

The universe is subject to an evolutionary development toward novel phenomena. Teilhard asks now whether the evolutionary process as we observe it has any direction. That is, whether evolution follows any deterministic law, and if so, what is the nature of that law. His answer is that, without leaving the scientific method, we can ascertain a general law of recurrence that defines and determines the course followed by the evolutionary processes—the law of complexity-consciousness. Let us examine briefly how Teilhard arrives at the existence of such a law, and how he defines it. We are at the center of Teilhard's cosmogony.[5]

If we examine the history of the world as a whole, we can ascertain that the evolutionary process is oriented toward greater levels of complexity. It always proceeds from less to more complex kinds of things—from the subatomic particles to the atom, from the atom to the molecule, from the molecule to the cell, from the cell to relatively simple multicellular organisms, then to more complex organisms, and finally to man. Man is the most complex being that exists in the world. In man we find all the previous levels of complexity, and all of them exceeded.

Complexity for Teilhard means not simple aggregation, as it occurs in a pile of sand, nor simple repetition, as in the phenomenon of crystallization. Complexity means "organized heterogeneity"; it implies an association of the component elements closed in itself. A pile of sand or a crystal are open systems in the sense that they are never externally finished. Complexity implies something which is externally completed. Moreover, it implies that the component elements are related according to a certain pattern, i.e. they are organized. An increase in complexity occurs when there are more component elements related by a higher level of organization. Higher complexity implies a greater concentration, or as Teilhard sometimes puts it, greater centreity. He considers it to be an undeniable fact that higher levels of complexity have been progressively reached in the evolutionary process.

The increase in material complexity represents only one aspect of the phenomenon of evolution. The scientific evidence of the evolution of the world shows that another development has also occurred, namely a gradual increase in the level of consciousness. The process of evolution is oriented both toward higher complexity and toward higher consciousness. Such orientation can be observed by the naturalist as a naturalist. The law of increasing complexity-consciousness belongs to the realm of natural science, since it is derived from direct observation; it is not a philosophical or metaphysical postulate.

In any kind of thing, Teilhard distinguishes two aspects—its external complexity and its interior consciousness. Complexity and consciousness are intimately related; they are two aspects of the same reality, like the two sides of a coin. A greater level of material complexity is always accompanied by a higher degree of consciousness.

> Whatever instance we may think of, we may be sure that every time a richer and better organized structure will correspond to the more developed consciousness. (*The Phenomenon of Man*, p. 60.)

Complexity and consciousness are, in fact, the two defining characteristics of matter.

> Spiritual perfection (or conscious "centreity") and material synthesis (or complexity) are but the two aspects or connected parts of one and the same phenomenon. (*The Phenomenon of Man*, pp. 60-61.)

For Teilhard there is no dichotomy between matter and spirit, or between inorganic and organic matter. The intrinsic relationship between material complexity and consciousness provides a tangible "parameter" which allows us:

> To connect both the internal and the external films of the world, not only *in their position* (point by point), but also . . . *in their motion*. (*The Phenomenon of man*, p. 60.)

Teilhard used the term consciousness in a very generic way by extending its meaning beyond its commonly accepted use.

> The term "consciousness" is taken in its widest sense to indicate every kind of psychism, from the most rudimentary forms of interior perception imaginable to the human phenomenon of reflective thought. (*The Phenomenon of Man*, p. 57.)

The question remains of how Teilhard justifies the predication of the term "consciousness" of all matter, inorganic as well as organic. According to him, the evolution of the animal world shows that the higher animal forms are characterized by a greater development of the nervous system, particularly of the brain. Greater morphological complexity is accompanied by a higher psychic level.

> Among the infinite modalities in which the complication of life is dispersed, the differentiation of nervous tissue stands out, as theory would lead us to expect, as a significant transformation. *It provides a direction*, and by its consequences *it proves that evolution has a direction*. (*The Phenomenon of Man*, p. 146. Italics as in the original.)

Teilhard finds the key to the direction followed by evolutionary processes in the gradual development of the nervous system which has accompanied the evolution of animals from the invertebrates to the higher vertebrates. The two developments, morphological complexity and psychism, which occur in the animal world, are taken as characteristic of all evolutionary processes. The parallel between external complexity and consciousness is extrapolated to the whole realm of matter. In all beings, organic or inorganic, there is a certain level of material complexity and a proportionate level of consciousness, even when we cannot directly observe the presence of psychism. The fundamental property of matter is its innate tendency to become organized in always more complex structures that for the same reason possess a more powerful psychism.

Teilhard's central idea is that the stuff of the universe possesses two dimensions or aspects—the external and the internal. The exterior side is constituted by the external dimensions of matter and the relationships among the material components. The internal side of reality is the phenomenon of psychism which is coextensive with matter. The internal side of things, i.e. consciousness, is also a cosmic phenomenon that must be incorporated in any phenomenological description of the universe.[6] The law of complexity-consciousness is at the core of Teilhard's vision. Being a necessitating law, it allows us not only to understand the evolutionary past but also to look into the future.

Omega: the Goal of Evolution

The history of the earth can, then, be seen as a gradual progress toward increasing complexity and consciousness. Many lines of evolutionary development have gone toward extinction or were lost in blind alleys. There is a "privileged axis," however, along which evolutionary progress continues; that is the evolutionary line of descent leading to man.

> Man is not the center of the universe as was naively believed in the past, but something much more wonderful—the ascending arrow of the great biological synthesis. Man is the last born, the keenest, the most complex, the most subtle of the successive layers of life. (*The Phenomenon of Man*, p. 223.)

Evolution has led to man. Evolution will continue following the necessitating law of complexity-consciousness. Higher levels of psychism will be reached. The goal toward which evolution is tending is called by Teilhard the "Point Omega," which is also Teilhard's symbol for God. Teilhard describes Omega as

> a harmonized collectivity of consciousness equivalent to a sort of superconsciousness. The idea is that of the earth not only being covered by myriads of grains of thought but becoming enclosed in a single thinking envelope so as to form, functionally, no more than a single vast grain of thought on the sidereal scale, the plurality of individual reflections grouping themselves together and reinforcing one another in the act of a single unanimous reflection. This is the general form in which, by analogy and in symmetry with the past, we are led scientifically to envisage the future of mankind. . . . In the direction of thought, could the universe terminate with anything less than the measureless . . . ? (*The Phenomenon of Man*, pp. 251-252.)

Teilhard is aware that his vision of Omega coincides with Saint Paul's notion of the final condition of the world, when all mankind will be united with Christ in a single Mystical Body and everything shall reach consummation in God. Again, it seems appropriate to point out that for Teilhard, cosmic history and salvation history are two perspectives of the same grand evolutionary development. In that sense at least, Teilhard's is a religious vision.

NOTES

[1] Pierre Teilhard de Chardin, *The Phenomenon of Man*, New York and Evanston: Harper Torchbook Edition, 1961, 318 pp. Like most of the significant works of Teilhard, *The Phenomenon of Man* was published posthumously. During his lifetime the publication of Teilhard's writings was largely suppressed by his superiors of the Jesuit Order. Unfortunately, the suppression of his writings still continues after his death in 1955. There is a volume of his letters to be published in the coming months by the New American Library, developed under the following four headings: (1) time: the fourth dimension; (2) universal evolution; (3) the parameter of complexity-consciousness; (4) Omega: the goal of evolution.

[2] E. S. Barghoorn and J. W. Schopf, "Microorganisms three billion years old from the Precambrian of South Africa," *Science*, v.152 (1966), pp. 758-763.

[3] Scientists have synthesized in the laboratory biologically active deoxyribonucleic acid (DNA), the chemical carrier of hereditary information. The synthesis of DNA, however, is accomplished by using a molecule of DNA as a template or primer. In any case, the artificial synthesis of an organic cell remains at best a remote possibility.

The classical discussion of the origin of life from inorganic matter is A. I. Oparin, *The Origin of Life*, New York and London: The MacMillan Company, 1938. For a modern discussion of the problem, see Gösta Ehrensvärd, *Life: Origin and Development*, Chicago and London: The University of Chicago Press, 1962, and Sydney W. Fox, ed., *The Origins of Prebiological Systems*, New York and London: Academic Press, 1965.

[4] As it is explained below, Teilhard assumes that traces of life, or as he puts it, of consciousness, are present in inorganic matter.

[5] To understand Teilhard's concept of the parameter of complexity-consciousness one must read, besides *The Phenomenon of Man*, one of his latest long essays, *Le Group Zoologique Humain*, Paris: Albin Michel, 1956, 172 pp.

[6] When Teilhard speaks of the "internal" and "external" side of things, he is obviously using a metaphor. The same idea is sometimes described by him by means of other metaphors, like the "within" and "without" of things, their "tangential" and "radial" energy, etc. It is not always clear whether

"material complexity" and "consciousness" are, for Teilhard, two different "principles of being," to use an expression from scholastic philosophy; or rather they represent two different ways of describing the same phenomenon. The idea that some traces of mind are universal has also been proposed by some philosophers, like A. N. Whitehead and C. Hartshorne, and by some biologists, such as B. Rensch, L. C. Birch, and Sewall Wright.

CHAPTER **XV**

The Use of Biological Concepts in the Writing of History

Allen D. Breck
University of Denver

Much of the controversy over the question whether historiography is an "art" or a "science" is derived from the encounter of nineteenth-century historians with biologists and with the popularizers of the Darwinian hypothesis. That it was a two-way encounter, with profit on both sides, there is no doubt. It should be possible, therefore, to assess the impact of the "grand style" of the historian's vision on the development of biological theory, to show how the ways in which historians took for granted such concepts as "growth," "development," "rise and fall," "emergence," and produced theories which accorded well with mid-Victorian optimism in a world becoming increasingly Europeanized. It would be fascinating to explore the fact that historical writings in the nineteenth century, with its frequent emphasis on the growth and development of institutions, races, and nations, was a considerable part of the air which biologists breathed, and that the impact of historical thought on biological reasoning was not inconsiderable.

The problem set for this paper, however, is to see this meeting of ideas from the other side. To what degree, and under what circumstances, it may be asked, were historians of the nineteenth—and early twentieth—centuries indebted to such discoveries as those of Charles Darwin? To what degree were narrative historians living in a variety of national societies influenced by the idea that human historical progress depended on natural laws, and, consequently, that the concepts and methods of the new evolutionary biology were adequate for the study of social processes? Whatever expropriation there was can be considered under two headings, that of *metaphor* and that of *formula*.

Among the Greeks, and in Europe since the Renaissance the metaphorical comparison was constantly made between the society of men and the appearance of natural phenomena. Thucydides, for example, was much influenced by medical practitioners, a fact which makes a careful reading of

217

The Peloponnesian War scientifically rewarding. To pick another example at random: typical of the use of terms concerning the natural world is a remark of Martin Luther's, when he speaks of monopolists as those who "oppress and ruin all the same merchants as the pike the little fish in the water, just as though they were lords over God's creatures and free from all the laws of faith and love."[1] Again, in the eighteenth century Johannis Nicolaus Funccius published five books in Latin entitled respectively (in translation) *The Childhood, Adolescence, Imminent Old Age, Vigorous Old Age,* and *Broken down Old Age of the Latin Language.*[2] But these instances (no matter how frequent in no way implied the adaptation of a world-view that considered "nature" to have a history of its own, and believed that the human species was part of that natural development. Such a revolution was to come only with the nineteenth century.

Biological metaphors indeed came easier to that century and its historians, but by the middle of the century it became difficult to disentangle the threads of three quite different theories of development: that of Hegel, that of the French and German Romantics, and that of natural scientists, whose eventual formulation was to be found in Darwinian evolutionary theory. Thus, when a historian of the time speaks of "progress," "evolution," or of the importance of "tradition," the reader must know something of the writer's background to see clearly from what theory the appeal has been taken. With or without the aid of scholars and thinkers in the field of "biology" (a term which was coined apparently about 1800, simultaneously in France and in England), enthusiasm for progress and the "wave of the future" led historians and other writers to trace the evolution of humanity through a definite progression. Thus the whole thrust of the Romantics was founded on an assurance of progress. Condorcet (1743-1794), with his discovery of three historical epochs (primitive past, progressive present, and irreversible future) and August Comte (1798-1857), with his law of the three stages (theological, metaphysical, and scientific) were part of the boundless optimism and heady sense of progress which made the future better than the present and the present superior to the past. The way to Darwinism was well and early laid.

I heard the other day of a student who said excitedly that "when the Renaissance came the Italians were ready for it!" It is less absurd to say that when Darwinism became generally known with the publication of *The Origin of Species* in 1859, the English and their continental contemporaries were ready for that. The almost instant success of the book was due perhaps to two factors: the felicitous phrasing of many paragraphs and the creation of usable metaphors. The title alone contained four most useful terms: On the *Origin of Species* by means of *Natural Selection*, or the Preservation of *Favored Races* in the *Struggle for Life.* Together or separately, they were unbeatable. And although Darwin might remark that he had only birds in mind with his sub-title, because "everyone is interested in pigeons," his popularizers and adapters knew better. The application of such metaphors to mankind could be made without straining the mind, or hopefully, the facts, too far. So much for metaphor.

A second expropriation from the biologists, and in particular Darwinian theories, is the development of formula in history writing. Beyond the use of metaphor lay the possibility of certain almost cosmic, universal processes, which by some sort of natural law saw that the fittest did indeed survive; hence it might lie within the realm of historical investigation to discover these overarching themes particularly if the evolution of institutions, states, or the like, could be made clearer. It is the contention of this paper that Darwinism (and genetic biology in general) affected practicing historians in their use both of metaphor *and* formula, but that a natural disrelish for far-reaching and demanding themes in the human past led many of them to disavow various formulas from time to time but unconsciously to accept the metaphors in their writing.

It is a truism, of course, that Darwin did not discover evolution, but was preceded by almost a century of men who had one insight or another into evolutionary processes.[3] Darwin's new idea was that "evolution" came about by natural selection, a process much like the breeding of animals. To the man in the street (and historians as well), however, Darwin was the innovator of evolutionary thinking in biology. To many historians, indeed, he gave fresh evidence of processes in the world of nature which they had come to conclude were taking place in the realm of history.

It is indeed curious that the popularity of biological science did not have an even greater influence on historiography, to the extent of causing more historians to see social laws in history which paralleled physical laws in evolution and genetics. The use of these hypotheses would indeed be "scientific history" of a sort, and a few did succumb to the lure. Perhaps, however, the development, particularly in Germany, of another sort of scientific history, coming from seminars which taught critical methods, and the development of internal criticism of the sources, produced something analogous to both science and art but which is sufficiently different from both so that history could be considered an independent third discipline.

What, then, *was* usable in Darwin and his predecessors? Surely not the whole work of any natural scientist, from Lyell and Lamarck to Darwin. Although any well-read man of the century could hope to "know" a science, and indeed many were inveterate readers and collectors of scientific books and journals, it was often the popularizers and interpreters of the classic thinkers whose works were better known to the public. To such readers out of the mass of scientific data offered them, certain propositions seemed worthy of defense or expropriation, or attack. First was the apparent destruction of the concept of design in the universe, or at least of the end of the concept of the Great Chain of Being which had assured man that this world was dark, the heavens filled with light, a controlling and deliberate providence on high. Now, on the contrary, one thought (despite the poetic and somewhat forced conclusion to the *Origin of Species*) that man moved upward, from primeval forms, from simple animal origins. It was the old order of nature reversed. Development was caused by variation and struggle, by survival and adaptation, by great purposive (but by no means determined) upward thrusts.

How well much of this new theory fitted with the spirit of progress, of nationalism, and the moving intellectual and physical frontier of the nineteenth century! In England, the spirit of the great Crystal Palace Exhibition of 1851 was still with people—England on the march, the Empire being won, progress everywhere! As for the continent, Carleton J. H. Hayes was right: "It would be difficult if not impossible to account for the immense vogue of sociological and philosophical Darwinism were it not for the spectacular wars which from 1859 to 1871 seemed to attest to its truth."[4] Weren't wars the testing-grounds for the survival of the fittest and the best? Darwin was, of course, popular for what he did *not* say: there is nothing here about the origins of life itself, the nature of the Creator, of the origins of races and nations. Struggle and elimination, yes:

> As many more individuals of each species are born than can possibly survive; and as, consequently, there is a frequently recurring struggle for existence, it follows that any being, if it vary however slightly in any manner profitable to itself under the complex and sometimes varying conditions of life, will have a better chance of surviving, and thus be naturally selected.[5]

But so far, only birds and beasts. What was needed for the historian was more explicit application to man and his social relations. This link was supplied by the popularizer and the interpreter. For the English and American historian, the mediating role of Herbert Spencer was spectacular. In the "System of Synthetic Philosophy," the reader could see straight progress from the here and now into ultimate perfection. What had in the past been depressingly simple was not promisingly complex; here was movement clearly discernable, from primitive savagery to present civilization. Where there had once been anarchy, he saw order, and knew law had succeeded brute force in the solution of human problems. How the words rolled off the tongue!— "survival of the fittest," "evolutionary stages," "recurrent social processes," "natural selection," "environmental adaptation."

James Harvey Robinson lyrically expressed the double sense of "scientific history":

> But the middle of the Nineteenth Century the muse of history, *semper mutabile*, began to fall under the potent spell of natural

science. She was no longer satisfied to celebrate the deeds of heroes and nations with the lyre and shrill flute on the breeze-swept slopes of Helicon; she no longer durst attempt to vindicate the ways of God to man. She had already come to recognize that she was ill-prepared for her undertakings and had begun to spend her mornings in the library, collating manuscripts and making out lists of variant readings. She aspired to do even more and began to talk of raising her chaotic mass of information to the rank of a science.[6]

The Spencerian *mystique* fell upon some who were convinced against their will. Thus Oliver Wendell Holmes wrote:

He is dull. He writes an ugly uncharming style, his ideals are those of a lower middle class British Philistine. And yet after all abatements I doubt if any writer of English except Darwin has done so much to affect our whole way of thinking about the universe. Of course he often is nor more than a *vulgarisateur*, but that was still more true of Sir H. Maine, and Sir H. M. covered much less ground.[7]

"Our whole way of thinking" could now include, thanks to such men as Spencer, the transposition of the metaphor and the conceptualization of biology to the study of human society: the transition was in many minds a small one. "A society is an organism," said Spencer, and one could easily think of "the state" or any other institution as a biological entity, possessed of a life history similar to that of the animals and the plants.

This life history was a *gradual* one the Victorian and his continental contemporaries believed, one emphasizing open-ended progress and only small increments of change. The old Latin tag, *natura non facit saltum*, nicely summed up the attitudes of a class which, since Burke, had felt considerable resistance to abrupt social change, an aversion to fiery revolutionaries and earnest rejecters of "the proper line of progress." Darwinism seemed to promise an understanding of long, slow, evolutionary changes, of the importance of the community of the present with the past, a searching for the long-range collective wisdom of western man, the Church, the Anglo-Saxon and Roman legal systems. For we should not be blinded by the pyrotechnics of the Wilberforce-Huxley debate over "Darwinism" into thinking that all historians rejected a major part of the new biological thesis: quite the contrary.

Such theories proved to be what lawyers call "an attractive nuisance," either confirming earlier conclusions of the narrative historians (as in the case of Edward Augustus Freeman) or leading them along the path of happy discovery that nations did indeed survive from a struggle of the unfit with the fittest. James Anthony Froude, for instance, completed in 1870 a stirring twelve volume epic of the "deserved triumph" of Protestant England over Roman Catholic Spain, and in 1874 treated the rise of English power in Ireland in similar terms. Was it not true, and could not history demonstrate the fact, that "the superior part has a natural right to govern; the inferior part has a right to be governed?" A good example can be taken from the writings of Sir Henry Maine (1822-1888), who, in the words of Sir Frederick Pollock "did nothing less than create the natural history of law." In his work (*Ancient Law*, 1861; *Village Communities in East and West*, 1871, and *The Early History of Institutions*, 1875) he did the same sort of thing with which Darwin was busying himself in studying the evolution of the species. Here were the origins of civilization in the west: Aryan, German, Latin and Greek beginnings of the rich development of law in Anglo-Saxon civilization. From this time onward in England (roughly, the last third of the nineteenth century) the tracing of any configuration of customs or institutions to their origins served as an adequate discussion of the *causes* of the institutions with which Englishmen were surrounded. The grand themes were all there, and a conservative historian could see that he did no violence to the narrative when he admitted that the facts all seemed to be on Darwin's side.

The approach of the Marxists to Darwinism in the third quarter of the century was, however, quite different. There was, it would appear, an early marriage of Marxian theory with Darwinism. Friedrich Engels read the *Origin* in 1859, and wrote Marx that Darwin had completely destroyed teleology in biology and had proven the historical-materialistic development of nature,[8] and Marx replied to another comment that "Darwin's book is very important and serves me as a basis in natural science for the class struggle." But the very element which was seen by many at the time to be the distinguishing mark of Darwinian as opposed to other biological conclusions was to prove to be the "unfortunate" concept of the survival of the fittest.

In a letter of June 18, 1862, Marx wrote ironically:

> I am amused by the statement of Darwin that he applies the "Malthusians" theory to plants and animals *also*, whereas the whole point of Mr. Malthus lies in the fact that he does *not* apply his theory to plants and animals, but *only* to men—with geometrical progressions—as opposed to plants and animals.
>
> It is splendid that Darwin again discovers among plants and animals his English society with its division of labor, competition, opening up new markets, "inventions" and Malthusian "struggle for existence." This is Hobbes's *bellum omnium contra omnes* and reminds one of Hegel in the *Phenomenology* in which civic society is expressed as the "spiritual animal kingdom," whereas with Darwin the animal kingdom represents civic society.[9]

For Marx there was indeed struggle, but it was the familiar one of economic classes pitted against one another. Another sort of social and historical process was also at work:

> Darwin did not know what a bitter satire he wrote on mankind, and especially on his countrymen, when he showed that free competition, the struggle for existence, which the economists celebrate as the highest historical achievement, is the normal state of the *animal kingdom*. Only conscious organization of social production, in which production and distribution are carried on in a planned way, can lift mankind above the rest of the animal world as regards the social aspect, in the same way that production in general has done this for men in their aspects as species.[10]

This new attitude marks the change in Marxist thought away from the use of biological idioms and insights and toward the use of metaphors and analogies from the physical sciences. The line of descent from Marx and Engels to Lenin and contemporary Russian historians seems to have been derived from Darwinism only in gross anatomy, and without the physiology of the system of the "survival of the fittest," which made it attractive to conservative historians.

It is true that Marx had later a higher opinion of Darwin and wrote unavailingly in 1880 to ask permission to dedicate a volume of *Das Kapital* to him. At the grave of Marx, Engels compared the two, but found their theories to be parallel rather than derivative:

> As Darwin discovered the law of evolution in organic nature, so Marx discovered the law of evolution in human history; the simple fact, previously hidden under ideological growths, that human beings must first of all eat, drink, shelter and clothe themselves before they can turn their attention to politics, science, art and religion.[11]

The reception of Darwin in pre-Marxist Tsarist Russia, however, is another story.[12] A study of the origins of social theory among the *intelligentsia* shows that young radicals welcomed with special sympathy the first translation of *The Origin of Species* in 1864 and found the whole work to constitute a sensible corollary to Newton's physical science. With true Russian enthusiasm, however, they read more into the story, perhaps, than the Darwinian traffic would bear. The final paragraph of the *Origin* had such an optimistic teleology that their conviction of a happy outcome to the human story was confirmed—although the inhumane struggle for existence did go on in nature, it did not of necessity happen in society.

Did not the laws of nature point rather to cooperation—to *Sobornost*, to centuries-old Russian feelings of "togetherness" and "purpose"? As the goal of individuals was surely cooperation and human solidarity, so the progress of society could now be measured by the index of the amount of "collectivity" it expressed. Prince Peter Kropotkin (1842-1921), the author of a popular work on the French Revolution, was cheered by Darwin's conception of progress, but deplored the degree to which Darwin's followers had reduced the struggle to one for survival—"to its narrowest limits." Russians, he believed, could do much better.[13]

Perhaps the most significant, or at least the most interesting, of those Tsarist historians who used the concepts and terminology of the biological sciences was Nicholas Y. Danilevsky (1822-1885). He had been trained in the natural sciences at St. Petersburg University, where he received a master's degree in botany in 1858. His later field-work in Siberia and his reading of the productions of the German historian Ruckert inspired him to develop in detail a theory of the "cultural-historical types" of mankind. In his celebrated study, *Russia and Europe* (*Rossiya i Evropa*, 1869), he rejected the possibility of a single "European civilization" and posited in its place a variety of racial groups, whose individual history he found to be guided by "natural" (read "biological") laws. Each people was thus an organism, which passed through different stages of development. The types were few.

For him, there were only ten varieties, each of which had gone through childhood, adolescence, maturity, old age, and eventual death, within an allotted number of years. We have thus the Egyptians, Chinese, "Ancient Semites," Indians, Iranians, Hebrews, Greeks, Romans, Arabs, and Europeans. As each was divisible into species, or races, we are brought around to the Slavs, a race both distinct and superior. With modifications, we can see the work of Oswald Spengler, who obviously had read Danilevsky, though perhaps not with great care.

Spengler (1880-1936) had studied mathematics and natural science as an undergraduate, but as a young teacher was already inveighing against the conclusions of Darwin, particularly those which he thought assumed that human progress was easy and biologically predictable. The publication of *Der Untergang des Abendlandes* at the end of the first World War in 1918, however, was revealed as a work laden with biological terms. It offered a complete model of the inevitable development of contemporary spiritual, cultural, and political epochs, strung along a chain of seasons: spring-summer, autumn-winter, from birth to death, from creativity to predictable oblivion. The work was a mixture of Nietzschean, Hegelian, Romantic, and other attitudes toward life and history, with an unusually heavy use of biological metaphors. Indeed, it is safe to assume that his discovery of "morphological" history was simply an enlargement of the biological and botanical metaphor that had intrigued historians of the last decades of the nineteenth century. Much as he might inveigh against the "trivial optimism of the Darwinian Age," his own conception of organic struggle and growth seems to be directly derivative from that of Darwin and his popularizers.[14] Yet it must become obvious that his use of terms is strangely different: "race," "morphology," "style," all are significantly different in the hands of this German *savant* from that of scientific language derived from the field and the laboratory.

In the United States there was both ready popular acceptance and scholarly rejection of Darwinism. Evolution had been popularly discussed in newspapers and periodicals since the publication of Charles Lyell's *Principles of Geology* in 1832 and Robert Chambers' *Vestiges of Creation* in 1845. Popular historians such as Bancroft, Parkman, and Motley had seen in the evolution of American people and their institutions an orderly development, in which the relative role of superior and inferior peoples was clearly evident. Like Irving and Prescott, they acclaimed Jefferson when he said that "history is the record of a nation advancing rapidly to destinies beyond the reach of mortal eye." All this narrative was, of course, metaphor, and acceptable equally to the prevailing Hegelianism, Romanticism, and scientific insight of the day. Americans, famous for their anti-ideological posture and their boundless optimism that their rapidly changing way of life would lead to an even brighter tomorrow, were ready for a Darwinism that validated their feelings with scientific support.

Thus, when the first copy of the *Origin of Species* appeared in the United States it seemed to affirm what historians had long been saying, if not in content, at least in style and metaphor. With the twin advent of Darwin-Spencer on the one hand and the scientific methodology of the German historical seminars on the other, a change in method, in content, and in style, became imperative. One has only to compare Francis Parkman's *Discovery of the Great West* (1869) with Turner's *The Rise of the New West* (1906) to see how profound the change from Romanticism to a form of genetic historiography had been. Romantic values—the individual, the irrational, the excited revelation of the remote past—were swept away in the impersonal tones of a "realistic" historiography, the methodology for which lay in the precise and impersonal research techniques of the biologists and other scientific investigators.

Justin Winsor, mirroring a common Romantic belief, had once said that history was "all shreds and patches," but now (certainly after 1870) the historian could begin to see a dynamism within that self-explanatory, self-contained model of the animal species *homo universalis* which was the American people. Institutions could (and should) be the object of historical thought, and the individual the exception, or, rather, the confirmation of universal presuppositions. There were, of course, by exception works such as those of George Bancroft, which conveyed, without any sense of incongruity the idea that the single, though perhaps not the immediate, causal factor in the great sweep of American history was the Creator Himself, Who spoke in the familiar words of Jacksonian democracy.

That word was spoken in 1860, but between then and the turn of the century a majority of historians had been won over to the new "scientific history." One element of Darwin's thought caught on rapidly with middle-class people and with a large part of the academic community. The application of the principle of natural selection to the development of human society, with special emphasis on competition and struggle, is familiar to us as "social Darwinism." Did not history "prove," one asked, and was it not self-evident that "the fittest" (that is, those who had survived in the grim economic struggle or in the competition of nations) were in the natural processes of human history so constituted that they survived?

Among the first American historians to be caught up in the net of biological necessity was the forceful and ebullient John Fiske (1842-1901), whose conversion to Spencerian Darwinism was almost complete. Of the theory of evolution, he commented:

> To have lived when this prodigious truth was advanced, debated, established, was a rare privilege in the centuries. The privilege of seeing the old isolated mists dissolve and reveal the convergence of all branches of knowledge is something that can hardly be known to the men of a later generation, inheritors of what this age has won.[15]

Fiske's admiration for Darwinism began when he was eighteen, and lasted through his honeymoon (when he took Spencer's works with him) and into the classroom, finally coming to rest in writing of history itself. For him, the Creator was still in command of the process, so an understanding of the goal of history could be suitably combined with a search for the origins of contemporary institutions—indeed, the one demanded the other. The discovery of evolution revealed that:

> the things of the world do not exist or occur blindly or irrelevantly, but that all, from the beginning to the end of time, are throughout the farthest illimitable space connected together as the orderly manifestations of a Divine Power.

As an historian, therefore, he sought the origins of American institutions in the European past—the New England town in the forests of Germany, the founding fathers in the signing of the Magna Carta, even the federal system in the leagues of Greek city states. The question must be raised, however, whether the basic narrative of Fiske's story of the American people was truly dependent on his reading of Spencer and of Spencer's rendition of Darwinism. A careful search in the twenty-two volumes of his collected works gives only a qualified support to the view of such dependence. His attitudes toward evolution are perhaps best seen in "American Political Ideas Viewed from the Standpoint of Universal History" (1885). Here, as in many of his speeches, which were immensely popular on the lecture circuit, he conveyed to

Americans ready to hear them, the concepts of the immutable laws of biological growth and the natural superiority of Anglo-Saxon institutions. There were biological impulses at work, his hearers were told, which made the English a people fit for populating and governing the whole world. If this were true, why, indeed, should that impulse be denied?

Here, however, we may discern a converging trend, wholly outside the ken of the scientist, but one which gave support to such ideas of inevitable development. German Romanticism was to furnish what may be called the "evolutionary germ theory" of institutional and "folkish" development. Brought to England in the Sixties, the methodology and focus of German scholars led such people as Bishop Stubbs and E. A. Freeman to concentrate on the origins of institutions, on the record of progress from simplicity to complexity. Whatever was, was not necessarily right, but further heady progress was possible *if only* the past were better known—so ran the argument.

Many of these ideas came with great force to American historians, so such a man as Herbert Baxter Adams at Johns Hopkins soon attempted to find the key to historical law in the doctrine of institutional evolution, much as did his contemporaries in Germany and England. But America was not Europe—here, on the contrary, was a long tradition of sturdy, independent, American "do-it-yourself" reform on the one side, and on the other the apparently almost limitless opportunities on the disordered western frontier. Both would seem to argue against any sort of determinism or a Darwinian or any other variety. What could the historian say to such a confrontation? Was a synthesis of systems possible?

We now come to two historians (one a professional, the other an amateur) who did see American development in biological, even Darwinian terms—Frederick Jackson Turner and Theodore Roosevelt. Somewhere between them is the optimist-turned-pessimist, Henry Adams. Born on the frontier in Wisconsin, Turner produced a masterful (though somewhat ambiguous) lecture on "The Significance of the Frontier in American History" in 1893, and thereafter a number of works on frontier and sectional history. Although he had been trained by Herbert Baxter Adams at Johns Hopkins, he reacted against Adams's respect for the European past and substituted for it a study of the magnificent sweep of the continuing American frontier. In 1888 he had identified his heroes as "Darwin, Spencer, and Lincoln." He was never to borrow *directly* from the Darwinians however; rather he seems to have had a growing conviction that the history of America could be explained wholly in terms familiar to readers of Spencer the sociologist—from simplicity to complexity, from primitivity to civilization, from agrarian America to the new frontier of the cities—in sum, the analogy of organismic change from youth to maturity. "The United States," he said, "lies like a huge page in the history of society. Line by line as we read this continental page from West to East we find the record of social evolution."

In biological terms, here was continuity, adaptation, improvement in terms of environment, the whole evolutionary process. On the frontier, out of the stimulus of the environment, were born such traits of bravery, independence, cooperation, the ability to survive, as could be transmitted socially, much as Lamarck had seen acquired characteristics transmitted biologically. With Turner, narrative history almost disappeared, the great movements and individuals of the Romantics have faded into the distance, to be replaced by types and movements more familiar to students of the natural sciences.

The full acceptance of scientific laws came to full flower from Henry Adams. Darwinism was in the air when he wrote of an early experience (1867-8):

> He felt, like nine men in ten, an instinctive belief in Evolu-
> tion . . . Natural Selection led back to Natural Evolution and at
> last to Natural Uniformity. This was a vast stride. Unbroken
> Evolution under uniform conditions pleased everyone—except
> curates and bishops; it was the very best substitute for religion; a
> sage, conservative, practical, thoroughly common-law deity. Such a
> working system for the universe suited a young man who had just
> helped waste five or ten thousand million dollars and a million
> lives, more or less, to force national unity and uniformity on
> people who objected to it; the idea was only too seductive in its
> perfection; it had the charm of art.[16]

He (Adams) was engagingly frank, and conceivably typical of many historians of his day when he said: "Henry Adams was Darwinist because it was easier than not, for his ignorance exceeded belief, and one must know something in order to contradict even such triflers as Tyndall and Huxley."

He had great hope for social evolution when he began teaching at Harvard; there, his course in the Middle Ages began with primitive forms of organization, from which moved like a great evolutionary stream the institutions of classical and medieval society. He was never to give up some sort of belief in this type of social evolution: we find him speaking in his presidential address to the American Historical Association (1894), "The Tendency of History," of hope that the historians would be able to do for society what Darwin had done for nature. But his essential pessimism led him eventually to finding truth in the Second Law of Thermodynamics, in the formulation that energy would finally disappear from the universe, so that devolution, rather than evolution, was the first order of nature.[17]

On the other hand, Darwinian concepts of natural selection and the survival of the fittest found a trenchant defender in Theodore Roosevelt (1858-1919), who noted a natural order of bravery, manliness, and self-improvement at work in the world, all of which made for progress. In the famous speech of 1899, "The Strenuous Life," he commented that "in this world a nation that has trained itself to a career of unwarlike and isolated ease is bound, in the end, to go down before other nations which have not lost the manly and adventurous qualities." Indeed, in the whole "Winning of the West," to which he devoted five volumes, could be found "the culminating event in the whole mighty history of racial growth from the German forest tribes to the present." The relationship between the civilized man who had reached a peak and the barbarian yet in the shadows below was clear in his mind:

> Such a barbarian conquest would mean endless war; and the fact that nowadays the reverse takes place, and that the barbarians recede or are conquered, with the attendant fact that peace follows their retrogression or conquest, is due solely to the power of the mighty, civilized races which have not lost their fighting instinct, and which by their expansion are gradually bringing peace into the red wastes where the barbarians hold sway.[18]

Greatly inspired by the drama of racial expansion, Roosevelt could identify the Turks, the Abyssinians, the Moraccans, as the barbarians, and the Greeks, the Italians, the Spaniards and other Europeans as the civilized peoples. We are a long way, in a straight line, from Darwin's finches! A profound change was, however, at the same time under way.

When, then, and under what circumstances, did historians, professional and otherwise, find thinking in Darwinian terms less useful? I would hazard as an approximate date the end of the first World War. This is not to say that here and there one does not find a historian who still thought in terms of evolutionary (read Darwinian) development. Robinson, himself, for instance, could still see all history in such terms—the struggle between privileged and unprivileged groups, the former of which had written almost all history down to his time, the latter mostly "written about." But his methods and ideas about physical and biological science were those of the nineteenth century— much faith, little scepticism, concentration on the large-scale model, the formulation of permanent laws, a preference for gradual change over catastrophe. The work of Hugo De Vries might never have taken place.

But great forces were at work. The end of the frontier, the catastrophic events of the 1914-1918 war, the recession that followed—all were upsetting to a theory of easy ascent and the perfectibility of the human animal. Consequently, historical and evolutionary interests lost their commanding positions in the social sciences.

The transition from an older acceptance of Darwinian assumptions and methodology as normative for the writing of history to the vastly different assumptions of our own day is consequently best represented by the work of

Charles Austin Beard, a man who confessed himself unable to be more than a "guesser in this vale of tears," who hoped that the guild historian would end his aping of the natural scientist and return to his own sphere, that of "historical actuality." For him, science was simply *one* of the realities which were the components of history—along with the state and other factors in the development of mankind.[19]

His presidential address to the American Historical Association in 1934 was entitled "History as an Act of Faith." For Beard events could be subsumed under three categories. He was a cautious exponent of the third alternative:

> The first is that history as total actuality is chaos, perhaps with little islands of congruous relativities floating on the surface, and that the human mind cannot bring them objectively into any all-embracing order of subjectivity, into any consistent system. The second is that history as actuality is a part of some order of nature and revolves in cycles eternally—spring, summer, autumn, and winter, democracy, aristocracy, and monarchy, or their variants, as imagined by Spengler. The third is that history as actuality is moving in some direction away from the low level of primitive beginnings, on any upward gradient toward a more ideal order—as imagined by Condorcet, Adam Smith, Karl Marx, or Herbert Spencer.[20]

Since Beard's day, the historical profession has encountered two significantly different interpretations of historical processes. The first would affirm the germ-developmental theory, but would insist that in our time a "great mutation" has separated the slower evolution of the centuries from our own rapid social and technological growth. The second group would replace "gradualism" of a Darwinian sort with a new "catastrophism" in human affairs. But that is another story. Let us now summarize our thinking about the earlier chapter in this intellectual adventure—the nineteenth century.

1. History writing is an inclusive, compositionist discipline, whose practitioners work with metaphor and formula, being often influenced by the prevailing world-view of people around them.

2. In the nineteenth century the prevailing mood of writers of history was one which saw vast evolutionary changes in human institutions—in the law, the church, the state, in custom and theory alike. Consequently, scientists such as Darwin and his predecessors found the way prepared in the popular reading (history had a *good* press) for ideas congenial to physical evolution.

3. In turn, the ideas of physical evolution (slow, inevitable, adoptive, irreversible) influenced greatly numerous historians who found comfort for ideas of nationalism, racism, progress in the upward sweep of their group, class, or whatever.

4. The reception of "Darwinian" concepts varied from country to country but was everywhere a significant part of the intellectual baggage of western man.

5. Biological terms and formulas were used decreasingly by historians after the 1920's, as economic, social, and political problems appeared to be no longer solvable in terms of the upwardly-inclined gradient of nineteenth-century social progressivism. Henry Adams's failure to reconcile Grant's administration with any theory of cosmic progress seemed truly apposite to the new historian.

One last point. We need to know much more about the reading, the libraries, the lectures attended by both historians and biologists in this period of a century and a quarter, roughly from 1800 to 1925. That the impact was almost equal on both sides, however, I would maintain, is demonstrable.

NOTES

[1] "On Trading and Usury," 1542, quoted in Richard Hofstadter, *Social Darwinism in American Thought* (New York: Knopf, 1944), 242.

[2] Walter J. Ong, *Darwin's Vision and Christian Perspectives* (New York: Macmillan, 1960), 134.

[3] Among general surveys, William Irvine's *Apes, Angels, and Victorians* (New York: McGraw-Hill, 1955) is useful.

[4] *A Generation of Materialism* (New York: Harpers, 1941). See particularly 12-13, 246, and 255 ff.

[5] *The Origin of Species* (New York: Appleton, 1897), 9. R. G. Collingwood, *The Idea of History* (New York: Oxford, 1946), especially 129 ff, 332.

[6] *History* (New York, 1908), 14. Morton White, *Foundations of Historical Knowledge* (New York: Harpers, 1965), 262. See also *The New History* (1912) *passim.*

[7] *The Holmes-Pollock Letters* (Cambridge: Harvard, 1941), I, 57. For Spencer, see *Principles of Sociology*, 3d ed., (New York, 1925), part 2 of Chapter II.

[8] K. Marx and F. Engels, *Gesamtausgabe* (Berlin, 1932), Vol. II, part 3, p. 447. See also Friedrich Engels, *The Dialectics of Nature* (London: Lawrence and Wishart, 1941).

[9] Marx and Engels, *Selected Correspondence* (New York, 1942), 125.

[10] *The Dialectics of Nature*, ed. 1941, 19. See the *Gesamtausgabe*, III, part e, 77-8. In 1908 Lenin in his *Materialism and Empiro-Criticism* relied on physics rather than on biology, astronomy, or geology for his analogies.

[11] Franz Mehring, *Karl Marx* (New York, 1935), 555. See also Isaiah Berlin, *Karl Marx: His Life and Environment* (New York: Oxford, 3d edition, 1963), 274.

[12] James Allen Rogers, "The Russian Populists' Response to Darwin," a perceptive essay in *The Slavic Review*, 22 (September 1963), 456-468.

[13] See *Mutual Aid* (New York: McClure, 1902), *passim.*

[14] H. Stuart Hughes, *Osward Spengler, A Critical Estimate* (New York: Scribner's 1954), 4. See also Charles Loring Brace, "Darwinism in Germany," *North American Review* (1870).

[15] E. N. Saveth, "Race and Nationalism in American Historiography: Late Nineteenth Century," *Political Science Quarterly*, 54 (September 1939).

[16] *The Education of Henry Adams: An Autobiography* (New York: Heritage Press, 1942) especially Chapter XV, "Darwinism."

[17] Merle E. Curti, *The Growth of American Thought* (New York: Harper, 1943), 554-5. William S. Jordy, *Henry Adams: Scientific Historian* (New Haven: Yale Univ. Press, 1952). Roy Nichols, "The Dynamic Interpretation of History," *New England Quarterly* 8 (June 1955).

[18] See not only "The Strenuous Life" but essays such as "How Not to Help Our Poorer Brother," "Social Evolution," and "The Law of Civilization and Decay."

[19] Henry F. May, *The End of American Innocence* (New York: Knopf, 1959). Howard G. Odum (ed.), *American Masters of Social Science* (New York: Holt, 1927).

[20] Boyd C. Shafer, "History, Not Art, Not Science, but History," *Pacific Historical Review*, 29 (May 1960). Compare Theodosius Dobzhansky, *The Biological Basis of Human Freedom* (New York: Columbia Univ. Press, 1956). See also Stephen Toulmin and June Goodfield, *The Discovery of Time* (New York: Harper, 1965), particularly Chapters 8 and 9 on Darwin's world.

CHAPTER XVI

What is a Historical System?

Hayden V. White
University of California, Los Angeles

The title of this paper is misleading. For it suggests that I have an answer to a problem which has been debated, with inconclusive issue, by historians, philosophers, and social theorists for over one hundred and fifty years. It is also misleading because I propose primarily to tell what biological systems are *not* rather than what historical systems consist of. And this might seem presumptuous, given the presence in this audience of so many better qualified to speak on this subject than I am. But it seemed better to err on the side of presumption in the interest of encouraging debate than to stifle discussion by re-rehearsing the trivia of my own discipline's internecine squabbles. In order to do our work, we historians frequently have to act *as if* we knew what a biological system was, the point at which the biological level of integration shades off into the historical level, and the ways that the two levels are related to one another. It would have been cowardly not to have admitted this at this gathering and to have avoided the issue altogether. And so, in the interest of possible clarification and at the risk of possible self-annihilation, I have decided to set forth what I believe to be some crucial distinctions between biological and historical systems, at least as they appear from the vantage point of the historian. If it turns out that these distinctions are not justified from the standpoint of biologists and philosophers of science, so much the better. We only discover the error of our ways by testing them in the presence of those best qualified to judge them.

This preliminary gesture to specialized competence having been made, I would like to begin by noting what seems to be a commonplace to investigators working at the transition points between the different levels of integration to be dealt with. It is true while historical systems can be profitably described in part by biological modes of representation and explanation, the reverse does not seem to be the case. That is to say, although we *often* talk about socio-cultural systems as if they were strictly analogous to biological systems, we only *occasionally* treat biological systems as if they were analogous to socio-cultural systems. For example, we may choose to treat the growth of a nation, a political institution, or even a socialized human being as if it had undergone the same kinds of processes of germination, maturation and development as a tree or a forest or an animal organism; but when we try to characterize the evolution of a plant or of an animal society, we do not ask *all* of the same questions about it that we do about nations, institutions, or human beings. This is because the life-history of a biological system is *in principle* exhaustively describable in terms of the laws of genetic inheritance, variation, and mutation that govern such systems plus the boundary conditions obtaining in that milieu in which the organism lives. We do not normally invoke the concepts of choice, purpose, or intent when describing the responses that biological organisms or entire species make to stimuli coming from their environments. Of course, we *can* invoke such concepts for narrative purposes, as ways of presenting our findings about such systems in common language; but they are not necessary for the scientific understanding of the systems in question. In the analysis of historical systems, by contrast, we cannot do without such terms; we need them to account for a whole range of socio-cultural data, especially those data which suggest that man *as a social being* can and often does choose self-destruction in service to some ideal value or culturally provided norm of comportment rather than yield to the laws of adaptation and survival that ought to govern him as mere *mammal.*

It is a characteristic of socio-cultural systems that they sometimes seem to choose, self-consciously and programmatically, *not* to survive, if survival entails abandonment of the life-style constituted by their dedication to ideal values, goals, norms, or aspirations. Thus, for example, the devotee of the Japanese code of Bushido responds to the culturally induced sense of shame which particular kinds of failure require him to feel by committing suicide. From the standpoint of his membership in a genetic line of descent, this act of suicide is non-adaptive. From the standpoint of his group membership, however, it is *adjustive*, as psychologists use that term, for it removes the stimulus causing the irritation felt as shame; and moreover it is adaptive in sociological terms because it affirms the values which make of the Samurai class a specific group among all the other groups to which the individual might possibly belong if he chose to do so. Thus, in this case, social adaptation seems to run counter to the imperatives of biological adaptation. And we can imagine a situation in which all of the adherents of the code of Bushido might feel compelled to commit suicide *en masse*, in which case an entire genetic population might be wiped out in the interest of affirming the ideal values which make of this population a distinct social grouping. I do not think that animal populations make such gestures by which, in choosing their own self-destruction, they at the same time affirm a set of values without which life itself would appear to be not worth living. Suicide in the interest of ideal values, then, would seem to be one possibility open to socio-cultural systems that is not open to most biological systems. And it would be interesting to know whether this is in fact the case.

But it could be argued that the behavior of such a social group is understandable on the analogy of the laws of natural selection. It could be said that the code of Bushido is analogous to a set of genetically inherited characteristics, that the code, even though provided by education and indoctrination, functions as a habit; that habits function in the same way that instinct does; and that the *Bushi* who conforms to the standards of behavior required by the code is *acting* in the same way that an animal *reacts*, which is to say automatically or mechanically. We might want to say that those *Bushi* who do *not* honor the code, but who, in a situation in which they might be expected to commit suicide, instead decide to go on living and to cease being *Bushi* by this decision, represent a kind of species variation. When a significant number of individuals who might be expected to act in conformity to the code, choose instead to improvise another code, in which shame is not adjusted to by suicide, we might liken this to a species mutation; and we could view the new group as a new species characterized by its treatment of survival as a value higher than mere conformity to a social code. The growth of this mutation could then be accounted for by pointing to changes in the *Bushis'* environment, to such socially significant factors as industrialization, urbanization, secularization, and the effects that these have on traditional social units, such as the family, religion, and traditional military castes. And then both the behavior of those adherents of the code who committed suicide and of those who chose to change their way of life would be strictly analogous to the behavior of animal populations adapting or failing to adapt to changes in their natural environment.

In fact, when historians are in their more scientific frames of mind, this is the way they account for such ruptures in socio-cultural systems. The socio-cultural system is treated as *a genetic endowment* circumscribing specific kinds of behavior for the individuals indoctrinated into them. When such systems dissolve, their dissolution is explained by the fact that they were not programmed for the kinds of responses required by their environments. The social units which split off from the original group, abandon its ideals, and constitute themselves as a new group by modifying their behavior in such a way as to permit survival and reproduction of their kind, are treated as types of species variations with mutational capacities. The old group is said to have "died" and the new group to have been "born."

But what gets obscured by this mode of representing historical transitions is the fact that socio-cultural systems do not actually have a "life" except in so far as individuals honor them as appropriate systems for living a distinctively *human* life. Socio-cultural systems are constituted and dissolved by the choices of individuals, which choices are conscious precisely in the degree to which indoctrination into them requires systematic repression of instinctive modes of behavior in order to create specific habits of social comportment. Whatever the genetically provided constitution of the individual, he is asked to alter it in some way and to merge his aspirations with those of the group; and he is *forced* to alter them by programs of education and indoctrination that work on the individual's consciousness more or less directly.

In the course of his socialization, then, the human individual is asked to accept as his own a set of ideal values which are determinative of the characteristics of the group's life-style. This life-style has to be *re*-presented and *re*-accepted by each new generation of individuals born into it. And there is no way of predicting in advance of the event, whether a new generation will accept the life-style offered to it as the sole appropriate way of achieving a distinctively human life by the previous generation. Even if changes in the environment could be predicted with perfect accuracy, it would be impossible to predict whether a given generation would abandon its inherited life-style or affirm it, even in the face of its own imminent self-destruction. When individuals who have been given the same education as their progenitors no longer confirm the values that gave the parent group its distinctive life-style, the socio-cultural system does not so much die as simply dissolve. Whether the choices were dictated by environmental pressures or by personality characteristics received by genetic endowment is irrelevant. Socio-cultural systems do not have lives of their own; they exist solely as a function of the choices of individuals to live their lives this way and not another, regardless of what the environment would seem to require for survival. And when individuals cease to choose a given way of life, this way of life ceases to exist.

Now I suspect that the kind of choices that I have been referring to here have their analogues in biological systems. And it might well be that the difference I am suggesting between historical and biological systems depends upon the vantage point taken for observing them, depends in short on whether we look at the matter from the standpoint of the *object chosen* or from the standpoint of the *choosers*. But it seems to me that there are good reasons for distinguishing between socio-cultural systems and biological systems by invoking the concept of *conscious* choice to describe the operations of the former. For the fact that socio-cultural systems always come equipped with programs of education and indoctrination suggests the extent to which they are implicitly recognized to be *un*nnatural by the individuals who are promoting them. Indoctrination into the group requires that the individual's genetic endowment be shaped and molded, augmented and channeled, so that patterns of behavior demanded of him as a social being be regarded as more desirable than those patterns of behavior which are natural to him. The process of socio-cultural indoctrination itself presupposes that the individual *could* choose not to accept the proffered system or code, that he might be inclined naturally to reject it, and that he might be able to find perfectly adequate reasons in the economy of his psychic existence to abandon it altogether. In short, the life of a socio-cultural system is only as strong as its power to convince its least inclined potential member that he *ought* to live his life as a human being *this way* and not another.

One of the reasons genetic explanations appeal to historians is that we are interested in processes of *socio-cultural change or continuity that extend over many generations.* When a given generation chooses to abandon the system or code proffered it by its parents, we are not inclined to say that the rebellious generation is exercising a freedom of choice, but that *the system itself* has entered a crisis. If in the course of the crisis the individuals in the system choose to reconfirm the system, we say that the *system* was healthy, that it had adequate survival capacities for that time and place. If the generation fails to reconfirm the system, we say that it was sick, that it was overripe or decadent, and that in the end it simply expired.

Thus, we are inclined to speak of the "death" of Roman civilization and the subsequent "birth" of Medieval Christian civilization, as though these systems were living organisms, with indeterminant life-spans to be sure, but as having theoretically calculable capacities for survival which made it possible for them to adapt to certain environmental conditions and not to others. We conventionally speak of Medieval Christian civilization as having taken shape in the "womb" of Roman culture, of having "grown to maturity" in its "bosom," and of then having freed itself from it when it reached "maturity," as if civilizations were organic individuals with received ontogenetic capabilities, phylogenetically affiliated with their predecessors. And in fact our interest in socio-cultural processes justifies our use of these terms, for the human individuals which comprised these systems did not have the temporal endurance which the systems themselves had.

But we are inclined to overlook the fact that the language which we borrow from the ontogenetic and phylogenetic processes of organic systems merely provide us with metaphors for characterizing *long-range* temporal processes, processes which occur only on the macro-temporal scale of historical evolution. If we shifted our perspective to the micro-temporal scale, to the scale marked by hours, days, and years in the life of the individual, rather than use that marked by centuries and millenia, we should not be able to discern any *process* at all. We should lose sight of both continuities and significant changes in the historical system, swamped as we would be by the plethora of atomic facts which have no pattern whatsoever. So we shift to the macro-temporal scale in order to see both long-range continuities and long-range transformations in socio-cultural systems and to discern a plan or pattern of growth and decline. This gives us a bit of comfort in the face of the possible meaninglessness of the whole socio-cultural enterprise.

But we gain meaning only by sacrificing awareness of the essential evanescence of every socio-cultural system, only by obscuring the extent to which any such system rests upon fictions that are turned into lived realities by the individual's choices of them as the sole possible way of living a human life, of realizing his personal aspirations, of achieving his goals as a *human* being. And we also lose sight of the fact that when a generation chooses *not* to honor its received socio-cultural endowment as *its* way of realizing a distinctively human life, we are in the presence, not of a crisis in the system, but of a crisis in the lives of human beings who must now choose another way of life. Revolutions in historical systems do not occur automatically; they are manifestations of widespread discontent with both the received social-system and the system of education and indoctrination by which men are made into social beings. And as thus envisaged, no socio-cultural system can be said to "die"; it is simply abandoned. Nor are new socio-cultural systems "born"; they are constituted by living men who have decided to structure their orientation toward their future in new ways.

The points at which old systems are abandoned and new ones are constituted are the most difficult for historians to deal with, precisely because they do not lend themselves to analysis by appeal to genetic models of explanation. The compulsive power of socio-cultural systems cannot be discerned at this time; and it cannot be discerned because the systems themselves are in process of being abandoned and constituted by the choices of men. Our mechanistic biases make us loath to recognize the constitutive power of these choices. And so we look for the "survivals" of the older system and the "nascent" forms of the emergent one. Because we do in fact know that Medieval Christian civilization was in process of being formed between the third and eighth centuries A.D., we treat its nascent forms as if they were self-generated, or as if they were generated by phylogenetic processes.

But the abandonment of Roman civilization and the constitution of Medieval Christian civilization were not genetic necessities. Roman society did not die because it had exhausted its genetic potential; for that survival potential was the same so long as men continued to act as if it were adequate to their needs. It might have lasted much longer than it did in Western Europe, as it did last in Byzantium. But in fact men abandoned it; and then chose another system which they not only *felt* was more adequate to their needs but which they *treated as if they had genetically descended from it.* In fact, the Christian past was a segment of the total historical past, which included a Roman pagan segment, of Western European men. What happened between the third and eighth centuries was that men *ceased to regard themselves as descendents of their Roman forebears and began to treat themselves as descendents of their Judaeo-Christian predecessors.* And it was the constitution of this *fictional* cultural ancestry which signalled the abandonment of the Roman socio-cultural system. When Western European men began to act *as if* they were descended from the Christian segment of the ancient world; when they began to structure their comportment *as if* they were *genetically* descended from their Christian predecessors; when, in short, they began to honor the Christian past as the most desirable model for creation of a future uniquely their own, and ceased to honor the Roman past as *their* past, the Roman socio-cultural system ceased to exist. Or at least it was reduced to the status of a recessive characteristic in the ideal ancestry of the new system, the dominant characteristics of which were provided by the fiction that the system was a genetically provided legacy of Christian antecedents.

What I am suggesting here provides an important qualification on my distinction between biological and historical systems on the basis of the choosing capacities of the latter. I am suggesting that historical systems differ from biological systems by their capacity to act *as if they could choose their own ancestors.* The historical past is plastic in a way that the genetic past is not. Men range over it and select from it models of comportment for structuring their movement into their future. They choose a set of *ideal ancestors* which they *treat* as *genetic progenitors.* This ideal ancestry may have no physical connection at all with the individuals doing the choosing. But their choice is made in such a way as to substitute this ideal ancestry for their actual, genetically provided but socially undetermined, modes of comportment. And they act as if they were more obligated to this ideal ancestry than they were to their actual progenitors.

Thus envisaged, the process of socialization can be characterized as a process of *ancestral substitution*, as a request for individuals to act *as if* they were actually descended from historical or mythical models in preference to any model that might be provided by genetic inheritance. That retroactive ancestral substitution is an essential ingredient in the constitution of historical systems is signalled by the fact that every society recognizes the kinds of conflicts it causes in the individual, and tries to provide ways of sublimating them. The conflict between the individual's obligations to his genetic progenitors and his obligations to his culturally provided ancestral models stands at the center of Greek tragedy, Judaeo-Christian ethics, and Roman law. In the end, every individual, before he is fully socialized, in whatever system, is forced to make at least a partial substitution of the culturally provided set of ideal ancestors for his actual progenitors; and to structure his comportment on models found in the former, in ways often detrimental to the health and even the survival of his genetic line.

No such choice is required in biological systems properly so-called, I should think. Biological systems are always functions of all or parts of their genetic endowment plus whatever forces have worked upon their immediate progenitors to increase or decrease their adaptive capacities in his present. Trees, forests, animal populations and cells *may be* conceived as the effect of selective *causes* operating in their pasts; but only human beings are asked to selectively choose their ancestry retrospectively. No father can require that his son treat him as his ideal father, even though he can demand that he treat him as his real father. Fathers bestow sonship on their progeny by the simple act of insemination; they become fathers in the biological sense thereby. But anyone who has had a son will know how tenuous is this merely biological claim on fatherhood, how difficult it is to turn it into the ideal fatherhood required by the son to act in his society as a putatively free agent, or to overturn it and reject it if he needs to do so.

In fact, social fatherhood is only bestowed by the sons, and it is bestowed by the choices men make of models of comportment offered by the socio-cultural system. When a whole generation fails to find in the culturally provided repertory of possible ideal ancestors any which sanction satisfaction of what they consider to be justifiable needs, revolutions are in the offing for that system. Through no fault of the system. The disaffected members of the rebellious generation will begin to ransack the system's historical record for possible ancestral models hitherto acknowledge as having only questionable worth as ideal models. Failing to find in the historical past of the system any ancestor to which the reverence that imitation indicates can be given, the generation may recur to the historical records of other, even contending systems, import models of comportment from these systems, and demand that their contemporaries honor them as their ideal ancestry. If they succeed in incorporating this alien ancestry into the officially sanctioned repertory of possible models for the system, a revolution or a reformation has in fact occurred. We might see these revolutions and reformations *as variations or mutations in the system*; but they are inconceivable without the act of retrospective ancestral constitution which give them their specific contents. Thus, Luther and his followers bestowed ancestry on a group which had lost it during the course of the Middle Ages, and set the Reformation in process thereby. Lenin, by contrast, imposed a completely new set of ancestral models on Russian society, and consolidated a revolution thereby.

The history of Russia written by a socially assimilated Russian historian prior to 1917 would not have had to contain any reference to Karl Marx and other European socialists to count as a full and adequate representation of Russian evolution. After 1917, however, any history of Russia that did not place Marx and the European socialists in the main line of ancestral descent of the society in process of formation, would be regarded as incomplete. And rightly so; for what the Russians did in 1917, among other things, was to reconstitute, retroactively, but in a degree that was mutationally significant in a socio-cultural sense, the historical ancestry from which they chose to act as if they had descended. Once this *ideal* ancestry was established, it could be treated *as if* it were the *real* ancestry of the Russian people; and it *was* such in so far as the Russian people structured their comportment in terms of their understanding of their presumed obligations to their adopted ancestral models.

This retrospectively provided ancestry *appears* as an actual genetic constitution to individuals fully indoctrinated into the system which has chosen it as a legitimizing agency. Once constituted *and accepted by a group as a genetically provided past*, this past *is* the past for that group as a socio-cultural entity. And no amount of "objective" historical work pointing out the extent to which this *chosen* ancestry is *not* the *real* ancestry can prevail against the choosing power of the individuals in the system. This is because as an adherent of the Russian socio-cultural system, the individual *is* a descendent of what he has chosen to be descended from. His behavior is understandable only if this choice is taken into account.

A recent disciple of Freud has said that in psychotherapy the problem is less to come to terms with our *real* fathers than to find our *true* fathers. The same can be said of socio-cultural systems. When a set of "true" ancestors are found to which many individuals can give the reverence of imitation for structuring their lives in their present, a socio-cultural system is formed. This system takes precedence over the genetically provided biological system without which, to be sure, even the capacity to indulge in this fiction would be unimaginable. But the process of retroactive ancestral constitution, of abandonment of culturally provided ancestral models, and the search for new models of comportment by which the satisfaction of secondary needs can be achieved, this process stands at the heart of the historical system. Eliminate it and historical systems would not exist at all.

As a kind of coda to these incoherent ramblings, I would like to say something about history-writing, and more specifically, about the relation of historians to the process of retrospective ancestral constitution which I have put at the heart of the historical system. If I understand Schrödinger correctly, he characterizes life as a reduction of positive entropy in particular time-space locations through a borrowing ("sucking") of negative entropy from the immediate environment. Historical accounts which establish the putative genetic connections between sets of chosen ideal ancestors and the socio-cultural systems using them to legitimize their goals or ends serve a similar function retroactively. Historians try to provide both the sufficient and the necessary reasons for any achieved socio-cultural complex being what it is and not something else. They put "life" (in the Schrödingerian sense) where formerly there was only chaos. They do this by establishing putative genetic connections between an achieved present and past socio-cultural systems. Men seem to require an ordered past as much as they require an ordered present. They want to believe that what they have in fact created could not have been otherwise. And historians assure them that this was so; out of the chaos of individual choices, the historian finds the order which even the choosers could not have seen. As they are *lived*, then, historical systems seem to move *forward*, into the future; as conceived and justified, they appear to *back* into it. Our anxiety in the face of the unknown drives us to embrace the fiction that what we have chosen was necessary, given our past. But the *historical*, unlike the biological past, is not given; it has to be constructed in the same way and in the same extent that we have to construct our socio-cultural present.

In choosing our past, we choose a present; and vice versa. We use the one to *justify* the other. By constructing our present, we assert our freedom; by seeking retroactive justification for it in our past, we silently strip ourselves of the freedom that has allowed us to become what we are.

CHAPTER XVII

On a Difference Between the Natural Sciences and History

Stephan Körner
University of Bristol

Convenience and tradition have long favored a division of the whole field of empirical enquiry into two mutually exclusive and jointly exhaustive parts, namely the natural sciences (briefly, science) and the historical disciplines (briefly, history). The division, which is based on a *prima facie* contrast between the aspirations and achievements of research in the natural sciences on the one hand and historical research on the other, gives rise to a question which once again engages the attention of philosophers: Is the contrast only apparent or does it rest on a structural feature which is characteristic of scientific but, at most, incidental to historical thinking? The structural features which are usually adduced to justify a deep theoretical opposition between science and history are not sufficient. For example, the search for causal and other predictively useful generalizations is not peculiar to science, just as the concern with unrepeatable events is not peculiar to history. Similarly, the description and explanation of human actions, as opposed to externally observed behavior, is a task not only for history but also for the incipient sciences of anthropology and sociology as well as for economic history, conceived as a branch of economics rather than as the history of economic institutions.

In this paper I propose to exhibit a structural feature of empirical thinking which is present in the aspirations of all and the practice of most scientific disciplines, but absent from both the aspirations and the practice of historical research. Roughly speaking, scientific thinking proceeds, or aims at proceeding, within "double-layered" (sometimes, as *e.g.* in quantum-mechanics, "multi-layered") conceptual systems, whereas historical thinking proceeds within single-layered systems. I shall first describe the structure of the double-layered systems which arise from the axiomatization of the results of empirical enquiries. I shall secondly argue that the tendency toward axiomatization, and thus toward the employment of double-layered systems, is characteristic of the natural sciences, but absent from the historical disciplines. Next, I shall try to show how the double-layered structure of scientific thinking affects scientific descriptions, general scientific laws, the application of these laws and scientific explanations. I shall conclude with some remarks on the single-layered structure of historical thinking.

The discussion will proceed on a level of generality on which differences between the physical and the biological sciences are irrelevant. It will in particular not depend on the truth or falsity—or even the heuristic value—of the reductivist thesis that the vocabulary of biology is in principle translatable into that of the physical sciences and that the postulates of biology are derivable as the theorems of some (perhaps not yet available) physical theory. If physics is double-layered, then so is biology, irrespective of whether biology is, or merely contains, physics. But though the double-layered structure of, say, classical mechanics is perhaps more immediately obvious than that of many other scientific theories, my distinction between the historical disciplines and the natural sciences does not depend on the privileged status of some specific scientific theory.

On the Axiomatization of Empirical Material

The network of concepts employed in common-sense thinking about our spatial and temporal environment has three facets which are relevant to my argument: (a) It contains a great variety of different concepts and is adaptable to a great variety of very different purposes; (b) the concepts are internally inexact in the sense that they or their sub-concepts admit of neutral (border-line) cases which can with equal correctness be judged to be or not to be instances of these concepts or sub-concepts; (c) the deductive relations between propositions containing internally inexact concepts as their constituents are not always capable of clear demarcations. Any systematic enquiry into some region of experience, whether scientific or historical, reduces the variety of available common-sense concepts by selecting a small number of them for application. Without completely eliminating them, it also reduces the inexactness of the selected common-sense concepts and the haziness of their deductive relations by judicious stipulative definitions. However, the axiomatization of empirical propositions involves further and more radical modifications, some of which I shall briefly and somewhat dogmatically indicate.[1]

In a full-fledged axiomatic theory, state-descriptions (descriptions of spatio-temporally separated states) are deducible from other state-descriptions in conjunction with the theory's axioms according to the well-known deductive schema: $s_1 \wedge A \vdash_L s_2$, where "$s_1$" and "$s_2$" are two (maximal) descriptions of states of affairs in the vocabulary of the theory, "A" the conjunctions of the theory's axioms, and "$\vdash_L$" the symbol for deducibility in accordance with the logico-mathematical framework into which the theory is embedded. This framework "L" comprises (in the case of all scientific theories known at present) elementary logic, $i.$ $e.$ the logic of propositions, quantification theory and the theory of equality and usually also some parts of mathematics ranging from elementary arithmetic to the theory of real numbers and beyond. If s_1 and s_2 were, as some philosophers of science seem to hold, straightforward empirical propositions describing the results of empirical investigations without modification, then axiomatization would be merely a kind of systematization. But this is not so.

Since L contains quantification theory which rests on the assumption that all predicates are exact, the internally inexact concepts of such empirical propositions as are to be incorporated into L must be replaced by internally exact ones. Since L contains the elementary theory of equality, which is a transitive relation, and since empirical (operational) equality or indistinguishability is non-transitive, the formulation of equality statements in L involves the replacement of the non-transitive by a transitive relation. If L contains the rules for the addition of quantities, multiplication of a quantity by a number, etc., which unlike the corresponding empirical operations are isomorphic with a part of arithmetic, then the expression of the results of empirical measurements and the reasoning about them in L involves a further idealization, namely the replacement of a set of relations which is isomorphic with a part of arithmetic by a set of relations which is not. Any additional extension of L by a mathematical theory enforces additional idealization of the incorporated empirical propositions.

Apart from the general modifications which L imposes on sets of empirical propositions incorporated into it, the axiomatization of empirical material imposes special modifications which vary from one axiomatic system to another. One group of them I have called "deductive abstraction." It consists in simplifying common-sense concepts by eliminating some of their deductive relations to other such concepts which are irrelevant both to the axiomatization of a particular region of experience and, consequently, to the axiomatization of its results. For example, the common-sense concept "x is a very small physical body" which entails "x has position, momentum, shape, color, etc." is replaced in classical mechanics by the concept "x is a physical particle" which for all x logically implies "x has position and momentum only."

Axiomatization of empirical material thus involves not mere systematization, but modification. It adjoins to the common-sense concepts and propositions which the theorist shares with his theoretically innocent helper a layer of ideal concepts and propositions and thus engenders a two-layered conceptual system, both layers of which are jointly employed in scientific thinking. It must, however, not be assumed that all two-layered conceptual systems employed in science have reached the perfection of axiomatic systems or that such systems contain only concepts which have obvious empirical counterparts.

The Tendency Toward Axiomatization and Some Criteria of Its Presence

Axiomatizations of empirical material are the exception rather than the rule in the natural sciences. Yet, if one were to send out an appropriate questionnaire to the practitioners of any established or aspiring science, they would all describe the axiomatization of the results of their enquiry as, at least, a distant goal. Historians would find this ideal unacceptable. An answer to the question: "Do you consider an axiomatic theory as a goal of your and your peers' intellectual endeavour?" could, I think, serve as the shibboleth by which the practitioners of the natural sciences could be distinguished from those of the historical disciplines. It would be sufficiently discriminatory even for such disciplines as anthropology, economic history or philology, by dividing them into a scientific and historical branch.

But there are other criteria for the presence of the tendency toward axiomatization. The incorporation of empirical propositions into a logico-mathematical framework and deductive abstraction are not only necessary (though not sufficient) conditions *for the success* of any axiomatization of empirical material. They are also (very nearly) sufficient conditions *for the presence* of the axiomatizing tendency. There is, for example, some truth in the adage that science is measurement, since to make measurements and to reason about their results, is to adopt implicitly not only elementary logic, but also a mathematical extension of it. It is moreover unlikely that whoever employs methods of measurement does not also aim at quantitative laws; and that whoever aims at quantitative laws does not also aim at their optimal formulation. But the optimal formulation of quantitative laws is their formulation in an axiomatic theory.

The same, although to a lesser extent, applies to deductive abstractions— at least when they are made in the knowledge that they involve some modification of empirical concepts. It is difficult to see why anybody should, for example, form the concept of *homo economicus* as a simplification of the common-sense concept of man, unless he wishes to use this concept in describing and predicting human and social phenomena in regions of experience in which disregarding the difference between a man and a *homo economicus* yields the advantage of more efficient predictions. But in this case it would be unreasonable not to strive for the type of conceptual system which guarantees the most efficient ones, that is to say, an axiomatic theory.

Axiomatization of empirical material and its necessary conditions, *i. e.* incorporation of empirical propositions into classical logic and its mathematical extensions on the one hand and deductive abstraction on the other, are characteristic of the actual or aspiring sciences, as opposed to the historical disciplines. They all result in the construction of two-layered conceptual systems which affect the structure of scientific, though not that of historical thinking. It is thus desirable to consider some features of scientific thinking within two-layered conceptual systems.

Scientific State-descriptions, Laws and Explanations

The interrelations between the layer of common-sense concepts and the layer of ideal concepts in a two-layered scientific conceptual system manifest themselves most clearly in their joint use for prediction. If my account of the schema: $s_1 \wedge A \vdash_L s_2$ is correct, than s_1, A and s_2 do not describe the world of common-sense experience but an ideal world. How then can we explain the enormous success of the use of theories in predicting not ideal, but actual events? The answer is implied in my earlier remarks: although the propositions s_1 and s_2 describe ideal states of affairs, they can within certain contexts and for certain purposes be treated *as if* they were respectively empirical propositions, say e_1 and e_2, describing real states of affairs. What these contexts and purposes are is an empirical question. Thus for certain purposes and within certain contexts physical bodies can be identified with particles of Newtonian mechanics, just as for certain other purposes members of the real species "man" can be identified with members of the ideal special *homo economicus* of, say, a marginal utility type of economic theory. It is by virtue of such identifications that the deductive reasoning on the ideal level from the ideal state-descriptions s_1 and the ideal laws A to another ideal state-description s_2 is made available for predicting that if e_1 which describes a real state of affairs is true, then e_2 which describes another, spatio-temporally separate real state of affairs is also true. A detailed analysis of the manner in which the limited identification of ideal with empirical concepts functions in testable scientific reasoning need not here be given (see *op. cit.*).

Scientific thinking by means of two-layered conceptual systems which lack the systematic character of axiomatic theories is less neat, but in principle not different. The reasoning on the ideal level can again be represented by the schema $s_1 \wedge A \vdash_L s_2$, but with a somewhat different interpretation of the symbols. The state-descriptions s_1 and s_2 are formulated in terms of ideal concepts, but not in the vocabulary of a specific axiomatic theory, since such a theory does not yet exist. The ideal law or conjunction of ideal laws A used in the deduction is similarly not the conjunction of the axioms of an axiomatic theory. But the symbol "$\vdash_L$" has the same meaning as before and the connection between the ideal level and the level of common-sense experience is again established by identifications. Pre-Newtonian physics and contemporary nuclear physics are examples of two-layered, non-axiomatic scientific conceptual systems.

Although some philosophers of science tend to conflate scientific prediction and scientific explanation, the two must be distinguished. They have in fact been distinguished by scientists who, whilst acknowledging the predictive power of a scientific theory, refuse to admit that it really explains the phenomena or makes them intelligible.[2] The various proposed principles of scientific explanation—of what makes a predictive theory also explanatory—are paradigmatic, programmatic, or both. They are paradigmatic in so far as they require that any scientific theory, in order to be counted as an explanation, be in some important respects like one or more existing theories. And they are programmatic in so far as they require any scientific theory to possess some general features, whether or not they are exemplified by existing theories.

Some philosophers and scientists locate the explanatory, as opposed to the predictive, power of a scientific theory in one or more of its concepts, *e. g.* various kinds of causal, probabilistic or teleological connection. If the concepts require a mathematical apparatus for their formulation (as is the case with some concepts of causality and most concepts of probability), then the insistence on employing these concepts in explanations strengthens the tendency toward the incorporation of empirical propositions into classical logic and its mathematical extensions. Again, ever since the pre-Socratics, some philosophers and scientists have held that all natural phenomena should be explained in terms of a few relations between ultimate, unobservable constituents, *e. g.* the atoms and the void of Democritus. Insistence on this kind of explanation strengthens the tendency toward deductive abstraction. But whatever principles of explanation have been advocated, they all imply the requirement of that systematic unity which is best implemented in axiomatic theories.

The Single-layered Structure of Historical Thinking

With some notable exceptions,[3] recent discussions about the difference between scientific and historical thinking have centered around the question of the extent to which the deductivist schema $s_1 \wedge A \vdash_L s_2$ is applied and applicable in historical thinking. These discussions have remained inconclusive because the parties engaged in them have failed to notice that the axiomatization of empirical propositions, their incorporation into a classical logico-mathematical framework and deductive abstraction, turn the single-layered structure of common-sense thinking into a double-layered structure, the empirical and the ideal layer of which are linked by identifications. The systematization of empirical propositions in historical thinking leaves the single-layered structure of the common-sense conceptual system intact. Although this, I think, clearly borne out by the practice of history, it might seem to conflict with some occasional methodological pronouncements of philosophically inclined historians. The conflict, however, is merely apparent. Before considering some examples, I should, perhaps, make the trivial observation that historians frequently make use of the results of scientific enquiry.

Plato, it might be objected, employed his Theory of Forms both to deal with natural and historical phenomena. But whereas his philosophy of science implies the limited identifiability of relations between the ideal entities of geometry with relations between physical objects, his philosophy of history is based on the assumed decay of historical societies from *one* ideal Form. It does not allow for relations between a plurality of ideal entities and the limited identification of such relations with actual historical processes. Augustine's two cities, whose interactions under divine providence account, in his view, for the historical process, represent a division of real men in one real world. Vico's "eternal history" which is "traversed in time by the histories of all nations"[4] is conceived not as an idealization, but as an exact description, because in giving it, "he, who creates the things, is talking about his own creations."

The empiricist historians, who conceive their methodology in the manner of Bacon and his empiricist successors, lay great stress on inductive generalizations. But, as I have emphasized, the search for, and the assertion of, empirical inductions and their systematization, does not by itself involve any tendency toward axiomatization or idealization. Again, the Hegelian dialectics has nothing to do with the deductive organization of empirical material. This was quite clear to Hegel himself, who assigned any attempt at apprehending the reality of nature and history to Reason—as opposed to the Understanding which, according to him, manifests itself in the misrepresentations of reality by classical logic and natural sciences.

Whenever somebody advocates the organization of historical phenomena after the fashion of the natural sciences—however imperfectly he may grasp their structure—he instinctively suggests a new name for the proposed type of enquiry: social physics, sociology, social anthropology, and the rest. And he does so with the full approval of those historians who do not wish to join him in his new venture. For they know that history is fundamentally different from the natural sciences, even if they cannot explain the difference. They rightly leave this problem to philosophers who—as my talk may have shown again—have so far not been notably successful in solving it.

NOTES

[1]For details, see *Theory and Experience*, London: 1966.

[2]For such views concerning orthodox quantum-mechanics, see, *e. g.* *Observation and Interpretation*, London: 1957.

[3]See in particular W. B. Gallie's *Philosophy and the Historical Understanding*, London: 1964.

[4]*Nuova Scienza*, §393.

CHAPTER XVIII

Historical Taxonomy

Robert E. Roeder
University of Denver

Do historians engage in taxonomical thinking? Should they? If so, how should they go about it?

Under these three interrogative headings, this paper considers some of the problems that arise because historians do, in fact, engage in ill-developed taxonomical thinking. It considers also some of the problems that would arise if they attempted to create for themselves a comprehensive and non-arbitrary taxonomy of human societies. The experience of biologists with taxonomy offers both cautions and suggestions, but the argument here is not that there can be any easy borrowing by historians of the biologists' methods and principles.

* * * * *

Do historians engage in taxonomical thinking?

The answer is yes, not only for the meta-historians, but also for a great many, if not all, others. In the case of non-system building historians this answer is perhaps obscured by the fact that their taxonomical thinking is in no sense comprehensive or systematic. It most often goes no further than the employment of ill-defined types derived from an unconscious or implicit typology of social processes or of social structures.

For instance, any effort at comparative history attempts to distinguish what is generic in a society's form or process from what is merely individual and idiosyncratic. Comparative history then depends upon the delineation of types, if not of typologies. And types produced by some process of comparison, whether it be a conscious and careful process or an impression- istic and unconscious one, are common coin in historical explanations, at least those of a certain order of generality. It is, for instance, impossible nowadays to read any general or textbook history of Europe without frequently coming across phrases such as "feudalism," or "nation-state," or "parliamentary

251

democracy;" or, referring to process rather than structure, "industrial revolution," or "democratic revolution," or "socialist revolution." Each of these is, I submit, a type or species of society or change.

It may very properly be objected that, while historians frequently use type-terms, they seldom set out a typology, much less a taxonomy, of which their type is a part. This is certainly true in the vast majority of cases. What is generally found in such general histories is a narrative or analytic explanation of how, in the circumstances extant in one particular society, a new general societal form, a new institution or societal sub-system, or a new process of change was produced. This new thing is given a name ("parlamentary democracy" or "bourgeois revolution") and that name, or type-term, is thereafter employed to create analogies in the mind of the reader when the histories of other particular societies are described (e.g. "In Russia, parliamentary democracy was slow in developing because ..."). Frequently, explicit comparative statements are made which in effect are attempts to identify what is typical and what is not in analogous institutions or processes in two or more societies (e.g. "The German industrial revolution, unlike the English ..."). Seldom does one find an *explicit* typology of which the types employed are deemed to be parts. Nevertheless, the very use of such type-terms implies that some sort, perhaps a very shadowy sort, of typology is in the author's mind. Unless the *Standestaat* of late-medieval and early modern Europe is believed to be a distinct species of political organization in a genus in which there are other species (feudal state, absolute monarchy) there would logically be very little sense in using the term: it would distinguish or delineate nothing. In the non-system-making general historian's usage then, while types are explicit, typologies are usually implicit only.

It is of course possible to find histories and biographies of particular events, persons, or short periods in the life of particular societies which scrupulously avoid the use of any explicit type-terms and thus of implicit taxonomies. Possible, but not easy: the influence of the historical system makers and of the model- (or type-) building social scientists is nowadays so pervasive that one much more often than not finds the narrowest monographs liberally sprinkled with taxonomical terminology. It is not that a desire to borrow the prestige of "social-science" afflicts even the narrative or "literary" historian; it is merely that such terms have become, ever since Marx, if not before, part of the virtually indispensable and highly convenient language of history. They are, in effect, shorthand symbols which save historians much space, time, and sometimes (all too often, it is to be feared) thought.

When one turns to the meta-historians, taxonomical thinking becomes evident.

Marx is an initial, principal, and perhaps sufficient example. Marx rushed on from a crude and unsatisfactory societal taxonomy to an explanatory law and explanations of particular historical changes. Yet that law and those specific explanations were, and had to be, stated in the terms of the typology and thus were limited by it. As early as his *German Ideology*, Marx produced the simple taxonomy upon which he therafter relied. Humanity is there described as having lived in societies of four major economic types: the primitive communitarian type, the ancient slave society type, the feudalistic, and the capitalistic. We need not comment upon Marx's deterministic explanation of how and why society evolves, and had to evolve, from type one through type four, it is sufficient to note here that the explanation was cast in terms of this taxonomy of societal types, and that it had to be: one cannot propound a general law of societal evolution or a specific explanation of some society's evolution without indicating, however imperfectly, what is changing to what.

Elsewhere and later Marx speculated about the possible existence of fifth—Asiatic—mode of production different from any he had previously discerned. The potential historiographical consequences, should this simple addition be admitted to the typology, were explored and bitterly debated by Marxist historians in the 1920's and 1930's as they considered how Chinese history should be treated. Even a cursory examination of this historiography indicates the importance of the taxonomical question. If one adheres to the original four-type taxonomy, the Chinese Empire must be described as a feudalistic society, or as one which was in the ancient slave society form and at some (rather vague) date became feudal. Despite the formidable obstacles created by some obvious facts of Chinese Imperial history, orthodox Marxist historians insisted upon their traditional taxonomy and thereby guaranteed a long, sterile and ludicrous debate over the periodization of Chinese history. It must be admitted that the acceptance of the notion of there being an Asiatic type of society, would not in itself have perfected understanding and explanation of Chinese history, but the whole affair is indicative of the utility and persuasive force of even simple and crude taxonomies of society. From one such Marx derived a law of societal evolution; that his explanation of history has been influential—for better or worse—is hardly news.

Many other examples of the taxonomy involved in the efforts of meta-historians could be given, perhaps ad nauseum. Time permits only very brief mention of two, which are representative of kinds of taxonomical effort somewhat different from Marx's. Max Weber's typology of ideal-types of authority is certainly an influential sociological taxonomic effort. One nowadays finds much use of such terms as traditional authority, charisma, etc., in historical writing. One also finds use of such categories as "traditional society," a somewhat different matter. The influence of Weber's typology is representative in two respects: first, in that historians have borrowed type-terms from a social scientist's typology of a societal sub-system, which is what, by the very nature of their discipline, social scientists most often attempt to classify; secondly, in that some historians take the leap of using that sub-system category to characterize a whole society. In doing so they rather infrequently state or even imply a theory such as Marx's which explains how the state of the sub-system determines the state of the whole social system.

The second representative example is constituted by the works of Arnold Toynbee. Despite his bias in favor of systematics and his use of a considerable number of biological metaphors, including taxonomical terms used metaphorically, Toynbee is not engaged in setting out a schema classifying morphological types of civilized social structures. In fact, about the only "systematic" morphology he propounds is the simple dichotomy between the species "primitive society" and the species "civilized society." This typology, stated at the beginning of the *Study*, is later slightly elaborated by the addition of two more types or species: "proto-civilized society" and "church of higher religion-society." The addition does not change the intent of his *Study*, which is to construct a model of the sequence of historical processes or changes through which civilizations have gone. He denies arguing that these sequences are the result of some kind of genetically-determined life-cycle of the species; he contends rather that they are merely the observed sequences. In *Reconsiderations* he admits that a single model will not do, that one needs two or three to describe the observable sequences of the individual societies of the species "Civilization." Now a model is not a taxonomy. Nevertheless, in contending that particular change-processes in particular societies are not unique, that they are members of a class of change-processes, and in arguing that there are systematic relations between *such classes*, Toynbee is engaging in a process of thinking which would be properly called taxonomical, if he restrained himself from taking the further step of model-building. His most enduring influence upon historical thought will derive, in my judgment, from his taxonomical effort, and not from his model-building, for, while some of his classes of processes are too loosely defined to be useful, some provide a basis for the gaining of real insight into the phenomena they group. As has already been observed, such grouping of particular processes of change and giving the group a name—making it a type in some usually hazy or implicit typology of change-process—is a quite common practice among historians. Toynbee is merely a fairly representative modern historian when he does this; he is unrepresentative in his comprehensiveness and explicitness, and, of course, also in going on to model-building, to saying that there is a common sequence in which these processes appear from one civilized society to the next.

* * * * *

The foregoing is sufficient—I trust—to establish that historians nowadays frequently engage in various kinds of taxonomical thinking.

The next question is: "Should they?"

It would be easy to argue that they should not, that instead they should ruthlessly exclude taxonomical language from their writings and taxonomical notions from their thought.

An obvious criticism—and one which scientists, logicians, sociologists, historians, indeed anyone possessed of common sense would agree in making—is that the taxonomies employed so far by meta-historians are altogether too simple, crude or vague. Marx's type "ancient slave society," for instance, causes almost hopeless confusion if one seeks to use it in analyzing ancient history. It does not differentiate between, say, the temple-state of Sumer and pre-Imperial China, or between either and the polis-based society of Greece, much less between these early forms and later imperial and bureaucratized forms of the ancient civilizations. Weber's "ideal-type" typology of authority is also too simple. His work does indicate the existence of two types of legal-rational authority systems—the democratic and the authoritarian, but these two are hardly sufficient for close analysis of modern history. The problem created by such taxonomy is that undue simplicity renders the classificatory scheme inapplicable or misleading for the historian seeking to use it to gain clues to the explanation of particular behavior of particular societies. The very general types of which the taxonomies are made are defined in terms of a few characteristics and each such type includes a large number of historic societies. Faced with the bewildering variety of *other* characteristics possessed by the particular societies within the type, the working historian either finds it of little use either in producing general "laws" or as an indication of what to investigate and explain in his inquiry into the behavior of a particular society at a particular time.

The loose usage of taxonomical terms by the non-meta-historians may be sufficiently criticized in a sentence. It leads them to so frequent inconsistencies, semantic quagmires, and time-wasting historiographic debates as to make attractive suggestions for the elimination of such terminology.

* * * * *

But there are good reasons to resist the suggestions and instead of eliminating taxonomy, really doing it.

The most simple reason is the convenience of the usage. Unless one possesses types, which are in effect shorthand symbols denoting complicated patterns or lengthy records of human behavior, it is impossible to talk about very much. Historians without their symbols would be in the position of mathematicians possessing only numbers with which to work. Placing types within a typology is one way of giving them more nearly precise, invariable and intelligible meanings, and therefore a taxonomical effort is indicated as one way in which historians may have their symbols and at the same time reduce their confusions.

Secondly, two of the most fashionable logics of historical explanation indicate the desirability of having a historical taxonomy. I refer to what Dray calls the covering law theory and Dray's own theory of the legitimate logical basis of most historical explanations which he calls "rational" explanation. A proper historical taxonomy is necessary for the first and would be useful to the second mode of explanation. If there are really to be found any such things as the covering law theorists' general empirical laws under which particular historical events or institutions must be subsumed if they are to be explained, these laws must in part be stated either in terms of classes of men or in terms of categories of societies or categories of social action. A valid taxonomy would provide such categories and thus might make possible the discovery and statement of such covering laws as well as of explanations in terms of them. (I realize this justification will not satisfy those who argue the covering-law thesis in order to prove that there can be no or no non-trivial empirical generalizations covering historical phenomena and that therefore historians can offer no satisfactory explanations. Despite the elegance of this argument, historians are going to continue explaining history to vast audiences.)

It is somewhat less easy, but nevertheless important, to see the potential utility of a valid historical taxonomy to Drayian "rationalists." Dray argues that the most characteristic (though not the only valid) form of historical explanation demonstrates how some historical action of an individual, group, or society could be possible, given their known (or discovered) "principles of action." By principles of action he means customary rationales for acting in similar or analogous instances. Now, as Dray points out, the historian may learn or guess what those principles are for a given actor in any of several ways or combinations of ways. He may have evidence showing what the actor (individual or group) said the rationale was. He may be, but rather infrequently is, aided by a law produced by experimental psychology which states all men or groups of similar character reason and act thus and so in similar circumstances. Or, to make a point Dray does not clearly make, the historian may, and very often does, infer that the actor has a rationale, a principle of action, for his activities which is like that of *similar* actors. On the basis of fragmentary evidence about the sayings and doings of Charles I of England, for instance, a historian guesses that he was a would-be divine right autocrat. The historian has available considerable information about what other European divine right monarchs had in the way of principles of action, so he infers that Charles had those too. The historian—or at least the good historian—does not stop there; he looks for, and offers in support of his explanation, evidence that Charles I in fact held such principles. He will admit contrary evidence, if he happens to stumble across it, but in general finds, at most, what he sets out to look for. If the evidence cannot be found to support such in inference, or is too contradictory, the historian may make another inference from other possible analogous (to Charles I) character-type's known principles of action.

The utility of a valid taxonomy would be to make the process of inference just described more feasible and less subject to error. If valid means of systematically categorizing societies and their sub-groups can be found, there will be less chance of historians inferring when they should not, and of failing to infer when they should.

A third major reason for considering as desirable the construction of a historical taxonomy is that valid taxonomy is in principle indispensable and would be in practice very useful to comparative history. One important variety of comparative history is the method whereby a historian seeks to test a hypothesis about the behavior of group "a" in society A by seeing whether that hypothesis also appears to hold when it is applied to a similar group "b" in a similar society B; obviously we need some means of saying groups "a" and "b" and societies A and B, *are* (validly for our analytic purposes) similar. We can make ad hoc judgements about similarity; it would be easier and less dangerous to make them if we had a systematic classification of societies, particularly since any such classification would probably depend on a good deal of "anatomical" and "physiological" work, as I shall argue later. We would certainly be much more likely to make better tests of hypothesis if we were in a position to conveniently compare not only A and B, but A to B, C, D . . . .Z. And this, taxonomy would make much more feasible. Other varieties of comparative history would also be facilitated, but time does not permit a demonstration.

Finally, if historians are ever to get beyond having a vague notion that societies evolve to having an even modestly verifiable theory—or, more likely, complicated set of theories—explaining how, a taxonomy is virtually indispensable. Social scientists can produce models suggesting or explaining how certain kinds of social changes might take place. The applicability of the theories thus produced can be determined only if the theory is tested against the cases that have existed where there obtained the conditions the theory assumes. This can most feasibly and least inaccurately be done if a systematic classification is made of what conditions obtained where and when.

These considerations are sufficiently persuasive—I hope—as to the desirability of having a historical taxonomy, and therefore of attempting to produce at least one.

* * * * *

How should historians go about it?

Desirability is one thing; possibility another. Is it possible to systematically classify, to create a taxonomy of the past? How would one go about it?

The experience of biologists in the systematic classification of life-systems may be the (common-sense) logical place to look for some suggestions; all sciences find it useful to engage in some sort of taxonomy, but the one which considers the organization of life on another level than the social sciences and history may provide the best guidance, although one should expect that the difference in level at which the organization of life is considered will require at least some differences in taxonomical procedure.

My no doubt too cursory examination of the history of biological taxonomy indicates to me that three requirements would have to be met if one were to succeed in constructing an even modestly useful historical taxonomy.

First, there must be definable units or "individuals" which are to be classified. Secondly, there must be a species concept. Thirdly, there must be a principle with which to organize the species into higher orders.

This is easy to conceive; how history might meet those requirements is not. The balance of my remarks will examine only some of the difficulties and make tentative proposals as to how they might be overcome.

The Need for a Definable Unit or "Individual"

For biology this is apparently no real problem. The utility of using the individual organism as the basic unit to be classified seems so obvious and necessary that when one says biological taxonomy, one means classification of organisms—not classifications of cells or ecological systems—though I would guess such classifications also exist for certain purposes. The organism has an observable and rather sharp beginning and end—both spatially and temporally—and thus constitutes a "natural" unit. (At least this is true of sexually reproducing organisms.)

For history, however, the finding of a unit is difficult. To take the individual human organism as the unit would be useless. We certainly know enough psychology, sociology and anthropology (not to mention history) to know that cultures, social structures, and social processes cannot be fully or even very much explained by attributing their character and change to biological races of man. This path has been followed far enough that we may be warned that it leads over intellectual as well as social precipices. It may, of course, be possible to do useful biological and psychological classifications of the genotypes of man—but that would not be historical classification.

So the unit must be a society of some sort. What sort? And how can spacial and temporal limits be found for whatever sort is chosen by other than purely arbitrary or conventionalistic means? Societies after all are networks of human relations, systems and sub-systems of human interaction, structures of human institutions, but not organisms. They end without the lives of the individual organisms that create them necessarily ending, and the other way around. Their spatial as well as their temporal boundaries are often fuzzy. Time does not permit a full discussion of the difficulties and alternatives so I must simply state what seems to me the best solution:

The unit of historical taxonomy should be a whole social system (not a sub-system) defined:

a) spatially, by determining what (at any given time) was the largest social network in which the human members acknowledged either the legitimate authority of a person or a societal sub-system over all human members and sub-systems or the de facto impossibility of resisting successfully what was considered (by some members or sub-systems if not by all) as usurped authority.

b) temporally, by the points in time at which *any* major sub-system of a society crystallizes a significant change, or at which sizable new groups of human beings are brought into the society.

The spatial delineation comes very close to being "the state," but it is not quite the same thing since it would cover cases where there is nothing that can be readily recognized as "a state" and also, and more importantly, because it does not imply any particular theory about the nature of "a state." It also conforms to what in my judgement should be a cardinal precept for historians: historical protagonists must be either individuals or *self-identifying* groups.

The temporal delineation principle offers the best way I can think of having a sufficient number of "individuals" to make classification worthwhile without resorting to purely arbitrary divisions. It is phrased to avoid implying any theory about the relationship between one major sub-system and others. I realize the concept "crystallizes" may give theoretical trouble, but I believe it would create fewer practical difficulties than any other. The other terms need some further comment.

"Major sub-system" is definable in either of two ways:

First, a major sub-system is any sub-system changes in the character of which can be observed to have been followed by changes in other, functionally different major sub-systems in *any* society.

Second, a major sub-system is a sub-system which clearly has several sub-systems as constituent parts (e.g. the economic sub-system includes sub-systems of agricultural and artifact production, distribution, saving and investment, etc., etc.).

"Significant" change is defined as being any kind of change which has been followed by changes in *other* major sub-systems in *any* society.

This temporal limitation principle would perhaps require borrowing a catalogue of sub-systems from the structural-functional sociologists. But neither limitation principle requires the use of the a-historic assumption that such sociologists often make in practice, if only sometimes in theory, that all sub-systems of a society function toward the continuance, the survival of that society.

Both delineation principles have the added practical advantage of being close to some, the best, of the ways in which historians normally do establish such limits when they set out to define what they are talking about.

The Need for a Species Concept

Here again biology is marvelously well served by nature. The principle of reproductive isolation unfortunately cannot be borrowed by the historical taxonomer. Not being organisms, societies obviously do not reproduce sexually; nor can any *category* of societies be considered isolated from other categories even metaphorically.

Evidence as to stimulation of substantial structural changes in one society by changes in another is so great that the kind of "cultural isolation" Spengler used as a metaphorical species principle is simply untenable today. Besides, one of the purposes creating a taxonomy would be to engender better theories as to relations between inheritance of the great traditions and structural forms as they combine in the production of new institutions or processes. To select a species concept which presumed a simplistic theory (e.g. "The Great Tradition" or "culture" determines all) would eliminate one of the major purposes of taxonomy. For the same reason one should not use a kind of Great Cultural phylogeny as the principle for classifying the species into higher taxons.

What then would serve as a useful species concept?

The prudent answer at this point in the development of the social sciences and history would certainly be something like this:

Lacking anything akin to "natural selection plus genetics" in the way of basic theory of societal evolution, we must do what the biologists did when they suffered from a similar lack. We must classify by constructing morphological typologies, fully realizing the dangers of analogy and the ultimate unsatisfactoriness of this procedure. However, until we have that basic morphological taxonomy, we are unlikely to produce the theory necessary for a better species concept. We can and should temper the dangers by doing careful morphology (social anatomy it might be called). We must place in the same species only those societal "individuals" which have similar structures in at least most of their major sub-structures, not just in one or two. How many points of comparison would be necessary to make a "good species" would have to be left to experience.

While prudent and probably correct at this stage of the game, this answer is not very satisfying on such a speculative occasion as this. May I therefore indulge in risk-taking by giving an imprudent, speculative answer in addition.

A more useful species concept might be arrived at by combining morphological and sequential considerations. One might call a species all those societal "individuals" which were both morphologically similar and preceded and/or followed on the territory they occupied by societal "individuals" which were also morphologically similar to the other members of their "generation."

The major risk in attempting to use this concept would be that you might get too many species with only one "individual" in them, thus rendering them useless for the purposes of comparative history. Also the problem of what to do when a new people came into and dominated a geographical area would create troublesome (but not very frequent) problems.

This would have the advantage of introducing some consideration of the origin and descent of the species into the taxonomy without inventing an untenable theory of evolution in order to do so. It would be as non-arbitrary and empirically verifiable as the simply morphological species concept. It would also provide a way to meet the third requirement of a taxonomy, viz:

The Need for a Principle to Group Species into Higher Orders

Unlike the first two requirements this might be dispensed with; given the relatively small number of societal individuals to be classified, species might be all you needed.

Certainly if the species concept used were either of the two suggested, one could not adopt a Great Culture phylogenetic principle (which in my judgement is a good thing, since to do so would be to reduce the taxonomy to tautology).

If one used the purely morphological species concept, one would have, in order to get genera more rather than less inclusive than the species, to abstract one of the structures used for species definition and group into genera all species having similarity in that one structure. It would of course then be perfectly possible for a species to be in two or more genera, unless only one structure was used throughout. If you did use only one, your genera would turn out to be types in a typology of a single societal sub-system. Whether this would be satisfactory, I don't know, but I suspect not, because of the circularity involved.

If the morphological-sequential species concept were used, the problem would be evaded in all probability. The genera could be constructed by grouping all morphologically similar species, regardless of sequentiality. If your taxonomy went beyond genera to a higher order it would of course be confronted with the same problem as the morphology-only species concept would create. But probably it would not be necessary or useful to go beyond genera.

If, in our discussion, there emerge better ways of meeting the requirements of a historical taxonomy, the paper will have served its purpose—which is primarily provocation. If, on the other hand, it provokes an irrefutable decision that historical taxonomy is an impossibility, it will also have served its more generalized purpose: that of driving ill-conceived taxonomical terms from the historian's vocabulary.

CHAPTER **XIX**

Theories of the Universe in the Late Eighteenth Century

Harry Woolf
The Johns Hopkins University

Together with D'Alembert we may take cosmology to mean "la science qui discourt sur le monde," that is to say in the language of the enlightenment, that science which reasons upon the actual universe. But to reason upon the universe by the eighteenth century was not, spider-like, to spin a web of fancy from within oneself, however lovely the final design or rational the rules of argument along the way. Properly practiced cosmology was to be considered a general physical science tied to the facts of the world, though loosely at time, and constrained to account for the world visible as well as invisible by the sum of things known. The metaphysical meaning behind an assembly of facts, the analogies and bonds among them and their actual interrelationships under the rule of the general laws governing the universe, this was the task to undertake, and to lay bare such things in a satisfying account of the world was the minimum demanded of any competent cosmologist.

Impressed into this enterprise, at least during the period under considera- tion, was the now familiar belief in the continuous connection between all of nature's parts. "All beings are held together by a chain," D'Alembert tells us, "of which we perceive several continuous parts, though in a very large number of places the continuity escapes us. . . . The art of the philosophy consists in adding new links to the separated parts so as to render them the least distant possible; but we should not deceive ourselves (into thinking) that there will not always remain some voids in many places. To create these links . . . it is necessary to pay attention to two things: to the observed facts which form the substance of the links, and the general laws of nature which establish their connections."[1] Thus epitomized in the great *Encyclopédie* of Diderot and D'Alembert, the traditional compelling force of the great chain of being together with the kind of scientific principles just enumerated were parts of that mold in which so many an enlightenment cosmology was cast. Philosophical cosmology in the seventeenth and eighteenth centuries reflected the attempt to supply broad metaphysical explanations of the new world revealed by the advancement of science. As such it dealt with many of the traditional elements of philosophy: substance, its essentials, modes, attributes; bodies, their nature and forms, forces and actions. The composed beings of the sensible and material universe thus analyzed found their place in the various dispositions of learning represented by the hegemony of Cartesian ideas in France, the capture of the academic enlightenment in Germany by the Leibnitzian-Wolffian system (at least until Kant) and the ascendency in England of the Newtonian system, whose practical operational power, if not always its metaphysics, soon superseded the others.

These traditional themes are not part of this undertaking. Rather what I propose to discuss is the shape and character of astronomical cosmology, which for all the grandeur of its concerns remains a lesser subject than its philosophical progenitor. Cosmological speculation in this vein took a new turn in the eighteenth century with the publication of Thomas Wright's *An Original Theory or New Hypothesis of the Universe*, in 1750.

Thomas Wright was born in 1711 and died in 1786. A modest origin did not stifle an early passion for mathematical and astronomical ideas and soon this was joined to a decent pedagogical talent to make of the humble carpenter's son a successful tutor to the children of the aristocracy. This not only defined his career, it virtually set the framework of his accomplishment, for nearly all that he wrote was descriptive, utilitarian, pedagogical—from schematic layouts of Euclidean geometry and texts on the use of globes to discussion of Irish antiquities and universal architecture. Thus, in his twentieth year, after teaching navigation to seamen laid up for the winter in the seasonal coal trade of the time, Wright tells us in his journal of doing a "General Representation of Euclid's Elements in one Large Sheet: and the Doctrine of Plain and Spherical Trigonometry all at one view. . . ."[2]

Teaching devices of this sort were not uncommon in the eighteenth century, providing ponies for a quick trot through the world of learning; their persistence and commercial success to this day suggest how many prefer the minimum in life. The important thing is that these schemata were part of Wright's normal way—solar eclipses were so represented, paths of comets, and (from his journal for 1736) in the same space in which he tells of his arrest for nonpayment of debt, he speaks of the "great expense in Copper Plates and Prints" for "his invention of the Theory of Existence and Represent ye Hypotheses in a Section of the Creation sixteen feet long."[3] But the high point of this progression to the *Original Theory or New Hypothesis of the Universe*, which concerns us, was his publication in 1742, the same year in which he was invited by the Czarina Elizabeth to become Professor of

Navigation at the Imperial Academy in Petersburg,[4] of a "synopsis of the Universe or the visual world epitomised," a chart of 4 ft. X 6 ft. accompanied by an explanatory text and subschemes, the *Clavis Coelestis.*

At this point, as we come to identify the contributions of Wright, a passing word on the astronomical background of the times is in order. The sixteenth and seventeenth centuries had borne witness to basic developments. That the fixed stars were at varying distances had been confirmed, thereby allegedly accounting for different apparent magnitudes. Astronomers had also come to believe that the Milky Way consisted of clusters of innumerable small stars—that stars were suns that might in turn possess attendant planetary systems and that a plurality of inhabited worlds was thus possible. Such ideas had received wide attention, especially as a result of the popularizations of Fontenelle, Huygens and others.

Gravitational astronomy advanced rapidly after the publication of the *Principia* in 1687. Euler, Clairaut, D'Alembert and others, exploiting the more felicitous mathematical language of the continent directed Newton's fundamental principles toward the details of applied practice, bringing a wide range of errant phenomena to heel at the command of the laws of gravitation and motion. Though it had lagged behind the achievements of theoretical astronomy at the end of the seventeenth century, observational astronomy also advanced apace as the quest for accuracy drove the manufacture of scientific apparatus into professional hands, and instruments of a precision hitherto unknown became a reality. The demonstrated periodicity of comets and the discovery of the proper motions of stars, essential elements of cosmological argument in the eighteenth century, are both direct results of these accomplishments. This, in brief, is the matrix within which Wright operates. What does he do?

Several features of his thought are worth considering in the earlier *Clavis Coelestis.* First, in dealing with comets, he not only joins Halley in recognizing their periodicity, but though he holds them to be durable bodies, he rejects their possible habitability. In his own words:

> 'Tis hardly possible that these Globes, from the great Extreams of Cold and Heat, they are subject to pass through, . . . should be inhabited Bodies, or that they should finally drop into the Sun; which Sir Isaac Newton intimates they must at last do; for tho' we are almost positive, he has an attractive Power, yet we have many Reasons to believe that he has also a repelling one near his Surface; and if he has Power to throw vast Masses of Matter from or out of his own Body into his Atmosphere, such as appear like Spots upon his Disk, . . . no other Body can be supposed to have Power to approach him.

> The only Cause of new Stars they cannot be. Vid. periodical appearances. But how all this may really be, whether they are

ordained to feed the other parts of Nature's Works with fresh
Supplies of solid or of fluid Matter, or whether they are barren
Bodies fram'd on purpose for a perpetual Change in Creation's
great Variety; the Cause of Conflagrations, Deluges or total
Catastrophes, or be they Worlds themselves not quite ...
destroy'd, or new creating from their old Remains, will never be
known, till some of the human Race have been Eye Witness of
such a Dissolution. ... [5]

Secondly, Wright emphasizes the importance of the observer's position in
reaching conclusions about the disposition of the heavens, beginning first with
the relativity of observations within the solar system. "To Venus, the Sun and
the Earth are the greatest apparent Bodies in the Universe. ... Mercury
appears to Venus, as Venus does to Earth, and the Sun appears to Venus near
thrice as big as to us."[6] But far more important is his extension of this same
truth, widely appreciated by then, to the stars, where on assuming a group of
equal size, he demonstrates that "Their various apparent magnitudes are
entirely the Effect of an unequal Distance, as will manifestly appear from the
Scheme of their Disposition, where the same stars D E F G H, altho' their
Globes be represented equal, appear respectively of different Magnitudes, as is
evident from the different Points of View A, B and C in the equal Arches of
Vision RS, TV, and OQ" (Figure 1).[7] Furthermore, Wright shows that clusters
of stars, and double stars may at times be no more than optical effects
derivable from the observer's station. Thus, he writes, "E F G H, as seen from
D, appear like part of the Pleiades, a bright Knot of Stars, and D K from C, as
the double Star of Castor."[8] This method of treating astronomical matters
"clearly and plainly", as Wright was to boast in the *Clavis Coelestis* is not,
unfortunately, carried through in the far more important *Original Theory or
New Hypothesis of the Universe, founded upon the Laws of Nature, and
solving by the Mathematical Principles the General Phoenomena of the Visible
Creation; and particularly the Via Lactea* of 1750.

Though it comes eight years after the pedagogically oriented *Clavis
Coelestis*, and is the last of his important published astronomical works, the
Original Theory of the Universe is a work rooted in Wright's youth, going
back to unpublished documents of 1734, loosely titled the *Elements of
Existence*. In fact, the development of Wright's cosmological view can best be
seen in the progression of ideas from the manuscripts of his twenty-second
year, through the *Original Theory* of 1750, to a recently discovered *Second or
Singular Thoughts upon the Theory of the Universe* which internal evidence
dates as after 1771.[9] In a word, we have in this mixture of manuscript and
published materials the complete cast of an eighteenth-century British
cosmology. Let us look at it.

In the earlier documents, we discover the key to Wright's entire
cosmological undertaking. It is nothing less than the complete integration of

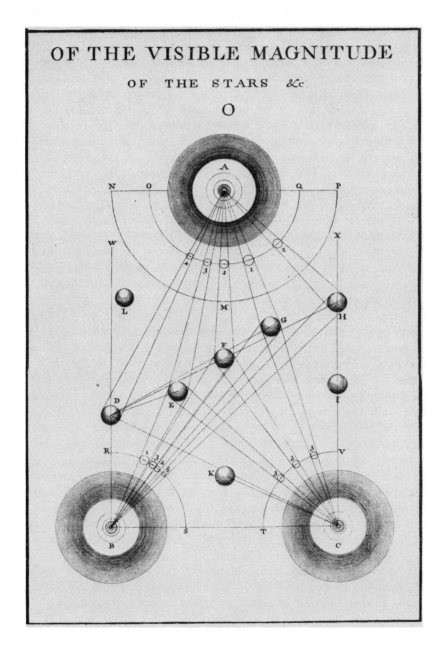

Figure 1.

the moral and physical universe. Assuming that the moral and physical order possess a common center in the Sacred Throne of God, he sets out, in his words:

> to solve, nearest to the level of ye human understanding, the Being of God and ye origin of Nature grounded upon visible effects and final causes. Demonstrating not only ye visible creation to (be only) a finite view (of ye general frame of Nature), but also the general disposition of the whole (is evinced from it). Together with ye several mansions of hapiness and misery in a future state and to be rationaly expected after death. All of which is represented . . . [as] extending from the Imperial Seat or *Sedes Beatorum* to ye verge of chaos bordering upon ye infinite abiss. Comprehending first ye Paradise of imortal spirits in their several Degrees of Glory surounding the Sacred Throne of Omnipotence. Secondly the Gulfe of Time or Region of Mortality, in which all sensible beings such as ye planetary bodies are imagined to circumvolve in all manner of direction round the Divine Presence, or ye eternal Eye of Providence. Thirdly the shades of Darkness & Dispare supposed to be the Desolate Regions of ye Damn'd.[10]

Thus, enveloping the common center of the universe are three zones, though the last may be thought of as open-ended; in the smaller, inner region the timeless world of Heaven, arrayed as he says in its "several degrees of Glory"; in the second layer, the mortal probationary world receding from but circling the Sacred Seat, and in the third, the old eternal night, chaos and the world of hell. The solar system is situated toward the center of the middle region and so shares by its geography, the moral position of man "equally open on all sides both to ye influence of Heaven and ye insults of Hell, . . . exposed for a trial of vertue. . . ."[11] That Wright's task is neither exclusively scientific nor theological is clear throughout his text, for the pendulum of his interest will swing from a discussion of the mansion of immortality where the guardian angels reside to the distribution of stars as actually seen from our small corner of the universe. Here is the crux of our problem, then, to understand the full amplitude of his oscillation, of course, but to deal mainly with a single direction of his movement, the scientific. So it is that we turn to his observation that,

> upon all sides the principal stars of the visible creation are exhibited in their natural order as seen from ye Earth by ye naked eye. Those of ye first magnitude nearest to our own system, and the rest proportionable removed according to their respective phenomena. Beyond these are others more remote crowned by a penumbral shaddow such as we call Telescopic stars and again without them more, supposed to be at immense distance & by no

means perceptible to ye human eye. At a certain distance from ye
sun equal to a vissual ray of ye smallest star is a faint circle of
light terminating the utmost extent of ye visible creation, in a
finite view from ye Earth. . . .[12]

In this, his first approach to the problem, the Milky Way is thus that "faint
circle of light" at the very edge of human visibility. But there is a curious
error here, for in the same document he tells us that there are "meriads of
systems in all manner of dispositions with an infinite number of worlds . . .
variously distributed round their . . . Suns or Centers." If this is so, it belies
the uniqueness of the Milky Way, for it is seen only in the plane drawn from
the observer to the phenomenon itself. But for an observer in a world where
myriads of systems exist in all manner of dispositions, a faint circle of light
would exist in every plane of his vision. That is to say, where the stars are
randomly distributed in every direction—a position Wright holds through the
later, 1742 publication of the *Clavis Coelestis*[13]—the observer's entire sky
should glow with a faint light. But this is not so, and the Milky Way appears
to be the result of a concentration of stars in a single direction. With the
publication of the *Original Theory* in 1750, Wright finally comes to grips with
this problem.

This major work is a rambling, theologically-oriented study aimed
primarily at an analysis and explanation of the Milky Way. Organized as nine
letters to a friend, the proper work of the book really begins with the fifth
letter, "Of the Order, Distance and Multiplicity of the Stars, the Via Lactea,
and the Extent of the Visible Creation," but the prefatory comments and the
scientists and poets to whom he turns for support set the tone and the
character of his argument. Newton is there, of course, but so are Huygens and
Bruno. There is an occasional genuflection to Milton and Dryden, but more
frequently the didactic lines from Edward Young's *Night Thoughts on Life,
Death and Immortality* compress an issue to its essentials, and there is always
the ready wit of Alexander Pope to set a mood. Anxious to set some limits to
his discourse, for all the grandeur of his theme, Wright informs us that "how
the heavenly Bodies were made, when they were made, and what they are
made of, and many other things relating to their Entity, Nature, and Utility,
seems in our present State not to be within the Reach of human Philosophy;
but then that they do exist, have final Causes, and were ordained for some
wise End, it evident beyond Doubt, and in this Light most worthy of our
Contemplation."[14] And here Pope comes to his aid:

> He who thro' vast Immensity can pierce,
> See Worlds on Worlds compose one Universe
> Observe how System into System runs,
> What other Planets, and what other Suns;
> What varied Being peoples ev'ry Star;
> May tell why Heav'n made all things as they are.

But let us return to the problem of the Milky Way. Having set up the alternative of whether or not stars are promiscuously distributed or follow a regular order, Wright now rejects his earlier argument of 1734 and says that they "are not only Bodies of the Nature of the Sun . . . performing like Offices of Heat and Gravity," but they do so "in a regular Order, throughout the visible Creation."[15] From this he proceeds to argue that the sun being no different in nature than other stars, it can have no superior claim to being "seated in the Center of the mundane Space."[16] With this as his first point, he invites us to think of a vast layer extended and bounded between two parallel planes or surfaces. Within this space, we are to "imagine all the Stars scattered promiscuously, but at such an adjusted Distance from one another, as fill up the whole Medium with a kind of regular Irregularity of Objects"[17] (Figure 2). An observer at position A, for example, looking in the direction H or D would first see the bright stars close to his position, then the more distant, fainter stars, the process continuing until the faintest and most distant stars would blend together into a single zone of light. If he looked the other way, of course, toward B or C, he would see a scattering of bright stars against a background of dark, empty space. This view along the line of sight AH or AD, Wright explains, "I take to be the real Case, and the True Nature of our *Milky Way*, and all the Irregularity we observe in it at the Earth I judge to be intirely owing to our Sun's Position in this great Firmament, and may easily be solved by his Eccentricity, and the Diversity of Motion that may naturally be conceived amongst the Stars themselves, which may here and there, in different Parts of the Heavens, occasion a cloudy Knot of stars, as perhaps at E."[18]

The sentence just quoted is the basis of Wright's major claim to fame, for the idea that the Milky Way is an optical effect explained by our position in a layer of stars was to have important consequences. It led directly to Immanuel Kant's own theory of the heavens and it may have influenced William Herschel as well. Had Wright stopped there we might have been content to leave him among the ranks of the enduring cosmologists, and interestingly enough, this is where most modern students of the subject keep him. But in the *Original Theory* as a whole and in the manuscript study *Second or Singular Thoughts Upon the Theory of the Universe* his explanation of the Milky Way occupies a lesser place, a minor incident almost on the road to a greater vision. And that greater aim remains the reconciliation of the moral with the physical universe. We have, he tells us, "to apply this Hypothesis . . . to our ideas of a circular Creation, and the known Laws of orbicular Motion, so as to make the Beauty and Harmony of the Whole consistent with the visible Order of its Parts. . . ."[19] The problem, in other words, is to make the idea of a flat layer of visible stars compatible with a universal scheme in which our own stellar system surrounds the invisible center required by

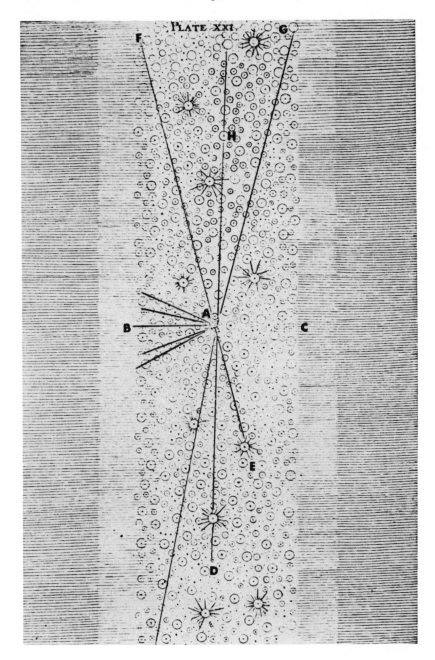

Figure 2.

Wright's physico-theology. Two possibilities present themselves, one in which the two surfaces that limit the layer of stars are themselves spherically concentric to the supernatural center, and the other in which the two parallel surfaces lie in the same plane but encircle the center, like the rings of Saturn. With Wright's own diagram let us examine these schemes.

This drawing is a representation of the entire creation. In Wright's words: "a universal Coalition of all the Stars consphered round one general Center"[20] (Figure 3).

This is meant to be a cross-section of the previous drawing with the "Eye of Providence" seated at the center, itself the "Agent of Creation" (Figure 4).

The double construction revealed in this diagram is Wright's way of showing how a hierarchy of worlds might be constructed. In his words: "a superior Order of Bodies C, may be imagined to be circumscribed by the former one A, as possessing a more eminent Seat, and nearer the supream Presence, and consequently of a more perfect Nature"[21] (Figure 5).

The increased detail of this drawing is meant to further explain the theory. By adopting a unit length for the distance BC in Part 2, the distances AD, AE, etc. can be calculated. The numbers need not concern us, for Wright is only anxious to demonstrate that the number of stars seen in the direction AD is much greater than those to be seen along AC or AB, thus adhering to his earlier account of the Milky Way. So much then for the first possibility (Figure 6).

In the second arrangement, the stars lie like the rings of Saturn in a thick plane which includes the possibility of additional worlds in subsequent rings such as those of Figure 7, Parts 2 and 3.

Worlds of visible stars, Wright's visible creations, may occur at B and C in Part 1 and with the profile revealed of one of these in Part 2 we can see how a Milky Way visible to each might be explained (Figure 8).

In both of these systems, the visible world, our own stellar system, occupies a disc-like region. In the case of the concentric spheres, the disc may be distorted by the gentle curvature of the spherical layer itself, giving it the shape of a shallow expanding dish, but this of course could not be visible to the observer. In the case of the ring-like universe, the disc would occupy one section of the ring, with its parallel surfaces as in the original 1734 proposal. Thus it is fair to speak of Thomas Wright as advancing the cause of a disc theory, and modern astronomers have looked to his rarely-read volume as to a precursor on this issue, but it is quite another matter to credit him with this view as a theory of the universe. His universe was far more restrictive, and meant to fuse, not to separate theology and astronomy. His explanation of the visible Milky Way thus deals only with a local universe so to speak—or, to use the looser language of a later epoch, with an "island universe."

Wright dealt with other problems in the *Original Theory* and in the subsequent, then unpublished *Second Thoughts* that are part of the

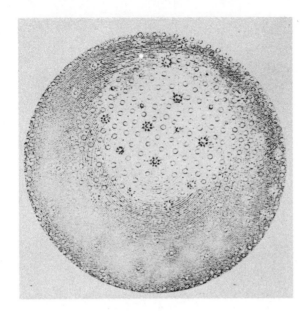

Figure 3.

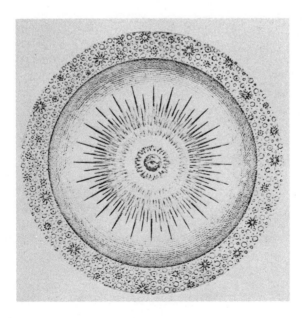

Figure 4.

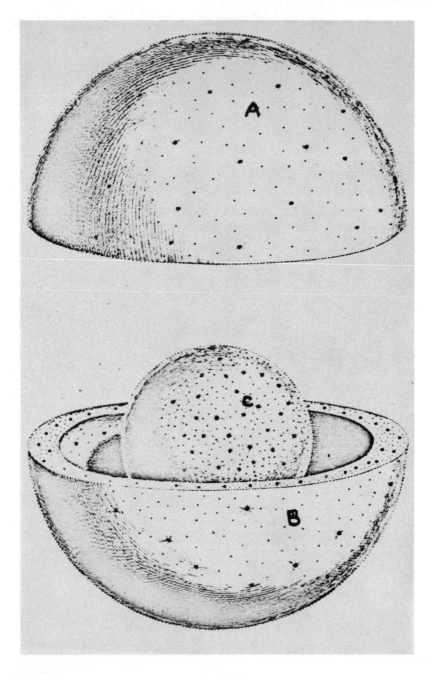

Figure 5.

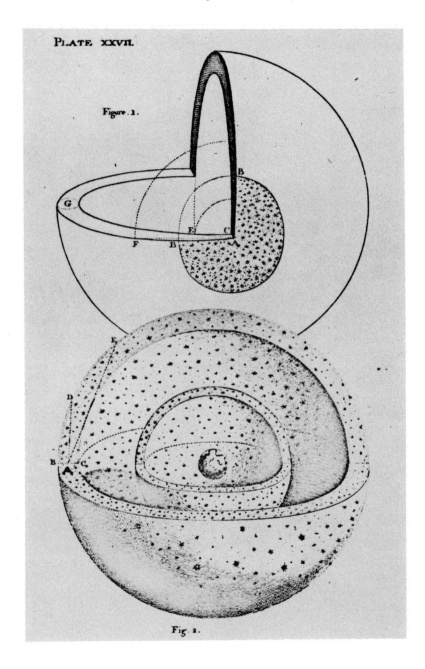

Figure 6.

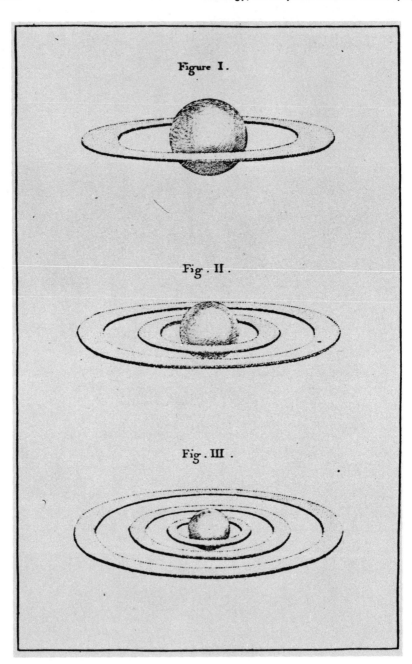

Figure 7.

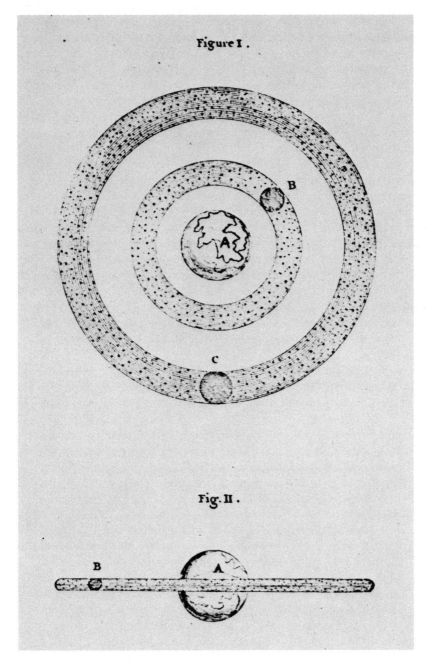

Figure 8.

cosmological debate of the enlightenment. Mostly, these are concerned with the stars and before going on with a more general, summary statement of that development in which we will hear frequently echoes of Wright's voice, it would be well to bring his views to a close. For Wright, the stars are clearly subject to gravitation and also move in response to a "projectile, or centrifugal Force, which not only preserves them in their Orbits, but prevents them from rushing all together, by the common universal Law of Gravity, which otherwise, as a finite Distribution of either, regular or irregular bodies, they must at length do by Necessity."[22]

The randomness of the stars and the irregularity of their orbits are apparent phenomena, explicable by the observer's position, for just as the planets appear orderly from the sun, so for the stars, "there may be one Place in the Universe to which their Order and primary Motions appear most regular and most beautiful."[23] The last two letters of the nine composing the *Original Theory* deal with the scale of the universe; the thickness and depth of the stellar shells, the period of a stellar orbit around the center of the universe and so forth. Once again he reiterates the sphericity of the creation and the unity of the natural with the supernatural: "In the general Center of the whole," he writes, there resides

> an intelligent Principle, from whence proceeds that mystick and paternal Power, productive of all Life, Light, and the Infinity of Things.
>
> Here the to-all extending Eye of Providence, within the Sphere of its Activity, and as omnipresently presiding, seated in the Center of Infinity, I would imagine views all the Objects of his Power at once, and every Thing immediately direct, dispensing instantaneously its enlivening Influence to the remotest Regions every where all round.... To this common Center of Gravitation, which may be supposed to attract all Virtues, and repel all Vice, all Beings as to Perfection may Tend; and from hence all Bodies first derive their spring of Action, and are directed in their various Motions.
>
> Thus in the Focus, or Center of Creation, I would willingly introduce a primitive Fountain, perpetually overflowing with divine Grace, from whence all the Laws of Nature have their Origin, and this ... would reduce the whole Universe into regular Order and just Harmony, and at the same time, inlarge our Ideas of the divine Indulgence, open our Prospect into Nature's fair Vineyard, the vast Field of all our future Inheritance.[24]

Wright's rapture could not prosaically close, and so he sings:

> Here Heav'ns wide Realms an endless scene displays,
> And Floods of Glory thro' its Portals blaze;

> The Sun himself lost in superior Light,
> No more renews the Days, or drives away the Night:
> The Moon, the Stars, and Planets disappear,
> And Nature fix't makes one eternal year.[25]

It would all be lovely if it ended here, an "unlimited plenum of Creation,"[26] but the culmination of his life's work came in the *Second or Singular Thoughts Upon the Theory of the Universe* (post 1771), which was meant to be a supplement to the *Original Theory* of 1750. An extended discussion of those thoughts is not warranted, for though he produces a complete physico-theology, the moral and the physical world finally reconciled, the astronomy is surprisingly regressive. Opening with the proverb "El sabio muda consejo, el necio no" (The wise alter their counsel, the foolish do not),[27] Wright only teaches us that in some cases it is better to follow English poetry than adhere to Spanish maxims, for the second look is a lesser one indeed. The earlier explanation of the Milky Way as an optical effect based on the location of the observer is completely abandoned. He now regards it as "a vast chain of burning mountains forming a flood of fire surrounding the whole starry regions, and no how different from other luminous spaces but in ye number of stars that compose them, or where there are none, in the vast floods of celestial lava that form it."[28] Moreover, these stellar volcanoes are now imbedded in a solid firmament that takes us back to the crystalline spheres of Aristotelian and medieval astronomy. Ostensibly this new structure provided Wright with a superior apparatus for harmonizing faith with fact, but the price was very high indeed.

Of general importance to enlightenment cosmology were certain questions concerning the stars. Most significant were those relating to their distances and their sizes, the causes of their occasional, changing aspects (fluctuating brightness, disappearances, etc.), their proper motion and its meaning, the nature of their constitution and finally, whether or not they were really subject to gravitational law.

The quest for a stellar parallax as a measure of distance was unsuccessful in the eighteenth century. It was not discovered until 1838 by Henderson and Bessel, and though few doubted its existence, astronomers varied in their beliefs about its possible detection. Some felt that the stars were so distant as to forever exclude the discovery. Others tried to estimate distances based on stellar brightness, using the sun as a standard and assuming similar values of size and brightness for another star (Sirius was a favorite). Calculations were then made, for example, to determine how far away the sun would have to be to shine with the same light as Sirius. The weakness of this method lay in the number of assumptions, for it was impossible to know whether differences in apparent brightness were due to distance itself or to the intrinsic brightness of a star, to its size or to any combination of these.

Changes in the stars themselves had been observed at least since the sixteenth century. Using the sun as a model again, some, such as Newton and John Keill, attributed the changing brightness or color of a star to a rash of spots of varying numbers revolving about it in such a way as to account for the irregularity of the light and even its total disappearance. Others, such as Hevelius, thought of it as due to clear or murky stellar atmospheres and their local meteorological variations. Still others, Maupertius for example, thought the variations due to the changing position of a rotating star, successively seen from on edge or in full face with all the gradations in between. Various theories of the sun's structure were also advanced, in part because of the belief that an explanation of the sun would also be one of the stars. Thus it was alternately a hot, burning body or a cold dark body with a luminous atmosphere. Sunspots might be the smoke from burning volcanoes (this was William Derham's idea, a man whom Thomas Wright read avidly), or openings in the glowing atmosphere to reveal the dark interior (Lalande and later William Herschel held this view) or the result of satellites revolving around the sun.

It was, however, the idea of universal gravitation that established the most critical problem for stellar astronomy. Why is it that the stars, if they are subject to gravitational forces, do not come together? And directly connected with this question: what is the spatial distribution of the stars? For a finite world, the stars would sooner or later run together, though Newton thought they were too far apart to be subject to gravitational attraction,[29] whereas in an infinite world this might not happen at all. In his *Opticks*, Newton had spoken of a repulsive force beginning where the attractive force ends, and though he had the microcosmic world in mind in that context, Roger Boscovich, taking his cue from this argument, advanced the astronomical analogy that gravitation ended in stellar space and repulsion began thus guaranteeing the separation of the stars and preventing the universe from collapsing.[30] We have also seen how Halley's discovery of the proper motion of the stars was employed by Wright not only to account for their distribution but also for their rotation in a harmonious system around a supernatural center.

But the most interesting contribution of sidereal astronomy to the eighteenth-century cosmological debate was that made by John Michell, one of the most fertile and imaginative scientists of the eighteenth century—a man well known to his peers and even to historians of science of the nineteenth century, but until our present day strangely beyond the attention of twentieth-century scholars.[31] Born in 1724, graduated from Cambridge in 1748 he remained there at least sixteen years in a succession of posts that carried him from the teaching of Hebrew, Arithmetic, Geometry and Greek to the Woodwardian Professorship of Geology from which, in order to get married, he resigned in 1764.[32] He remained active until his death in 1793, advancing some of the most suggestive ideas in science of the epoch. He was the first to publish the inverse square law for magnetic attractions (*Treatise of Artificial Magnets*, 1750). His analysis of the great Lisbon earthquake of 1755, published in 1760, led to his election to the Royal Society that same year. More importantly, though the thesis of the study proved to be in error, his method for locating the epicenter of an earthquake and large parts of his explanation of the phenomenon itself have become essential elements of modern seismic theory. He invented the torsion balance used by Henry Cavendish (after Michell's death) in the famous direct measurement of the force of gravitation, and was an early English advocate of Roger Boscovich's atomic theory. But our catalogue must cease; our interest in him is astronomical.

The two major articles which John Michell published in the *Philosophical Transactions of the Royal Society* for 1767 and 1784 contain some of the most profound astronomical thought of the entire century. Michell's first paper, "An Inquiry into the probable Parallax, and magnitude of the fixed stars, from the quantity of light which they afford us, and the particular circumstances of their situation," was aimed at obtaining the distance of Sirius by the application of the common photometric techniques I have already described, that is, the establishment of a relationship between comparative brightness and relative distance. It is quite likely that Michell was drawn to the problem of determining sidereal dimensions by the enormous international attention (a sort of international geodetic year was under way) then being given to the attempt to settle the actual scale of the solar system by means of the rare transits of Venus of 1761 and 1769.[33] The motive, naturally, need not concern us, but some of the conclusions should. At one point in a discussion of stellar luminosity, Michell appends an extraordinary, unfortunately lengthy footnote:

> We have at present no means of judging the comparative brightness of the Sun and of the fixed stars, in proportion to their respective sizes, excepting from the comparison of the Sun's brightness with that of our common fires; but the Sun's light exceeds the light of our brightest fires in so very great a proportion (viz. of some thousands to one), that we want some middle terms to be able to form any analogy, which might serve to carry us farther. We find however in general, that those fires, which produce the whitest light, are much the brightest, and that the Sun, which produces a whiter light than any fires we commonly make, vastly exceeds them all in brightness; it is not therefore improbable, from this general analogy, that those stars, which exceed the Sun in the whiteness of their light, may also

exceed him in their native brightness; now this is the case with regard to many of them; and, on the contrary, there are some that are of a redder colour.[34]

Here Michell has indicated that a relationship exists between color and intrinsic brightness, probably for the first time, though others had of course discussed stellar light in terms analogous to earthly fires. Michell goes on to discuss the likelihood of double stars and the possible calculation of their masses:

If however it should hereafter be found, that any of the stars have others revolving about them (for no satellites shining by a borrowed light could possibly be visible), we should then have the means of discovering the proportion between the light of the Sun, and the light of those stars, relatively to their respective quantities of matter; for in this case, the times of the revolutions, and the greatest apparent elongations of those stars, that revolved about the others as satellites, being known, the relation between the apparent diameters and the densities of the central stars would be given, whatever was their distance from us: and the actual quantity of matter which they contained would be known, whenever their distance was known.... Hence, supposing them to be of the same density with the Sun, the proportion of the brightness of their surfaces, compared with that of the Sun, would be known from the comparison of the whole of the light which we receive from them, with that which we receive from the Sun; but, if they should happen to be either of greater or less density than the Sun, the whole of their light not being affected by these suppositions, their surfaces would indeed be more or less luminous, accordingly as they were, upon this account, less or greater; but the quantity of light, corresponding to the same quantity of matter, would still remain the same.[35]

This is an amazing combination of ideas, considering the era. Not until the late nineteenth century was it demonstrated that there was a relationship both between luminosity and color, and luminosity and mass. In addition to these prescient assumptions, Michell's reasoning about double stars is most compelling. By 1767 a relatively small number of double stars had been recorded, and these were usually explained as optical doubles, that is, as effects attributable to the observer's perspective, much as Thomas Wright had explained (see Figure 1). Between 1779 and 1781, however, when William Herschel undertook his second major sweep of the heavens, stellar pairs were on his mind, and his efforts produced a catalogue of two hundred sixty-nine double stars. Yet even before this impressive evidence was at hand—and Michell was to use it quite effectively, when it appeared, for his second astronomical paper of 1784—he argued that a statistical analysis of the distribution of the stars would reveal certain basic truths. We need not re-examine the fullness of his argument, but at the end of his chain of reasoning, Michell was able to say: "We may from hence, . . . with the highest probability conclude (the odds against the contrary opinion being many million millions to one) that the stars are really collected together in clusters in some places, where they form a kind of system, whilst in others there are either few or none of them, to whatever cause this may be owing, whether to their mutual gravitation, or to some other law or appointment of the Creator. And the natural conclusion from hence is, that it is highly probable in particular, and next to a certainty in general, that such double stars . . . as appear to consist of two or more stars placed very near together, do really consist of stars placed near together, and under the influence of some general law. . . ."[36] To my knowledge this is the first application of probabilistic thinking to the problems of stellar distribution.[37] It does not matter that there were assumptions in his argument and elements in his analysis that would not hold up under the careful scrutiny of the developing theory of proabilities; the principle was certainly sound. Indeed, toward the end of his career (1803-1804) and some twenty years after producing his first catalogue of binary stars, Herschel went back to the same work and discovered that some of the double stars had actually altered their relative positions in such a way as could only have been produced by their rotation around each other, thus fulfilling Michell's predictions and incidentally demonstrating the extension of the law of gravitation to the system of the stars.[38]

Yet more must be said before we leave the good Reverend Michell, for he brings us back to the cosmological question by noting, in that same brilliant paper of 1767, that, "after what has been said, it will be natural to inquire, whether, if the stars are in general collected into systems, the Sun does not likewise make one of some system; and which are those, amongst the . . . stars, that belong to the same system with himself."[39] That the sun was a member of a larger stellar system was of course an essential part of Thomas Wright's cosmology. But by bringing this forward now I am not trying to argue that there was any direct influence of Wright upon Michell. There is no evidence for it whatsoever, yet Michell's scientific reasoning certainly comes to support some of Wright's fancies.

Thomas Wright's books were rare almost from their first publication, but there was a surprising durability to his views in the history of cosmology—or rather to certain parts of his judgements. This was mainly due to the odd circumstances by which they came to be associated with Immanuel Kant. Within a year of the appearance of Wright's *Original Theory* in 1750 it was well reviewed in a Hamburg periodical.[40] Each of the nine letters was summarized with extensive quotations from some, but none of the fine plates were reproduced. There Kant discovered Wright and gratefully acknowledged his debt when he came to publish his own *Allgemeine Naturgeschichte und Theorie des Himmels* . . . of 1755. In this, his *Universal Natural History and Theory of the Heavens, or an Essay on the Constitution and Mechanical Origin of the Whole Universe, treated according to Newtonian Principles*, Kant writes of his work: "The First Part deals with a new system of the universe generally. Mr. Wright of Durham, whose treatise I have come to know from the *Freie Urteile* [The Free Judgements] of 1751, first suggested ideas that led me to regard the fixed stars not as a mere swarm scattered without visible order, but as a system which has the greatest resemblance with that of the planets; so that just as the planets in their system are found very nearly in a common plane, the fixed stars are also related in their positions, as nearly as possible, to a certain plane which must be conceived as drawn through the whole heavens, and by their being very closely massed in it they present that streak of light which is called the Milky Way."[41] For all the brevity of the review—Kant never did see the original work itself—Kant was clearly aware that there were two possible stellar systems to be found in the *Original Theory*, whatever Wright's own final conclusion may have been. Of the two, one a system of concentric spheres and the other a succession of flat rings resembling those of Saturn's, Kant obviously chose the planetary model. "I cannot exactly define the boundaries," he writes, "which lie between Mr. Wright's system and my own; nor can I point out in what details I have merely imitated his sketch or have carried it out further. Nevertheless, I found . . . valid reasons for considerably expanding it on one side."[42] Thus, it is Kant's great fame which, for a long time, perpetuated Wright's name, though curiously enough Kant's own book was also to become a rare work rather quickly, for the publisher went bankrupt and only a few copies reached the public.

A biographical sketch of Kant is certainly uncalled for in this place, save to note that he was born in 1724 and died in 1804, and that his main scientific work is confined to the theory of the heavens of 1755 and to an earlier, shorter piece on the retardation of the rotation of the earth published in 1754. Other than to point to the initial stimulus provided by Wright, I do not intend to discuss the origin of Kant's cosmological theory, except to suggest that it was motivated by a desire to extend Newton's theories beyond their author's own limitations, and that is also had its beginnings in the continuous debate over the nature and development of the planets. One need only cite Descartes' theory of other dead suns that had been drawn into the vortex of our sun, from *Les principes de la philosophie* of 1644, Swedenborg's ejected solar matter from his *Principia* of 1734 and Buffon's tidal theory from the *Histoire naturelle* of 1745—to name only a few—to see into what company the young Kant entered with the *Allgemeine Naturgeschichte*.

The "one side" of Wright that Kant expanded in the first of the two parts into which his theory of the heavens is divided, was his elaboration of the Milky Way as a flat, stellar system, whose innumerable stars obeyed the laws of gravitation. Before him was the model of the solar system and so, by analogy, Kant argued for a rotation of the galaxy around an undiscovered center. We see the sky crowded in one direction and not another, he reasoned, because our entire solar system lies in the plane of the galaxy. Moreover, to an observer external to the system "this world will appear under a small angle as a patch of space whose figure will be circular if its plane is presented directly to the eye, and elliptical if it is seen from the side or obliquely."[43] This phenomenon would always be distinctly different from the observation of a single star, because of the character of its light (faint and diffuse), its shape and its enormous diameter. Such nebulosities had of course already been seen by astronomers and Maupertuis in particular had described their occasional ellipticity. Where Maupertuis had thought of them as single stars of enormous size, and William Derham had seen them as doorways to Heaven, for Kant "these elliptical figures are just universes, . . . Milky Ways, like those whose constitution we have just unfolded."[44]

The incipient general philosopher in the young scientist also emerges when Kant warns us that "the Nebulous stars . . . must be examined and tested under the guidance of . . . theory. When the parts of nature are considered according to their design and a discovered plan, there emerge certain properties in it which are otherwise overlooked and which remain concealed when observation is scattered without guidance over all sorts of objects."[45]

The heart of Kant's book, however, is the second part in which, like Lucretius, but more aptly for the eighteenth century with its Alexander Pope, he takes us from the first state of nature to the planetary and celestial system:

> See plastic nature working to this end,
> The single atoms each to other tend
> Attract, attracted to, the next in place
> Form'd and impell'd its neighbour to embrace.
> See matter next, with various life endued,
> Press to one center still, the gen'ral Good.[46]

The common direction of planetary motions and their distribution in the "prolonged equatorial plane of the sun" point to a primary cause exercising its influence in the universe. For Kant, this cause was mainly, but not exclusively, the principle of gravitation.[47] And so, though he recognized that space is virtually empty, "bereft of all matter that could cause a community of influence,"[48] he wrote; he reasoned that at its origins the matter of the universe was in a gaseous state, almost uniformly distributed. "But the variety in the kinds of elements," he wrote, "is what chiefly contributes to the stirring of nature and to its formative modifications of chaos, as it is by it that the repose which would prevail in a universal equality among the . . . elements is done away, so that the chaos begins to take form at the points where the more strongly attracting particles are."[49] These high density regions then become centers of condensation for stars or planets.

Nature has other forces in store, however, to guarantee the particular dynamics of the universe and these are the forces of repulsion which are manifested in "the elasticity of vapours, the effluences of strong-smelling bodies, and the diffusion of all spiritous matter."[50] It is through a combination of these attractions and repulsions among the basic elements of the universe and by a chain of reasoning which we need not reiterate here that Kant brings us to the "free circular movement" of the solar system.[51] This is, perhaps, the weakest part of his argument, for he is telling us that the solar system by its inner evolving interactions could generate its own angular momentum, a violation of the principle of conservation of angular momentum that modern science cannot accept. But this was not known to Kant in 1755, for it received its first complete formulation by Euler in 1775.[52]

The question of the origin of rotational motion apart from most modern theories of planetary formation favor a Kantian view. They argue for a condensation or collisional accretion of relatively cold matter in the region or nebula of the sun, with additional forces at play such as those of the pervasive, weak magnetic field of interstellar space, the temperature effects of radioactivity and the role of turbulence in differentially rotating gaseous clouds. Clearly these are not issues for present discussion.

In retrospect, there is error enough in Kant, perhaps because he was so specific about so many aspects of stellar and planetary formation. For example, we reject his arguments about the distribution of planetary densities in the solar system, and the pattern of orbital eccentricities, among others. But Kant is really important for the evolutionary principle which permeates his universe, and for his vision of a cosmos gradually emerging from chaos in accordance with natural law.[53]

Though we have dealt with unimaginable dimensions of time and space in our cosmological tour, the terrestrial trip only took us from Thomas Wright to Immanuel Kant, in brief, from an optical account of the Milky Way to a theory of cosmic evolution. Many who shared in this disucssion in the eighteenth century, certainly not without importance, were set aside— Lambert, Buffon, Laplace, and William Herschel—to name only a few of the major figures. Lack of space is obviously sufficient reason, but for those omitted there was also the problem of choosing to summarize the unoriginal for the sake of historical completeness (Lambert, for example), or, with unjustified brevity, of dealing with the truly great (Laplace and Herschel, for instance). I have taken the easiest path and avoided both.

Thomas Wright of course is grist for the historian's mill—a village Hampden who moves the mighty Kant—as is John Michell, whose quiet, unsung genius, encased for posterity in only a quartet of great papers, demonstrates the superiority of having a small output from a great man rather than the reverse.

Finally, I have avoided the cosmogonic question as much as possible. Whether it all began with a big bang, with one atom or many, or because the demiurge thrust his hand into the stuff of the world and gave it a twist or because it all reflects God's great plan:—these were and are issues of continuing cosmological debate. But the joys of such speculation are not ours today; besides, Lucretius reminds us,

> ... who can build with puissant breast a song
> Worthy the majesty of these great finds? [Book V]

Perhaps the Latin poet of the nature of things spoke for all cosmologists when he wrote that

> ... all terrors of the mind
> Vanish, are gone; the barriers
> on the world
> Dissolve before me, and I see
> things happen
> All through the void of
> empty space. ...
> I feel
> A more than mortal pleasure
> in all this.

NOTES

[1] J. L. D'Alembert, "Cosmologie," *Encyclopedie* (Geneva, 1777), Vol. 9, p. 593.

[2] E. Hughes, "The early journal of Thomas Wright of Durham," *Annals of Science* (1951), Vol. 7, No. 1, p. 7.

[3] *Ibid.*, p. 13.

[4] *Ibid.*, p. 17.

[5] T. Wright, *Clavis Coelestis, being the explication of a diagram entitled a synopsis of the universe or, the visible world epitomized.* First published, London: 1742. Reprinted London, Dawsons, 1967, pp. 49-50. Students of the history of astronomy of the eighteenth century are deeply indebted to the studies and sumptuous editions of Thomas Wright's works by Dr. Michael Hoskin, of which this is one.

[6] *Ibid.*, p. 18.

[7] *Ibid.* This diagram appears opposite p. 42.

[8] *Ibid.*, pp. 75-76.

[9] T. Wright, *Second or Singular Thoughts upon the Theory of the Universe* (London: Dawsons, 1968). The date is established by Hoskin, p. 8.

[10] T. Wright, *The Elements of Existence of a Theory of the Universe* [Wright MSS. Vol. VII, Central Library, Newcastle-upon Tyne], p. 1, as cited by M. Hoskin in the introductory matter to his forthcoming edition of Wright's *An Original Theory or New Hypothesis of the Universe. . . .* I wish to express my thanks to Dr. Hoskin for permission to draw upon this material.

[11] *Ibid.*, p. [C 2]. This reference is given to the manuscript designations as cited by Hoskin in note 10 above.

[12] *Ibid.*, p. [C 3]-[C 4].

[13] Wright, *Clavis Coelestis*, p. 25.

[14] Wright, *Original Theory*, pp. v-vi.

[15] *Ibid.*, p. 30.

[16] *Ibid.*, p. 59.

[17] *Ibid.*, p. 62.

[18] *Ibid.*, pp. 62-63.

[19] *Ibid.*, p. 63.

[20] *Ibid.*, p. 64.

[21] *Ibid.*

[22] *Ibid.*, p. 57.

[23] *Ibid.*, p. 62.

[24] *Ibid.*, pp. 78-79.

[25] *Ibid.*, p. 80.

[26] *Ibid.*, p. 83.

[27] Wright, *Second Thoughts*, p. 25.

[28] *Ibid.*, p. 79.

[29] I. Newton, *Principia* (Cajori edition), p. 544.

[30] R. Boscovich, *Theory of Natural Philosophy* (English ed., Cambridge, Mass., 1966), p. 146.

[31] The best essay on John Michell written to date, and one to which I am greatly indebted is Russell McCormmach's "John Michell and Henry Cavendish: Weighing the Stars," *British Journal for the History of Science*, Vol. 4, No. 14 (1968), pp. 126-155.

[32] C. L. Hardin, "The scientific work of the Reverend John Michell," *Annals of Science* (1966), Vol. 20, p. 27.

[33] See H. Woolf, *The Transits of Venus. . .* (Princeton, N.J.: Princeton Univ. Press, 1959).

[34] J. Michell, "An Inquiry into the probably parallax, and magnitude of the fixed stars, from the quantity of light while they afford us, and the particular circumstances of their situation," *Phil. Trans. of the Royal Society of London*, Vol. 57, Part 1 (1767), p. 238.

[35] *Ibid.*, pp. 238-239.

[36] *Ibid.*, p. 249.

[37] See R. McCormmach, *op. cit.*, pp. 139-140.

[38] W. Herschel, *Phil. Trans.* (1803), 93, pp. 339-382, and *Phil. Trans.* (1804), 94, pp. 353-384.

[39] Michell, *Phil. Trans.* (1767), 57, p. 250.

[40] W. Hastie, *Kant's Cosmogony*, p. 180.

[41] *Ibid.*, p. 30.

[42] *Ibid.*, p. 32.

[43] *Ibid.*, p. 62.

[44] *Ibid.*, p. 63.

[45] *Ibid.*, p. 64.

[46] *Ibid.*, p. 69.

[47] *Ibid.*, p. 71.

[48] *Ibid.*, p. 72.

[49] *Ibid.*, p. 74.

[50] *Ibid.*, p. 76.

[51] *Ibid.*, p. 78.

[52] C. Truesdell, "Whence the Law of Moment of Momentum?" *Mélanges A. Koyré*, Vol. I, p. 607.

[53] The opportunity of reading an unpublished, forthcoming essay by Gerald J. Whitrow on Kant's cosmogony has helped me to formulate my own views, and I wish here to acknowledge my indebtedness to Dr. Whitrow.

CHAPTER **XX**

"Must a Machine be an Automaton?"

J. O. Wisdom
York University, Toronto

Some of you fellow-robots may have been brainwashed into a doctrine of reductionism; at any rate let us suppose that this particular Colloquium implies that we must reduce everything to something else, biology to physics, sociology to biology, mind to body, and so on. I happen to have been brainwashed the other way. So there is some kind of dichotomy of robots.

Reductionism and Epiphenomenalism

The question I am really raising is an old one which has proved wholly intractable. In fact, the question I am really asking has to do with the theme that human activity is in some degree controlled by or produced by states of mind, or alternatively that it is not controlled in any degree by states of mind.

Otherwise expressed, this is the question whether human action is "reducible" to physiological state; or, in technical terms, it is the old hackneyed question of epiphenomenalism. This claims that states of mind, though they exist, really have no power to operate upon anything, and exercise no control over the action of the body, which is entirely determined by its physiological states. The psychological states are in fact superfluous.

Epiphenomenalism is a central issue raised by the question, "Must a Machine be an Automaton?" I should like to be bold enough to step outside the shoes of a philosopher for a moment and give a straight answer to this question, hoping that this will be off the record because no self-respecting philosopher ever gives a straight answer to a question like this.

The answer is "No." A machine need not be an automaton, it could be something much more elaborate. But, after all, the philosopher is quite right. He comes back at me straightaway and says, "This won't do, you have to explain a little bit more about the idea of a machine and of an automaton before we can see whether your answer is even clear." This is perfectly correct. So one has to do a certain amount of examination of these concepts in order to reach an answer.

Such an examination brings out one of the curiosities about apparently fixed concepts—how they change down the ages, a thing that we lose sight of rather easily. I want to spell this out a little bit in connection with the concept of a machine.

Various Conceptions of a Machine

Strangely enough there are different ideas of what machines are: The classical model of a machine is a clock ("Machine equals clock"). And a clock is mechanical. I would rather not go into what mechanical is, what constitutes being mechanical; I think one might spell that out, but intuitively we do know.

That is the overall idea that has held sway for a number of centuries until fairly recently. I do not think it is the correct picture of a machine today.

Let us consider a little bit more fully what a clockwork consists of. Predictability is one of the characteristics you expect of a clock; and you expect it to be deterministic.

Then we come to something more interesting: the absence of autonomous control. I have looked around for some sort of a simple expression for this without finding one. A clock has no control over its own activity, but we robots think we have.

Then all-important: absence of all mental attributes. I do not propose to go into what they are; we all know. When we were still human, before we received our university education and became robots, we all knew what mental attributes, feelings, and things of that kind were. If you have been brainwashed by your education so that you can no longer remember, you will not be able to follow the rest of this paper. However, some of you robots will remember from days gone by what mental attributes are.

We need not spend long on predictability and determinism, because in statistical and quantum mechanics you can have machines which do not embody these characteristics—at least predictability is vastly modified, and in some cases there is literally no predictability. So the concept of machine has changed now considerably.[1]

As a result of the fairly recent development of cybernetics, centering on the mechanism of feedback, autonomous control enters the scene, while the old-fashioned machine re-enters it. Fifteen years ago this was virtually unknown, but now I take it to be commonplace knowledge. Thus, with feedbacks and mechanisms like thermostats, machines can control their own activity. Hence there is a great change in the conception of the machine.

Machines and Mental Attributes

We are now left with the absence of mental attributes as the only item that seems to distinguish machines from non-machines in contemporary days.

So much so that I think I can say it has become for some people a tautology that if a thing is a machine it cannot be human, if a body is a machine, it cannot have mental attributes. In other words, at the present time the concepts of machine and human being are used to exclude one another. (Though this does not mean they have to be so interpreted.)

Now to turn from that point to those who believe that minds count, that is to say, humanists. If you are a humanist, you think that minds count, and therefore you don't like the idea of thinking of a human being as a machine. Maybe that is a prejudice, but I think this is the way we tend to look at things.

However, there are those who think that humans are machines. By this would be meant that a person would be in all respects human despite having no transnatural (or spiritual) factor in his make-up.

The curious thing about those who think that humans are machines is that mostly they give the impression of feeling that they have to whittle away the notion of minds when they are discussing this kind of topic, making it thinner and thinner. On the other hand, those who think that minds count take the line of denying that humans are machines. Thus, there are two opposite tendencies, and doubtless others in between.

Put slightly differently, there is a tendency to assume that if minds count, human beings cannot be machines, and if human beings are machines, then minds have to be got rid of (because they would interfere with the conception of a machine). Then the whittling down process of the conception of mind proceeds like this. There are great advances in producing machines or artifacts which can reproduce all sorts of things that human beings do. The first obvious one is calculation. Calculation is far more efficiently done nowadays by computers. Secondly, artifacts can be built to reproduce appropriate reactions, that is to say, if they are moving about and bump into something, they will take action to avoid bumping into it. Thirdly, as the second feature implies, they will exert autonomous control.

Rudimentary Mentality: Perceptual Consciousness

I think it might be possible to suppose that you could reproduce in a machine something like perceptual consciousness, and certainly the idea is not unknown, because Descartes certainly held it, and a great many in television production seem to hold it also. But I do not in the least see why some elaborate technological discovery should not be made such that consciousness could be built in.

Thus an artifact might be able to calculate, to exert automatic control, to possess perceptual consciousness, including sensations of "physical" pleasure and pain.

The whittling-down of the conception of mind by people who want to reproduce human activity in machines is steadily pursued. What does not appear in the program is reproduction of more developed feelings. That is handled in a somewhat different way.

On the other hand, for humanists the conception of a machine should exclude mentality, or include it only in the most rudimentary form.

Interpretative Mentality

So the human being and the automaton, or the machine, differ for the humanist as regards the status of consciousness. But I do not think that is the main point. The main point concerns feeling.

By feeling, in this context, I do not mean sensations, e.g. of touch, or pleasure and pain, but experiences such as wanting or fearing, and things of that sort. (And here again are things that those who are operationally minded do not understand after they have become brainwashed by a university, but they did understand them once.)

Now, along with these processes there are one or two others that seem basic. Imagining is such a one. On a different kind of level is the all-important one of interpreting. You see, an animal automaton (if there is such a thing, which I do not believe), when confronted with a lion, is supposed to react according to the principle of stimulus and response. The stimulus of a lion shape will produce some nervous system reaction of fear and flight. One is inclined to slide in the word "automatic" at this point, but this would be question-begging. What I want to bring out is that if you are not an automaton, interpreting is one of the factors that plays a part in determining the outcome of your behavior. In other words, the determinant is not simply a stimulus picture of the lion; the lion is *interpreted* as a lion. Further, interpreting involves anticipating—anticipating what sort of thing a lion would do, as opposed to a sheep and anticipating that the sheep would do something quite different.[2]

Those are the sorts of thing, it seems to me, that make up the difference, or at least a great part of it, between what I regard as a human being and an automaton. There are other factors too. On a previous occasion I have put it forward that the difference between a robot and a human being could be described in terms of such factors as those mentioned. To amplify, you could have a wrinkled forehead or you could have a frown; they may look the same but they are in fact quite different things. Now, what cybernetics has so successfully taught us is that we can imitate or reproduce in artifacts things like wrinkled foreheads. But we cannot reproduce frowns.

In short, what we have succeeded in doing is to reproduce *physiological behavior*. Wrinkled foreheads are movements of the body and therefore can be called behavior of the body; whereas what is wanted, if you want to go the whole hog, is to reproduce *psychological behavior*, as typified by, say, a frown as opposed to a wrinkled forehead.

Interpretative Mentality as Part of Control System

The next question that arises concerns the possibility of building these things (the capacity to feel, imagine, want, interpret), into machines. We assume the cybernetics experts have been very successful in building in rudimentary quasi-purposive movements; can they do the rest?

It is the usual scientific belief, though it is not actually fully proved, that life evolved from inanimate matter, and then that human beings evolved from non-human living cells. This is part of the usual evolutionary view of the world. The current version is that originally there was just hydrogen dust, which by attractive forces became lumps of something different; high pressure led to chain reactions, atomic explosions, and thus produced other elements, leading to other kinds of atoms; eventually after cooling down molecules were formed, then complex molecules, then living cells, then mobile bodies, eventually the cortex; and somehow or other these remarkable things called feelings and other mental phenomena.

I know of no alternative account unless you adopt the theory that somehow life and thinking and feeling were breathed in upon dead matter. So I am going to assume that the conjectured evolution is virtually the only possible account.

Now if such an evolution has happened once, all we have to do is find out how nature did it, so that we can do it again. Nature is rather slow; we might be able to find a quicker alternative way of doing it.

I fail to see why, with technological luck, this could not be done. That is just a little hurdle on the way.

If we assume that we can hope to build in all these things in the course of time, then we are in a position to raise a fundamental question. What do we get when we have done it? What sort of creature is this body that we built all these things so successfully into? It is able to want, to fear, to imagine, to interpret. Is it a sophisticated robot or is it human? (I should perhaps mention on the side, because the matter is sometimes dragged into the argument, that it is quite irrelevant whether the creature comes out of the laboratory instead of out of the womb.)

The answer seems to me to depend entirely on whether the factors thus built-in play an effective part in governing the behavior of the artifact,[3] as opposed to be being present as a figure-head going through the motions but not relevant to its workings, like the "decisions" taken by a parliament under an absolute dictatorship. In the latter situation what we should have would be a robot with imitation human behavior.

I shall try to bring out the problem by putting two alternative questions.

Could autonomous control and capacity for having feelings be built into the same robot as *independent* systems? In such an artifact the mechanism for autonomous control and the mechanism for having feelings would not interlock at any point. It seems to me perfectly conceivable that this sort of artifact could be built. But it would be a rather curious kind of set-up.

Now, the alternative question is whether you could build these in as *linked* systems. The answer is not known. I raise the question, mainly with the object of trying to bring out what the question means. If these two capacities are linked systems, this would mean that a feeling (or interpreting or anything of this sort) would be a factor in the feedback control-mechanism. If the feedback control-box contains only thermostats and gas flames and things of that sort, that is one thing. But if feelings are part of the machinery, then they would be linked, and thus play a real part in producing whatever behavior ensues. This is the central point I wished to bring out.

Thus, on the first alternative, what we have in fact is a new model of the old-fashioned theory of epiphenomenalism. Epiphenomenalism allows autonomous control; but consciousness or feelings, though they exist, play no role. Contrasted with this is the opposite process—a change in your ideas effecting a change in your body. Thus when you get information on the telephone that your pay has gone up, your body reacts, your knees and legs move and carry you to a place of celebration. The reductionist thesis precludes this process.

I think that some of us have been brainwashed to suppose that human beings fall under the first category, epiphenomenalism, while for others they fell under the second category, where the feeling is part of the machinery controlling the control-box. My aim has been not to try to answer this question, which is still pretty intractable, but to try to sharpen up just what is involved in these two alternatives.

Conclusion

I shall conclude by trying to draw together the various threads of the paper, for the problem I am really concerned with may easily get lost sight of amid questions that, important as they may seem, turn out to be peripheral.

Part of the twentieth-century *Weltanschauung* is the adherence to reductionism. Applied to mind and body, it means explaining all mental activity by means of physiological processes; thus it is equivalent to epiphenomenalism. This is the central issue in discussion machines and automata.

There are, and have been, widely different conceptions of a machine. (i) Classically machine equals clock; a clock is mechanical. This includes (*a*) predictability, (*b*) determinism, (*c*) absence of autonomous control, (*d*) absence of all mental attributes. (ii) Post-classical physics discards (*a*) and (*b*), and gives us the notion of a statistical machine. (iii) Cybernetics discards (*c*), and introduces the notion of a self-regulated or *autonomous machine*.

Hence the contemporary conception of a machine may seem to be a statistically mechanical, autonomous system lacking all mental attributes.

Some regard it as part of the very notion of a machine to exclude mental attributes. But some try to build rudimentary mental attributes into machines.

The outcome is that the first group denies that human beings can be machines; but the second group, subject to the same pressures, admit humans as a species of machine only by whittling down the conception of mental attributes until there is nothing human left in it.

With an effort, many might admit the possibility of reproducing in an artifact rudimentary attributes such as perceptual consciousness, sensations of touch or pleasure and pain.

Even so, I contend, the artifact would still be an automaton.

The next step would be to build in "interpretative mentality," including imagining, wanting, anticipating. (Our conjectured history of the earth and of evolution provides a reason why this should be physically possible.) We need to build in not merely the capacity for purposive *movements,* after the manner of cybernetics, but also the capacity for *psychological activity* (frowns as opposed to wrinkled foreheads).

But, though necessary, this would be not quite sufficient; for, if the interpretative capacity were isolated from the control-system, the autonomous control of an artifact would operate independently of it—we should have a robot, possessing interpretative capacity indeed but only as an epiphenomenon.

Hence for an artifact to be human, the interpretative capacity would have to be reproducible but also to be built in as part of the autonomous control system.

That is why epiphenomenalism is still a fundamental question.

It should now be clear that the question of a machine's being restricted to automata can in principle be answered in the negative: this requires broadening the conception of a machine to cover artifacts that would include just those systems (of feelings, interpretative capacities) that the automaton lacks.

Thus, the most fundamental question is whether we can find a way of deciding whether feelings, imagination, and interpretation are elements in the control-box, or whether they are not.

And that is a material question, one of great significance and interest, and not a question concerning definitions.

It should now be clear that the issue is not the question whether there are the grounds for asserting or for denying that we could ever reproduce the whole of human behavior. And if we could, what could the control system consist of? For, there is no known way of saying reproduction in an artifact is impossible. Anyone who maintains it cannot be done in principle, is doing a piece of uncheckable speculation. The significant questions concern the *kinds* of attributes that may or may not, be reproducible in an artifact; and what *kinds* are to be found in the control-box. Further logical possibility of reproduction is not the point; *physical* possibility is what is at stake.

To end with some general comments. It is vital to realize that *there is no essential conception of a machine.* I am now saying, in terms of my viewpoint, that humans *are* machines in a significant sense, but also that machines can have these human characteristics which do not have to be explained away because of being non-mechanical. What I think is important is to bring out what the human characteristics are that can be represented in a machine, and then it is totally uninteresting whether you decide to call the end-product a machine or not. The general question of constructing a human being involves not just the achievements of current technology and its extension in the near future; it involves establishing the desiderata, the *kinds* of characteristics that would be required to be reproducible in an artifact, needed to solve the strategic problem. The practical, tactical, problem is subordinate to this. In finding the desiderata, it is necessary to eliminate all sorts of irrelevant considerations, e.g. kinds of materials used, origin of artifact, limitations of current technology, operational misrepresentations of mental processes, parascientific domination by epiphenomenalism, and essentialistic conceptions of machine and human being. Some of these irrelevancies have now become generally known, but some have not. In this paper I have been concerned to expose the essentialistic conceptions, and to bring out the desideratum on linking interpretative capacity, contra epiphenomenalism, in with the autonomous control system.

NOTES

[1]If you go back to Cartesian days, animals were considered to be machines, yet they had mental attributes of a sort; an animal had perceptual consciousness, recognized smells and foods and things of that sort. So before the tradition I am considering, there was yet a different conception of a machine.

[2]J. O. Wisdom, "Mentality and Machines," *Proc. Arist. Soc.*, Sup. Vol. 26, London, 1952, 1-26.

[3]We have, no doubt, the difficulty to face of understanding how thinking and other mental processes might have evolved out of material systems, and secondly to understand how they might have evolved out of such systems in such a way as to develop a measure of autonomy. But there are parallels that make such a model conceivable. (J. O. Wisdom, "Some Main Mind-Body Problems," *Proc. Arist. Soc.*, London, 1960, 60.)

CHAPTER **XXI**

Epistemology, the Mind and the Computer *

Henryk Skolimowski
University of Michigan

Epistemology and the Brain

The three great traditions in the philosophy of mind are: Cartesian, LaMettriean, Kantian. These three traditions have been continued in the twentieth century in a variety of forms, and often under a range of disguises.

The Cartesian tradition is primarily concerned with the mind-body problem. It attempts to ascertain what kind of entities are mental entities. It investigates their status from an ontological point of view. Because of this, the Cartesian tradition could be called ontological. In the twentieth century it culminated in a number of doctrines which interpret mental entities in entirely corporeal terms, which in other words attempt to demonstrate: that there is nothing "mental" in so-called mental events; that we need not postulate a special realm of non-physical entities to explain the phenomena of inner life, sometimes called psychic life; that the language and categories of the physical world suffice to give an acount of the inner workings of the mind.

Thus, the mind-body dualism is here resolved in favor of the body. The result is a materialistic monism. Some might claim that the problem of dualism in such circumstances is not resolved but dissolved. The ghost has been chased out of the machine, to paraphrase Ryle. It should be noted that the doctrines which were most emphatic in chasing the ghost from the machine were developed with the aid of the refined tools of modern semantics; they were couched in semantic or linguistic idioms. Perhaps the

---

*By "epistemology" we shall understand the theory of scientific knowledge, particularly as related to the growth of this knowledge. By the "mind" we shall understand the *cognitive* organ which participates in the growth of this knowledge. No definition of the "computer" is offered here in order to avoid such inadequate definitions as: "The computer is a high speed moron."

two most consistent doctrines of "semantic materialism" are Ryle's, as expressed in *The Concept of Mind*, and the Polish philosopher, T. Kotarbinski's, as expressed in his *Gnosiology* and a number of other publications. Kotarbinski's doctrine is known as concretism or reism.*

Theories like Ryle's and Kotarbinski's could be called the materialist theories of mind. They have been continued in recent times by Australian philosophers such as J.J.C. Smart (*Philosophy and Scientific Realism*) and D. M. Armstrong. Armstrong's impressive *A Materialist Theory of Mind* is a critical analysis of reasons for and objections against a materialist theory of mind.

Ontological theories of mind, as we have mentioned, and particularly materialist theories, focus their attention on "mental entities" (events, phenomena). They attempt to account for these entities (events, phenomena) from an ontological point of view; they attempt to answer the question: what are these entities, events, phenomena?

The physiological or Lamettrian tradition, although often arriving at a position very much like that of Ryle or Kotarbinski, is primarily interested in a different problem. Its focus is on the brain conceived as a physiological organ which coordinates our inner life. Thus, the main problem here is *to explain the function of the mind in terms of the function of the brain*, and the latter in neurophysiological terms. LaMettrie's *L'homme-machine* provided a paradigm (as imperfect as it was) for this tradition. Contemporary neurophysiologists and behaviorists continue the same line of inquiry.

The important point to notice here is that, although ontological and physiological traditions often converge, particularly in their conclusions, they should not be confused. They are two distinct approaches. The ontological tradition, particularly as exemplified by Ryle and Kotarbinski, does not resort to physiology and science at all; its arguments are conceptual and semantic; its objective is an ontological reduction of one kind of entity to another.

The physiological tradition, on the other hand, derives its arguments primarily from science: from experimental psychology, physiology, neurology, chemistry. Its scope of problems varies. The minimal scope or the minimal programme is to demonstrate that since the brain is a neurophysiological organ, its entire function must be explained in neuro-chemical physiological terms alone. Such a claim cannot be questioned because under these conditions the programme is a tautology. However, and this must be firmly borne in mind, the minimal programme of the physiological approach is hardly a theory of mind. It *becomes* one when it is claimed that not only the physiological functions of the brain, but also the nature of the mental

---

*For a connoisseur it will be interesting to observe that Kotarbinski and Ryle quite independently arrived at a number of *identical* results. (See my *Polish Analytical Philosophy*.)

phenomena and especially the nature of conceptual thinking are explicable in neurophysiological terms.

Now, both the ontological and physiological traditions are concerned only marginally with the acquisition of knowledge by the mind and with cognitive processes. If one insists that the human mind distinguishes itself from the animal mind by the power of its conceptualization, then one must realize that the theories of mind just discussed sadly neglect the peculiarities of the *human* mind. They neglect the epistemological problems involved in the study of the mind. For the epistemologist the problems of first magnitude are related to the questions: how do we acquire knowledge? What is the role of the mind in this acquisition? And particularly: what is the function of the mind when new knowledge and specifically new scientific knowledge is generated?

Theories of mind which address themselves to these questions are by and large of Kantian origin. We can call them epistemological theories of mind as their main focus is on cognitive knowledge, and the relationships of this knowledge to the mind. Kant is considered to be the initiator of this tradition. However, this tradition has been already pursued more than a hundred years before Kant. In Herbert of Cherbury's *De Veritate* (1624) there is the suggestion that there are certain "principles of notions implanted in the mind" that "we bring to objects from ourselves . . . [as] . . . a direct gift of Nature, precept of natural instinct" (p. 133).[1] It is only, according to Herbert of Cherbury, by the application of intellectual truths which are "imprinted on the soul by the dictates of nature itself" that we can compare and integrate individual sensations and make sense of our experience in terms of objects (pp. 105, 106).[2]

Now, before we examine epistemological theories which concentrate on the cognitive structure of the mind, a few words should be said about those physiological theories that make claims about the cognitive structure of the mind which, in other words, attempt to give an account of conceptual thinking and particularly the acquisition of new knowledge by means of sole references to the function of the brain.

The Reduction of the Mind to the Brain

The literature concerning the relation of human cognition to the function of the brain is voluminous. Various aspects of the problem have been analyzed. Hardly ever, however, is the problem confronted directly, that is by asking the question: how does man's conceptual thinking be rendered in terms of neurophysiology? Let us consider the requirements which a satisfactory neurophysiological theory of the mind would have to meet. Let us also consider an example of such a theory as provided by one of the luminaries of the physio-neurological approach, D. O. Hebb.

Now, Hebb himself notices that the relation between the brain and the mind is by no means simple or directly proportional. *"Even while the brain is losing cells (in middle age) its level of function is still rising."*[3] Let us observe the first anomaly in the very formulation of this otherwise interesting statement. As Hebb asserts, "The brain begins its course of slow disintegration when man is about twenty-five." From a neurophysiological point of view it would seem incorrect, if not contradictory, to talk at the same time about slow *disintegration* (losing cells) and the *improvement* of its function. If the brain in question is solely a neurophysiological organ, then its degeneration in neurophysiological terms cannot mean its improvement in any other terms, because neurophysiological terms are ultimate. If we assert, however, that "its [the brain's] level of function is still rising," we immediately assess this "rising" in terms which are beyond the neurophysiological; we judge its function in *cognitive* terms. We should therefore say that, *although* the brain degenerates from a neurophysiological point of view, it does nevertheless improve from a cognitive point of view. Thus, even if we attempt to make a seemingly innocent statement about the function of the brain, we are drawn inadvertently but inevitably into the assessment of this function in cognitive terms.

Another difficulty of the same sort can be seen in the statement which asserts: "A large brain injury need not reduce intelligence in proportion to its size, because of the conceptual development that has occurred before the injury."[4] Hebb is perceptive enough to ask immediately: "But *how*? What is a 'concept,' neurophysiologically? All this did was to replace one nagging problem with another . . . ."[5] Quite so. And this epitomizes the present situation concerning physiological theories of the mind.

Let us return to Hebb's question. "What is a 'concept,' neurophysiologically?" Most neurophysiologists avoid such questions as awkward. Some might even suggest that the question is ill formulated. But it *must* be answered if a consistent neurophysiological theory of the mind (understood as an instrument for the acquisition and development of knowledge) is to be provided. As it might be expected, when a neurophysiological account of conceptual thinking is given without evasion and equivocation, or replacing one problem with another, it becomes a highly speculative venture, very difficult for direct empirical testing. Such an account was in fact given by Hebb himself in his book *Organization of Behaviour*. We shall briefly examine his hypothesis, not in order to exhibit its weaknesses which are quite apparent, but in order to demonstrate that even hard-headed scientists, when confronted with *real* problems, must resort to imaginative hypotheses. Of this Hebb is perfectly aware as he insists that "research that breaks new ground demands the guidance of imaginative speculation."[6]

Hebb's theory of mind, or more precisely that part of the theory which is concerned with the conceptual activity of the mind, is called the theory of "spinning electrons" or "cell-assembly." Hebb writes:

The long and short of the theory is that an elementary idea consists of activity in a complex closed loop called a "cell-assembly" developed as the result of repeated stimulation during childhood; that one or more of these, simultaneously active, can excite another, in a series called a "phase sequence"—that is, a train of thought— . . . in the operation of these cell-assemblies, accounts for the selectivity and directedness of the train of thought.[7]

In other words, thoughts are generated by "phase sequences of spinning electrons." A new thought, a new idea, a discovery, according to this theory means that some electrons have departed from their phase sequences in order to form new phase sequences. We shall not discuss in detail the difficulties this theory entails. Let us notice just one. If a new idea is a new spinning route of electrons, the question is: why do new ideas occur to minds thoroughly versed in a given field which are altogether prepared for this new phase? Why does this new spinning route not occur in unenlightened peasants, for example? By a piece of sophistry, we could say that creative people are just those in whom electrons have chosen to trace new phase sequences. This would be ingenious but not really satisfactory.

The upshot of these arguments is as follows. In order to be a "good" scientist, when attempting to give a theoretical explanation of complex brain processes, one has to resort to imaginative hypotheses which some contemptuously but unjustly call "speculation." To be a mediocre scientist is to avoid the problem altogether. Thus we face this dilemma: either to remain strictly scientific and factual but to explain little about the relation of conceptual thinking to the neurophysiological basis, or to be bold and imaginative and propose theories which account for the complexity and intricacy of phenomena but which are less susceptible to direct empirical scrutiny.

Now, in asserting that neurophysiological explanations of the mind do not yet begin to explain, we are not suggesting that there is no relation between the mind (which is meant to be a cognitive organ) and the brain (as the bio-neurological basis of the mind). What we are suggesting is that the model for the explanation of cognitive phenomena has to be far more subtle and intricate than neurophysiologists are at present prepared to entertain. Perhaps an analogy will help us here.

No doubt the brain is a necessary basis for the mind and its knowledge. This is obvious but trivial. No one would deny that the brain is the necessary hardware, so to speak, for the existence of cognitive processes. But the hardware is one thing, and the knowledge grafted on it or contained in it is another. It is quite apparent at a moment's reflection that regardless of how detailed is the inspection of this hardware, how detailed is the scrutiny of the components that produce intellectual products, the nature of these products is beyond any physical description of the hardware.

Thus, the relation between knowledge and the brain is like the one between electricity and the electric generator. There is no electricity without a generator. But the generator itself does not explain the nature of electricity. The structure of the generator and the function of its various parts in relation to each other tell us nothing about the production of electricity. However, if we know *beforehand* that electricity is originated by a given assembly of machinery, we can then, and only then, make sense of the various parts of the generator. In other words, the prior notion of electricity is necessary for our understanding the mechanical assembly called an electric generator. The mechanical and magnetic phenomena which are involved in the process of generating electricity begin to make sense *if* we know that electricity arises from these processes. These processes can be fully comprehended only when we know what their final product is, and specifically when we know that this product is different in nature from the structure of the generator and the processes which generate electricity. Electricity itself is, in other words, not a part either of the mechanical machinery of which the generator is made, nor of the processes which are necessary for its production.

The same holds for the relation between knowledge and the brain. We can best make sense of them if we assume that the brain is an electric generator and knowledge is electricity which, although produced by the generator, markedly differs in nature from the stuff of which the generator is made.

But this analogy should not be carried too far. We should not mistake knowledge for actual electricity or for some spinning electrons. The nature of both is so different that even if we discover all possible patterns of the firing of electrons and all the variety of possible configurations of electrons which accompany cognitive processes, they could never be *identified* with knowledge. It can also be argued that the brain-mind problem will never be completely resolved in *scientific* terms. To resolve it completely is to establish the meeting point of the brain and the mind in such a way as to make it the subject of direct empirical testing.[8] It should also be remembered that the more we learn about the brain, the more we know what *is to be known* about its extraordinary complexity and intricacy. A solution to one problem usually opens the way to many new ones. And there is no reason to suppose that our further probing will not result in further multiplication of problems.

It has been contended that if we did not have the brain, we could hardly talk about the mind. This is quite so. But we can reverse the tables. If we did not have a mind, we could hardly have practiced neurophysiology, nor could we examine the intricacies of the brain. The tough-mindedness of scientists always demanding empirical evidence is frequently a disguise for the poverty of their imagination. If neurophysiologists insist that the mind is nothing but the brain, the onus is on *their* shoulders to prove *empirically* that this is the case. Epistemologists do not have to prove empirically that this is *not* so.

Evidence for the cognitive structure of the mind does not lie in the realm of empirical testing, but in the cognitive history of mankind, in the history of science, in the history of knowledge and culture. To explain the phenomena of knowledge in cognitive terms is all the epistemologist is concerned with. As long as such terms as "meaning," "concept," "content," "rationality," "intelligibility" and the like remain specific epistemological terms, irreducible to the terms of other sciences, the reductivist programmes of neurophysiologists will remain in the sphere of unactualized possibility.

The Linguistic Aspect of the Mind

As we have mentioned, the most distinctive feature of the mind is that it is a cognitive organ. Perhaps the best definition of the mind is to call it the central coordinating cognitive agency. The mind thus must be designed in terms of the knowledge it possesses and the cognitive processes it is capable of. Without straining the issue, we might say that the mind is the sum total of knowledge contained in it and organized in conceptual patterns plus intellectual dispositions with which it is equipped to transform existing knowledge and to originate new knowledge.

The mind without knowledge is like an unprogrammed computer. The analogy with the computer enables us to see another peculiar characteristic of the mind. While the computer can operate only with the knowledge programmed into it, the mind, on the contrary, can so to speak "up-programme" itself. While it functions it can arrive at results which transcend its original programming. This process of up-programming during problem solving is absolutely essential for both, the growth of knowledge and the growth of the mind. If the mind did not possess this ability, it would be exactly like a computer, retrieving and transforming knowledge according to the patterns built into it, according to its strictly formalized programming. We may carry the parallels between the two a step further and say that the mind is a very peculiar kind of computer, a computer which has programmed into it the ability to go beyond the limits of its programming, beyond the limits of itself, which is in a way a contradiction in terms. This is why the mind is *not* a computer.

We must not assume, however, that the distinction between the mind and the computer is easily made; it is becoming increasingly difficult. One of the most curious facets of the invasion of the human world by increasingly sophisticated machines is our desperate search for a definition of man that would distinguish him from the machine. The machine has traditionally been conceived as a kind of clock, intricate maybe, but very limited in its function. Man has often been defined by contrast to the machine. When the machine became sophisticated, traditional contrasts became inadequate. The conceptual difficulties in defining man in the age of computers are only a part of the problem. The other part lies in our difficulties in defining the machine. We still cannot free ourselves from the conception of the machine as a clock-like mechanism. Thus, we have a dual problem here: how to redefine both man and the machine without circularity, that is in such a way that one would not imply the conception of the other.

This problem has already led to some extraordinary definitions. Some authors are prepared to grant the name of "human being" not only to "objects produced by biological reproduction from human beings," but also to objects recognized as such "upon the election by a majority vote of human beings of a court with jurisdiction in the designation of human beings; any objects declared as such by this court are human beings."[9]

Now, the increase of our knowledge does not consist merely in the accumulation of more and more facts of the same kind, which is the quantitative growth of knowledge, but consists also in the comprehension of these facts with increasing depth; the latter is achieved through the construction of more powerful theories which give a more penetrating account of phenomena and which reveal relationships not dreamt of before—and this is the qualitative growth of knowledge. The explanation of the growth of knowledge in depth is inconceivable if the mind were just a passive receptor, or what comes to the same, an ordinary computer. On the other hand, it is entirely conceivable once we think of the mind as possessing this peculiar ability to up-programme itself as it continues to arrive at new knowledge.

The mind is pregnant with half-programmes, fourth-programmes and hints of programmes which, given the opportunity, become fully articulated, and thus new extensions of knowledge. This process of articulation we call creativeness. How does this articulation occur? There is no articulation without elucidations of language. Thus, the growth of knowledge is inseparable from the growth of language which means introducing new concepts, splitting existing concepts, discovering in language concealed ambiguities, clarifying the multitude of meanings compressed in one term, refining the penumbra of uncertainty surrounding concepts. These elucidations of language—the discovery of new concepts for new contents, and the endowment of old concepts with new meanings—are essential for that function of the mind which is called creativeness, or, in other words, up-programming. The computer does not possess this ability because it is an intrinsic feature of the computer that it must work with fixed, unambiguous, precise, unalterable, formalized concepts. How these new conceptual elucidations could be explained in neurophysiological terms, that is, through the same electro-chemical constituents, remains an enigma if not a mystery.

My aim here is not to celebrate ambiguities, but rather to recognize what actually occurs when science advances. Practising scientists are most distressed when the concepts which they treated as solid rocks turn out to be wobbly and ambiguous. But the history of science without fresh distinctions, without new concepts, without making firm old established concepts wobbly, would be a catalogue of things, not a progression in the stages of understanding. With an eye to the new content which science discovers and explicates, Martin Gilbert wrote, some four centuries ago, in his famous treatise, *De Magnete*:

> We sometimes employ words new and unheard of, not . . . in order
> to veil things with a pedantic terminology and to make them dark
> and obscure, but in order that hidden things which have no name
> and that have never come into notice, may be plainly and fully
> published.

Let us take the concept of *flying*, for example. It is obvious that we do not fly in the same sense in which birds fly; yet, we do fly. Thus, after we discovered that it is possible for us to fly, but in a different sense from the flying of birds, we may conclude that the original concept of flying contained some ambiguity.

In a nutshell, the growth of science means the increasing content of scientific theories and the enrichment of the language of science. This growth has neither been smooth nor linear, and its convulsions are reflected in the twists of language with its conceptual shifts, changes in meaning, metamorphoses of concepts. Changes in language follow like a shadow the changes in the content of science. But it must also be remembered that new content of science can only be expressed through new concepts. Concepts are epistemological categories. To talk about concepts "neurophysiologically" or "biologically," or in any sense "physically," is to make a category mistake.

The human mind is a linguistic mind. Human knowledge is linguistic knowledge. Knowledge, particularly scientific knowledge, must possess content. Content must be expressible through symbols. Symbols organized in coherent wholes are languages. Language is thus understood here in a broad sense. Any coherent system of symbols may be considered to be language. It is a condition *sine qua non* of knowledge that it can be explicated and expressed by means of intersubjective symbols. It is thus in this sense that human knowledge is linguistic knowledge. And this reinforces our conclusion about the unique epistemic status of knowledge, of concepts, of conceptual thinking.

Now, in order to distinguish even more sharply the uniqueness of the mind as contrasted with the computer—which is taken by many "physiologists" to be the analog for the mind, thus an "empirical" evidence that it is only a matter of time when the mind will be shown to be but a computer—we shall analyze the structure of rationality in human beings and in the computer.

Rationality: Human and in Computers

It is well to remember that rationality is not an invention of modern science. It existed before modern science did. Aristotle has defined man as a rational animal. "Rational" for him meant reasonable, in possession of his reason, being able to judge on intellectual and non-emotional grounds. We must be quite clear that Aristotle did not define man as a *logical* animal but rather as a rational one. And we also must be clear that rationality was for him not identified with the capacity for computation but rather with the ability for making judicious judgements.

For the ancient Greeks, the philosopher was no doubt the epitome of the rational man. The essential characteristic of the philosopher, as we all know, was Sophia—wisdom. In other words, it was not logical dexterity which was associated with the sophists but a well balanced, judicious and wise judgement. Now, I am emphasizing this element of judgement in the structure of traditional rationality in order to make it clear that as originally conceived, rationality was not so much the ability to think logically but rather the ability to reason, to arrive at pertinent and relevant conclusions.

There is a peculiar difficulty in discussing and analyzing rationality. Rationality is embodied in all our cognitive systems. And I mean here cognitive systems produced in the post-Renaissance period in the Occidental world. The analysis of rationality must be performed within a cognitive system. Cognitive systems to count as cognitive and as intelligible must obey the criteria of rationality. Thus analysis of rationality is partly a self-referential process. It is very difficult to show the limitations of certain forms of rationality if we have to obey the criteria of rationality inherent in these forms. I shall try to exhibit the limitations of those forms of rationality which are the by-products of modern science. Computer rationality is the final stage in the development of scientific rationality; it is the embodiment of certain abstract features of this rationality.

We must realize it very clearly that one of the tacit assumptions of the Western intellectual tradition is that science is rational, and that the criteria of rationality are to be found in the characteristic features of the most advanced forms of scientific knowledge. In the 18th and 19th centuries it was physics that provided the basis for these criteria. In the 20th century it is logic rather than physics that provides this basis. It is again tacitly assumed that science and rationality are one: the articulation of one is at the same time the articulation of the other. The rational agent in science furthers the progress of science. In furthering the progress of science and in exhibiting its new features, it (the rational agent of science) furthers itself by articulating its own features.

Quite understandably therefore the crisis of contemporary science is at the same time the crisis of rationality. According to the established view there is only one direction for science and rationality to proceed: toward further refinement, further abstraction, further articulation. We have recently discovered, however, that this trend is unlikely to continue because of the inherent difficulties that are latent in the mechanistic model of the universe: reality seems to tell us through negative feedback that the idiom we have accepted was fruitful up to a point and becomes less fruitful after we have passed this point. There is still another reason. We do not want to pursue science along its traditional line because of human and social considerations. Clearly our decision to question the direction of science, because of some human and social objections, if such a decision is judged within the model of scientific rationality, is simply irrational. And yet we think that we ought to control the development of science and not *vice versa*. Furthermore, we think that our decision to exert a control over science is a rational decision though it may be incompatible with the direction rationality has pursued during the last two centuries.

It is evident therefore that even with regard to such issues as which direction science ought to follow, our rational decisions do not have to be based on the idea of rationality which is latent in science. Such decisions are based rather on what we consider to be judicious judgements very much in the manner of Aristotle and the Greek idea of Sophia. It is thus clear that human rationality is not wholly logical, while scientific rationality, particularly as exemplified by computers, is entirely logical.

To reiterate, in the second half of the 19th century and in the first half of the 20th century mathematical logic came to be viewed as a new paradigm for all knowledge. And consequently rationality became imperceptibly identified with logicality, that is, with the features characteristic of the systems of formal logic. Rationality and logicality became one. Those systems began to be considered rational which resembled the features of the systems of formal logic. In this context computers are no doubt the most rational creatures.

The Philosophical Roots of Computer Universalism

Before we proceed with a detailed analysis, let us make it quite clear that the dispute about the nature of the computer, its role, its function, and its future is really not about this or that aspect of the computer's workings, not about the division of labor between man and machine, but is a continuation of the fundamental dispute about the nature and structure of the universe. Computer theorists, at least the most ambitious of them, are the twentieth-century prophets of the mechanistic *Weltanschauung*. They seem to be arrogantly confident (as was LaMettrie two centuries ago when he developed his idea of *L'homme machine*), that the basic assumptions on which their claims are founded are unquestionable. And these assumptions are:

1. That the world is a closed system;
2. That its behavior is regular and is describable in scientific laws;
3. That phenomena of various kinds can be reduced to one kind;
4. That the whole world and its particular sub-systems are wholly determined.

These are not strictly scientific assumptions. They are of a metaphysical variety. Computer theorists thus form a neo-mechanistic school of philosophy. Their tenacious defense of some grossly exaggerated claims of what computers can and will do is more understandable if we realize that they represent a school of metaphysics.

I shall call computer universalism (or neo-mechanism) the thesis which claims that the computer will ultimately be able to perform all the functions that human beings perform. This universalism is based on a reductive programme. It reduces the human being to the functions of the mind; the mind to the brain; the brain to its neurophysiological structure; and this structure is then isomorphized by the structure of the computer.

What ultimately matters is not the legitimacy of these reductions but the demonstration that there are no activities which man can do and computers cannot. An ingenious argument demonstrating that some things cannot be done by the computer was put forth by J. Lucas in his "Minds, Machines, and Gödel." Lucas has argued that machines are definite; whatever is indefinite or infinite is not a machine. Machines are thus capable of performing only a limited, finite, definite number of *types* of operations. Any sequence of operations that is produced by the machine can be represented by the logician on paper. If we know the operations the machine can perform (rules of inference), and if we know the initial axioms and schemata, then every step, every operation of the computer can be recast on paper as a step of a formal proof. The machine cannot do anything but produce steps of formal proofs according to the rules of inference, that is, according to the operations it was programmed for. The machine can only reach the conclusion which follows the rules of a formal proof. In Lucas' words: "The conclusions it is possible for the machine to produce as being true will therefore correspond to the theorems that can be proved in the corresponding formal system."

Now, a Gödelian formula is a formula which we know to be true in a given formal system (if we stay outside the system), but which we cannot prove in the system. "The formula is unprovable in the system," but it is nevertheless true. Lucas rightly argues that any rational being could follow Gödel's argument and convince himself that the Gödelian formula, although unprovable in the given formal system, is nevertheless true.

As we have said, the computer can only produce provable formulae. If the formula is unprovable in the system, then the machine cannot produce it. A true formula that is unprovable in the system is out of the reach of the computer. The Gödelian formula cannot thus be proved in the system, whatever computer we take. The machine *cannot* produce the formula as true. The mind can.[10]

The standard reply would be to say that the Gödelian formula is unprovable by the computer in the system, but it can be proved in a meta-system. But this would not do. For by this operation we only remove the problem one step; and it will reappear on the meta-system level.

The conclusion is as follows: If any mechanical model of the mind must include a mechanism that can enunciate truths of mathematics (since this is what minds can do), then the machine cannot be an adequate model for the mind because it cannot do everything the mind can do. And this is not because of some limitations of existing machines, but because of the intrinsic nature of the machine.

I shall attempt to advance here another argument demonstrating the inherent limitations of the computer to perform certain human activities. I shall not treat these activities individually, as separate self-contained actions that can be treated as discrete units. But I shall rather exhibit the roots of these actions and demonstrate that without understanding the roots there is no understanding of the actions themselves. Indeed, the "behavior" and the outcome of this behavior may be identical in man and in machine and yet, as I shall argue, we have very good reasons to insist that this is not the *same* kind of activity. What then makes the difference? Human intentions, human rationality and everything else that produces behavior of a certain kind. "Behavior," it may be argued, stands for self-correcting activity, for "correct" responses to the environment, characteristic of entities endowed with consciousness, or at least with awareness of the outside world. From this point of view, we can talk about the "behavior" of machines only in a derivative sense. In the strict sense of the term the doings of the machine cannot really be called behavior at all. And consequently, since they cannot even start behaving or misbehaving, the *results* of their doings may resemble the results of human behavior, but in no sense can we say that their *doings* are identical with human *behavior*.

Now, for tough-minded machine theorists to talk about intentions of human behavior is to evoke metaphysical ghosts. Why? Because intentions are unobservable and unmeasurable, while "behavior," that is, the results of the computer's doings, *is*. But why should we accept the behaviorist criterion of validity? Computer theorists so often, in fact too often, take the behaviorist criterion for granted. But this in no way compels us to oblige them and accept this criterion as the only criterion of *human* behavior. Let us take an elementary book of English grammar. On the first pages we shall find the conjugation of the verb "to be." We shall also learn that its negation is "not to be." Let us combine these two phrases, which every foreign student learns

during his first few lessons of English. We shall obtain "To be, or not to be."
Well, this is a rather well known phrase. It can be uttered by a foreign student
learning English. It can be announced by an actor on the stage playing
Hamlet. It can be emitted by the computer "learning" to speak. And finally it
can be uttered by a person who is about to commit suicide. These utterances
are identical in sound, syntax and the meaning of particular words. But do
they stand for the same things? Clearly not. Consequently, verbal behavior
alone is not sufficient for the understanding of the particular kind of activity
this verbal behavior represents.

We may thus conclude that in order to understand an utterance of the
phrase "To be or not to be" unequivocally, we must reconstruct the *context*
to which it belongs. And the reconstruction of this context requires much
more than simply the rules of behaviorist psychology. To appeal to
context—Hubert Dreyfus rightly argues—is more fundamental than the appeal
to facts for the context determines the *significance* of the facts. And he
ingeniously illustrates this by analyzing the meaning of the statement: "The
book is in the pen." From our knowledge of the relative size of pens and
books, we infer that the book cannot be in a fountain pen and therefore that
the expression means: that the book is in a playpen or a pigpen. But the
expression: "The book is in the pen," when uttered in a James Bond movie,
can mean just the opposite of what it means at home or on the farm.[11] Can
the computer ever guess this latter kind of meaning without having seen a
James Bond movie?

Rationality as a Faculty

Now, I shall not pursue particular cases as John Lucas and Herbert
Dreyfus have done, but rather I shall focus my attention on one issue—on the
structure of human rationality versus the structure of computer rationality. In
describing their respective characteristics, it will become clear why the
machine will never be able to perform certain human activities and behave like
we do.

I shall argue that the structure of human rationality is a unique
organization of our cognitive faculties. Human behavior is distinct from both
animal and computer behavior (if it makes any sense at all to talk about
computer "behavior") in virtue of the structure of human rationality. Unless
the computer acquires the structure of rationality identical with our
rationality, it will never be able to "behave" in the way we do. But unless it
has gone through the same biological evolution it will never acquire the same
structure of rationality.

It may be argued that the term "structure," when used in the phrase "the
structure of human rationality," is used figuratively rather than precisely. No
scrutiny of our body or our brain will reveal a structure which is the structure

of rationality. The objection is valid. The structure of rationality is not the kind of structure which we find in structural engineering. It is not, in other words, a physical object. The term is employed rather in the sense in which we speak about the structure of language. There is no *physical* structure to be found in language, yet we maintain that the fundamental features of language are such that it is quite legitimate to talk about the structure of language. It is in this sense that I use the term "structure" with regard to rationality.

As I have mentioned, human rationality is grounded in biology. Computer rationality, on the other hand, is grounded in electronics. The consequences following from this simple observation are far-reaching and usually unnoticed by computer theorists. Because human rationality is grounded in biology it is, though this may sound paradoxical, artificial, whereas computer rationality not tied to biological roots in natural. What do I mean by these terms: "artificial," "natural"? Human rationality is artificial because it emerged so to speak *against* the biological system, and in the course of its evolution alienated itself *from* this system. The grounding of our rationality in biology is the source of all kinds of imperfections of our rationality. In human beings rationality is an attempt to overcome the natural limitations of the biological organism. Rationality is for biological organisms quite unique, and is therefore "unnatural." J. J. Rousseau put it so well when he remarked that "thinking is against human nature," and "the thinking human being is a degenerate animal."

To reiterate, although grounded in biology, human rationality transcends this biology. Therefore human rationality is not an extension of our biology but is in a sense its denial; not a refinement of biological necessities but a confinement of these necessities. Rationality could thus be defined as a liberation from biological necessity. Rationality is thus a faculty which enables the animal who possesses it to even turn against the biological forces which sustain the life of this animal. The only animals who possess this faculty are human beings. No possible analog can be found among other animals or plants; their biology must be unconditionally obeyed.

To be rational for a human being is not necessarily to be logical. Quite often our behavior is illogical but nevertheless rational. Such a situation does not and cannot occur with the computer. What does make us rational is the realization of our inconsistencies and our ability to overcome them. To put it in stronger terms: human rationality is a reconciliation of contradictions. It must be emphasized that this is not a resolution of contradictions, but rather our ability to contain them and live with them. There cannot be any algorithm for a reconciliation of contradictions. So long as human life is what it is and has been, one of the vital functions of this capacity which we call rationality is to reconcile us to the necessity of contradictions. No computer, that is to say, no formal system, can "live" with contradictions.

Contradictions at the foundation of a formal system mean the non-function of the system.

In order to exist, the computer must be perfectly logical. If some illogicalities are to be admitted in the system, they must be algorithm-built illogicalities, thus *logical* in nature. Rationality in the computer is thus perfectly natural. Why? Because it is the consequence of the computer's perfectly logical structure. Rationality, as grounded in electronics and as based on logical algorithm, is the only possible state of being of the computer. Computers do not overcome biological conditioning because they are not subject to it. And this is their great strength which is obvious to everybody. But this is also the source of their great weakness which is not obvious at all. I shall thus advance a seemingly paradoxical thesis: *the greatest weakness of the computer is its perfect logicality*. This trait alone makes the computer incapable, and not only temporarily but in principle, of simulating and performing many actions and forms of behavior whose validity and meaning is essentially rooted in our imperfections.

Marvin Minsky is one of the foremost advocates of the thesis that in essence the computer can do anything that human beings can. His main argument appears to be this: "There is no reason to suppose machines have any limitations not shared by man."[12] This is his motto and this is his fundamental mistake. Why? Because we ought to look at the situation the other way around. As paradoxical as it is, it is in our limitations that we must see our strength, and it is in the lack of these limitations in the computer that we must see its weakness. *Our* (biological) *limitations are responsible for the superior structure of our rationality*. It is because the computer does not share these limitations that it is such a poor problem-solver.

Experts in artificial intelligence seem to be oblivious of the fundamental fact that the greatest obstacle in making the machine do what we can do is not the machine's still limited ability to simulate and repeat the logical characteristics of our behavior, but rather its inability to inherit our non-rational and illogical traits of our behavior and our rationality.

We may inquire whether it might not be possible in some distant future to simulate the structure of human rationality with all its imperfections, weaknesses and aberrations. Can we not imagine a machine so "perfect" that it embodies all human imperfections? After all, Fred Hoyle has described such a creature which he called Andromeda in his science fiction novel *A for Andromeda*. Logically, this is indeed conceivable. However, it must be borne in mind that these perfect replicas of human beings which would entail all our imperfections would be indistinguishable from human beings. If such replicas are ever made, or even had been made in the past (which, after all, is a *logical* possibility), they would be indistinguishable from human beings. They would simply *be* human beings. And then the problem: man versus the machine no longer exists.

In emphasizing the superiority of the structure of human rationality over computer rationality we have not suggested that human rationality is superior in every respect, but rather that it is superior in some *vital* respects: superior in its comprehensiveness, superior in its ability to solve problems. Human beings, viewed from the standpoint of their rationality, are problem-solvers. In solving countless kinds of problems during our long evolution, we have developed the fantastically complex structure of our rationality. This structure is not as efficient or as infallible as computers are, but instead we are capable of solving *new* problems or solving old problems with increasing depth.

Minsky and other artificial intelligence experts should like to reduce man to a problem-solver of the kind of problems which are formalized or at least programmable. But this is a travesty of human problems. It is a fundamental characteristic of genuine problems that they are unprecedented, thus neither programmable, nor formalizable before they are solved in a heuristic manner. Selfridge and Neisser made a similar observation in their classical study "Pattern Recognition by Machine." ". . . A man is continually exposed to a welter of data from his senses, and abstracts from it the patterns relevant to his activity at the moment. His ability to solve problems, prove theorems and generally run his life depends on this type of perception. We suspect," they continue, "that until programs to perceive patterns can be developed, achievements in mechanical problem-solving will remain isolated technical triumphs."[13] Furthermore, if various modes or perception, to be found among different animals are, as recent studies demonstrate, initial strategies for problem-solving in the process of survival. For instance, the frog does not perceive the light spot when it is stationary, and begins to perceive it immediately when it starts to move, then our conclusion is reinforced that problem-solving, pattern-recognition, etc. are intimately connected with the evolutionary life of the animal, and are not a matter of clever programming.

From the fundamental fact that human rationality is grounded in biology and that it was developed in man, the problem-solver, follows a number of important consequences. We shall now contrast the characteristic of human rationality with the characteristic of computer rationality:

	the structure of rationality	
	human rationality	**computer rationality**
(i)	qualitative	quantitative
(ii)	continuous	discrete
(iii)	multi-layered	linear
(iv)	hierarchical	operational

(i) Human rationality is both qualitative and selective. It does not only compute logically but also selects and evaluates. Its selection and evaluation is not guided by any explicit set of criteria. The set of criteria may vary from situation to situation. Whether a situation requires the same, or a new set of criteria is a matter of judgement. In the final analysis, the criteria for selection and evaluation to be applied in a given situation must be seen as the result of the whole experience of mankind in problem-solving. No such experience can be conceivably stored in the computer unless we assume *a priori* that it can be stored. Then of course we *assume* that the computer may one day become indistinguishable from human beings, which is question-begging and which makes the whole problem: man versus the machine, non-existent.

(ii) Human rationality is continuous and is composed of multifunctional "parts." If some links fail, others carry on the process. The weakening of some or even all links causes the weakening of the rational process but not a complete breakdown. Computer rationality is discrete and is composed of one-functional parts. A break in some links results in a total breakdown. The whole system is rigid, while in human beings the system is flexible and self-adjusting.

(iii) Human rationality is multi-layered. Which layers respond to a given situation and how various layers are integrated together is for us a bit of a mystery. Computer rationality is linear. Given a problem, the computer simply follows the programming. The computer cannot by itself switch to another programming in order to probe a problem with increasing depth, in order to find its new aspects as it happens with problem-solving by human beings. If the computer does switch to another programming, this switch must be rule-governed, thus pre-programmed.

(iv) Human rationality is hierarchical. We can distinguish important from trivial problems and can concentrate on the former at the expense of the latter. Computer rationality is indiscriminate. Given certain premises it inevitably proceeds to conclusions, even if these are patently absurd. In the jargon of computer theorists, this lack of discrimination is summed up as: "garbage in; garbage out." To take a concrete case. During the visit of Khrushchev to the United Nations in New York in 1961, a computer spotted an oncoming wave of what appeared to be Soviet ballistic missiles. Its advice was: "retaliate." However, its decision was overruled by a human judgement. The general in charge decided that it would be too absurd to suppose that the Russians would start a nuclear war while their leader was in the United States. And he was right.

Our discussion of the characteristics of human and computer rationality can be summarized by means of a table taken from a special issue of *Science Journal* devoted to the topic "Machines Like Men." Here it is:

BRAIN/COMPUTER COMPARISON	BRAIN	COMPUTER
1 'reset' time of elements	10^{-2} second	10^{-7} second
2 rate of transfer of information	10-30 bits/second (typing at 200 words/minute)	6000 bits/second (higher with magnetic tape)
3 storage rate	1 bit/second in long term memory	10^{6} bits/second
4 storage capacity	theoretical maximum 10^{9} bits during lifetime	currently 3x10
5 processing	parrallel	serial
6 interconnections	rich	poor
7 filtering	very efficient	receives only pre-digested information
8 effects of component failure	rarely produces 'nonsense'	usually produces 'nonsense'
9 type of problems that can be tackled	very general	rather limited

Figures are only rough approximations; computer data, particularly of capacity, are still changing very rapidly   *

*N. S. Sutherland, "Machines Like Men," *Science Journal*, Vol. 4, no. 10, October 1968, p. 47.

The first three rubrics extol the glory of the computer and by comparison emphasize the deficiencies of the human mind. But the picture changes when we go on to examine the remaining rubrics. And these are really the important ones. Perhaps the most interesting is rubric 8 ("The Effects of Component Failure"); if anything goes wrong with the computer, if any element becomes faulty, then—to use Sutherland's words—"catastrophic errors occur and the print out may become complete nonsense." In contrast, the human mind hardly ever breaks down completely except in pathological cases.

It is interesting to note that even the protagonists of the computer admit explicitly that man being the result of an evolutionary process acquires, handles, and processes his information according to survival value. It is further admitted that this survival and self-maintenance are achieved because of man's genetic make-up and because of his built-in goals and drives. We are told, however, that "attempts are made to program computers to exhibit curiosity as a goal."[14] This is really an interesting idea. But are we not deluding ourselves by supposing that if we use the term "curiosity" and link it to some kind of process in the computer, we have thereby taught the computer to be genuinely curious? Is it not simply a linguistic perversion to talk about "curiosity" and "goals" with regard to a piece of electronic equipment which shares none of our biological predicaments?

We may therefore observe that such concepts as "choice," "evaluation," "selection," "curiosity," "drive," etc. when used to describe the "behavior" of computers and that of human beings, are altogether different concepts. "Choice," "evaluation" and "selection" for computers are equivalent to following the rules as determined by the programming, and if this is the case there is no choice, no selection, no evaluation in the ordinary sense of the words, but only logical determination.

The whole "syndrome" of the computer's thinking, intelligence and rationality is based on the confusion of rationality with logicality. There is only one form of rationality in computers and this is logicality. It is because the computer has no choice but to be logical that its rationality (judgement, thinking) is so limited. If intelligent thinking means computation plus judgement, then the conclusion follows that computers are not and *cannot* be intelligent. If rationality means an intricate ability to solve problems in the light of the experience of the species, then computers are not and cannot be rational. They are only logical. There is a large leap between logicality and rationality. This leap means 600 million years of organic evolution.

The difference between the structure of human and non-human rationality nowhere manifests itself more strikingly than in the use of language. After the initial rather successful attempts to teach the computer the rudiments of human speech, we have become more sober of late. Many think that the whole programme of the computer's acquisition of language has broken down.

Some suggest that "the machine that can understand normal fluent human speech may never be built. . . ." The difficulties of the computer with human language is not a temporary technical problem which can easily be resolved with bigger computers, faster computers, more expensive computers, more complex computers.

The structure of human language is as multi-dimensional, stratified, hierarchical, rich and open and thereby in a sense as the structure of rationality itself. *Indeed, human language mirrors the structure of human rationality.* Neither human rationality nor human language are entirely consistent or governed by logical principles alone. They are open systems and the process of change and refinement will go on forever. Both rationality and language are results of biological and socio-historical processes.

Lev Vygotsky, in his classical work *Thought and Language*, argued forcefully and convincingly that the development of speech and intellect is not only determined by our biological endowment but also by our socio-historical habitat. This socio-historical development—peculiar to human beings—has had an enormous influence on the structure of our language and thus on the structure of human rationality. Vygotsky observed that the later stages of the intellectual and linguistic development is not a simple continuation of the earlier stages. *"The nature of development itself changes,* from biological to socio-historical. Verbal thought is not an innate, natural form of behavior but is determined by a historical-cultural process and has specific properties and laws that cannot be found in the natural forms of thought and speech."[15]

The computer cannot yet simulate our most rudimentary biological responses which we share with other animals. Is it reasonable to assume therefore that in time it will be able to recapture our socio-historical development as well? We can assume this of course, but it will be a *very* optimistic assumption.

Jacob Bronowski has put forth the thesis that "Human language is unique not in any one of its unitary characteristics, but in their totality."[16] *Mutatis mutandis* we can say: human rationality is not unique in any of its particular characteristics, but in its total structure which acts as a faculty.

The Behavior of the Mind

The growth of language, its structure, its fascinating complexity are a revealing evidence of the behavior of the mind. When discussing the problems of epistemology, the mind versus the brain, the phenomenon of language, our concern is, of course, with the *conceptual* behavior of the mind. What is the best way to study this behavior? The nature of the products of this behavior. What are the essential products of the mind's conceptual behavior? Knowledge, particularly knowledge organized in recognizable cognitive patterns, and of course language itself.

We should therefore sharply separate two lines of inquiry: (i) The investigation of the neurophysiological processes accompanying the workings of the mind. (ii) The investigation of the conceptual behavior of the mind. In the former case we seek and try to establish empirical evidence for neurophysiological underpinnings of our thinking and conceptualizing. In the latter case we seek an explanation of the function of the "factor" which can produce conceptual structures in which knowledge is organized. Thus the first is an empirical inquiry; while the second is an epistemological one.

Our problem is thus: what kind of function must be attributed to the mind, what are the features of its behavior, what is its structure most likely to be in order to make the process of the acquisition and extension of knowledge intelligible and rational? We are confronted here with a dilemma. Given the consequent (q), i.e. knowledge organized in conceptual patterns —what kind of antecedent (p), i.e. the structure and the function of the mind, must we assume from which this consequent will follow in virtue of the features of the antecedents? Our problem clearly is not a formal logical problem to be solved by deductive reasoning alone. It belongs to the realm of inquiry which Charles S. Peirce called "abduction." In the function p→q, we are given q and try to reconstruct p from which q will follow. As we know, q may follow from many different p's. Our task then is to discover that p from which q follows *most compellingly*. The term "follows" is not here used in the formal logical sense. The relation of the structure of the mind to the knowledge it produces is contingent, not logical. However, given our knowledge about the growth of knowledge, and given *some* understanding (which we take here to be unproblematic for the sake of argument) about what the mind can and what it cannot do, we may reflect on various models of the mind from the point of view of their plausibility and comprehensiveness.

We shall require that a valid cognitive model of the mind explains the relation between the mind and its knowledge and in particular explains the "mechanism" of the acquisition of new knowledge. Any model of this sort must be tentative, open-ended, and subject to modification; if it is made of definitive and closed, it will cease to be adequate for the open system of knowledge.

In the first half of the twentieth century the study of the mind so conceived was systematically neglected. In most recent years however there can be observed a renaissance of this approach. It is, first of all, due to the inquiries of cognitive psychologists and psycholinguists.

Chomsky's Theory

In the philosophical world the works of Noam Chomsky deserve special attention. Although Chomsky is specifically concerned with the problem of the acquisition of language, he really confronts major epistemological issues concerning the structure of the mind. His main question is: What kind of structure must our mind possess to make the acquisition of language, especially by the child, possible?

What is primarily at stake is to determine the so-called "deep structures" which are innate structures. These structures are responsible for the diversity of linguistic forms characteristic of the "surface structures" dealt with by ordinary grammars. But "a consideration of the nature of linguistic structure can shed some light on certain classical questions concerning the origin of ideas."[17] "This investigation may reveal to us some of the specific mechanisms that enable us to acquire knowledge from experience, specific mechanisms that provide a certain structure and organization, and no doubt certain limits and constraints on human knowledge and systems of belief ... we can hope to learn [from this investigation] some important things about the nature of human intelligence and the products of human intelligence."[18]

Thus Chomsky's theory of language, which he calls *generative grammar* (as it is concerned with the structures that enable a mind to generate new linguistic forms and new knowledge), embraces a large field and implies a number of important consequences.

One of them is a doctrine of innate ideas, and another is psychologism. In Chomsky the two are related to each other: psychologism, which he calls "mentalism," seems to "follow" from the doctrine of innate ideas. Now, Chomsky's merit in revealing the bewildering complexity of the structure of language cannot be overrated. Chomsky's mentalism and his doctrine concerning innate ideas, on the other hand, are questionable, if not entirely spurious.

Chomsky says: ". . . It is natural to expect a close relation between innate properties of the mind and features of linguistic structure: *for language*, after all, *has no existence apart from its mental representation*" [Italics added].[19] Thus, mentalism implies the reduction of cognitive entities to mental entities. The doctrine of innateness, on the other hand, if it is to include a hypothesis about linguistic universals, implies the denial of the growth of scientific concepts. We shall discuss these two issues in some detail.

Chomsky's mentalism, if it is meant to be a resurrection of nineteenth-century psychologism, is open to all the objections to which psychologism is vulnerable; particularly to the objections related to the distinction between the *act* of cognition being mental and thus subjective, and the *result* of cognition being intersubjective content meaningful in virtue of the language in which it has been expressed. Thus, all the epistemological perils inherent in psychologism: its subjectivization of science and, in the final analysis, its subjective idealism, are hidden in the concept of mentalism. If mentalism, on the other hand, is not meant to be a resurrection of psychologism, then it is a misnamed doctrine whose obscurity will have to be considerably clarified.

What are the linguistic universals allegedly planted in the structure of our mind? Chomsky is tantalizingly evasive in describing them in detail.[20] Their general characteristics follow from a rationalist conception of the nature of language and a rationalist conception of the acquisition of knowledge. The essence of both is the view that "the general character of knowledge, the *categories* in which it is expressed or internally represented, and the basic principles that underlie it, are determined by the nature of the mind." From this follows that "The role of experience is only to cause the innate schematicism to be activated and then to be differentiated and specified in a particular manner."[21] Thus, "Experience serves to elicit, not to form, these innate structures." Or in still another manner, the stimulation "provides the occasion for the mind to apply certain innate interpretive principles, certain *concepts* that proceed from 'the power of understanding' itself. . . ."[22] Chomsky thus postulates innate "categories," "concepts," "linguistic universals" (". . . a characteristic feature of current work in linguistics is its concern for linguistic universals. . . .").

We can discern two distinctive theses in Chomsky's contentions: (i) that there are cognitive structures of the mind; and (ii) that these structures are innate. Clearly, we can assert (i) without asserting (ii); in other words, (ii) does not follow logically or epistemologically from (i).

I shall now argue that "linguistic universals," "innate concepts," "inborn categories," in sum all those phrases which purport to denote "inborn"—specifically, linguistic structures—are either used in a figurative sense, or, if used in the literal sense, are refuted by the actual growth of scientific knowledge.

The shortest possible form of our argument can be presented as follows: if there are innate *concepts* (categories, linguistic universals), then there is no *growth* of concepts; if there *is* growth of concepts, then there are no *innate* concepts; but in fact there *is* growth of concepts, therefore there are no innate concepts. This growth of concepts is exemplified particularly in the history of scientific knowledge; these concepts, it must be emphatically stressed, are not trivial or marginal, but essential in our edifice of knowledge.

We must, of course, disallow that *innate* concepts (categories) can change their nature, can, in other words, "grow"; if they are innate, they do not change; if they do change, they are not innate. Chomsky might argue at this point that what is innate are *ideas*, not concepts or categories. We may then retort: how are these ideas related to concepts and categories? What part or aspect of the concept is the "idea" supposed to be? How does an idea become a concept? And furthermore, how are these "innate ideas" related to the growth of concepts? Unless the relation of "innate ideas" to concepts and the categories of science is made clear, we shall have to regard them as mysterious or devoid of meaning.

Now, what kind of arguments are left for rescuing the doctrine of innate concepts? Can we question the notion of the *growth* of concepts? Hardly, for the history of science is the history of the growth of concepts. The extensions of knowledge and the refinements of scientific theories are inseparably linked with the growth of concepts. Concepts thus grow, change, undergo metamorphosis. It is sufficient to mention the evolution of such concepts as "force" and "gravity" to realize at once that their pre-Newtonian meaning was different from the meaning they acquired in Newtonian mechanics, and still different within the system of Einstein's physics: extensions and refinements of scientific knowledge are responsible for these consecutive metamorphoses. If this is so, then there are no innate concepts of "force" or "gravity," for if there were, which ought to be considered innate: pre-Newtonian, Newtonian, or Einsteinian? Or consider the extraordinary evolution of the concept of "matter." Which of its successive embodiments (and there were dozens[23] usually reflecting the state of scientific knowledge at a given time) is to be considered the innate one? It must be observed at the same time that the concept of *matter* is not a marginal one. If there were innate concepts, this concept ought to be among them. Or finally, let us take the concept of "innate ideas" itself. It means something different now from what it meant in Locke's time. We do not attach different meanings to concepts because we have become bored with their existing meaning. They acquire different meanings because they grow and evolve. And this is so in relation to all important concepts making up the corpus of scientific knowledge. Thus, if we admit that concepts do grow and change, we *cannot* uphold the thesis of innate concepts, in its classical sense at least.

Now, is it still possible to maintain any sort of doctrine of innate concepts (ideas) *after* we have recognized the evolution of scientific knowledge? It is indeed, if we are prepared to bear the consequences. There are at least two courses left to us.

We may postulate an innate, permanent and unalterable programming in our mind (involving specific concepts and categories) which is compatible with the changes in scientific knowledge. But then we shall have to assume, as Plato did, that various stages in the evolution of science are not really the result of the *evolution* of science, but rather they represent the process of the unfolding of constant, unalterable categories. Evolution does not concern the growth of concepts in this case because we are committed to the universe of constant concepts which do not change. This is a logical possibility, but a rather far-fetched one.

Another possibility is to assume that the innate programming, innate concepts and categories *change* during the process of the growth of knowledge. Such a doctrine no doubt would be burdened with perplexing enigmas. We would have to explain *why* the changes in the innate programming fit the changes in scientific knowledge so remarkably. In order to maintain that this cognitive programming is innate, that is to say, is not caused by external circumstances such as the growth of science, we would have to assume that this "fit" is coincidental. If we were to assume that it is not, i.e. that it might have been caused by external circumstances, then we would in fact renounce the doctrine of innate ideas.

To conclude our discussion. The growth of scientific knowledge forces upon us the following alternatives regarding the doctrine of innate ideas:

(i) to abandon the doctrine of innate ideas, and to expel any form of innateness from the structure of the mind;

(ii) to build into our innate doctrine a stipulation that innate concepts can evolve—a rather peculiar doctrine of innateness;

(iii) to assume that there are innate elements in the cognitive structure of the mind, cognitive dispositions of the mind which facilitate the emergence of concepts and ideas.

It is the third possibility that is the only tenable one. I shall explore the tenets and the consequences of this position in the remaining part of this paper.

It is clear that Chomsky's impetuous anti-behaviorist campaign led him to an untenable position in regard to the concept of mind. One can uphold a rationalist conception of mind, in the traditional sense of the term, that is, one can maintain that the mind is an active organ in the acquisition of language and of knowledge, and specifically that the cognitive structure of mind is a linguistic one, *without* at the same time being committed to the doctrine of innate ideas.

In the pages that follow I shall outline a model for science and the mind that combines the rationalist conception of mind with an evolutionary conception of science. This model stems from the conviction that the rationalist conception of mind cannot be systematically developed if the evolutionary conception of science is totally disregarded. On the other hand, the evolutionary conception of science cannot be consistently maintained without assuming a rationalist conception of mind. Furthermore, it appears that only if we combine the two can we arrive at a satisfactory justification of the objectivity of scientific knowledge.

The Interaction Between Knowledge and the Mind

The growth of the language of science reflects the growth of science. But at the same time the growth of the language of science is a reflection of our mental growth. Thus in the second sense, the growth of the language of science reflects the growth of our mind—that is, the cognitive structure of mind. In language we witness the culmination and crystallization of two aspects of the same cognitive development: one aspect related to the content of science; the other aspect related to our *acts* of comprehension of this content. Since it is inconceivable to express any of these two aspects without language—because language unites them both—we are bound to conclude that there is a parallel conceptual development of the content of science (as expressed through the outer language of science) and the inner mental structures of the mind (as expressed through inner acts of understanding). Since fundamental concepts through which we grasp and express our knowledge of the world have changed and evolved, it is fair to conclude that the structure of our mind has changed and evolved as well. The conclusion that we are led to is that the structure of the mind, the conceptual arrangement of the mind mirrors the structure and the limitations of the knowledge by which they have been shaped. Mental structures depend for what they are on the corresponding development of science and of all knowledge. Thus *the conceptual structure of the mind changes with shifts and developments in the structure of our knowledge.*

At this point we must assume that there is a parallel conceptual development of our knowledge and of the mind. Knowledge forms the mind. The mind formed by knowledge develops and extends knowledge still further which in turn continues to develop the mind. Thus there is a continuous process of interaction between the two. Although they are independent categories as far as their meaning is concerned, viewed in the overall cognitive development, knowledge and mind are functionally dependent on each other and indeed inseparable from each other. They are two sides of the same coin; two representations of the same cognitive order. The concept of mind must include knowledge which has formed it and which it possesses.

Bearing this in mind, we can easily see the strengths and shortcomings of Kant's theory of knowledge. It is fairly obvious that the mind was made by Kant to reflect Newtonian physics; the conceptual order of the mind was to mirror the conceptual framework of the physics of his time. Kant's crucial epistemological mistake was not in the construction of a defective model of the mind—given the knowledge of his time, Kant's model was superlative—but was rather in the assumption that the growth of scientific knowledge has ended, that science would progress no further.

Kant's main achievement thus consisted not so much in the *construction* of the universal mind, but rather in the *recognition* of the structure of the mind as formed by Newtonian physics. The distinction between recognizing the mind as shaped by the science of a given epoch and constructing the universal mind to reflect all science at all times must be very clearly made.

The changing state of science means the changing content of science. The total content of science is expressed by the totality of concepts and their relationships. This totality of concepts is sometimes called the conceptual apparatus of science. We shall call this totality of concepts *the conceptual net of science.* We can talk about the net of science *in toto* as well as about the conceptual net of a particular science, for example physics. The net of

Newtonian physics includes Newton's laws and other laws following from them, as well as the philosophical presuppositions concerning the absoluteness of space and time and the constancy of matter. But it also includes terms and concepts which belong basically to other sciences, such as chemistry and biology. And in addition it includes many terms and expressions of ordinary language.

The three expressions "the conceptual net," "the conceptual framework," and "the conceptual apparatus" are akin in their meanings and can often be used interchangeably. The conceptual framework usually describes the outline, the skeleton within which other concepts are located and to which they are related. By "the conceptual apparatus" we usually mean an aggregate of all technical terms specific to a given science. "The conceptual net" of a science, as the term is used here, comprises them both and also includes many expressions which are, strictly speaking, not expressions of this science, but which nevertheless are necessary to account for its content. For example, the logical particles (connectives), such as "or," "and," "if ... then," "not," are included in the language of every science. And so are many expressions of ordinary language such as "object," "outside," "inside," "related," which only on the surface are philosophically innocent. The term "outside" for example presupposes spatial relationships, presupposed a three-dimensional world. The term "object" presupposes a semantic and ontological commitment. No doubt, we could redefine both these terms ("object" and "outside") and make them technical terms. But this cannot be done with all terms and concepts which are comprised within the conceptual net.

Why is this not possible? Because our conceptual net would then become a formal language. In introducing our formalism, however, in describing its features, we would have to go beyond the limits of this formalism as we always do in such cases. While explicating our formalism, we relate it to a large conceptual framework. This conceptual framework cannot be formalized. To formalize it would require explicating its formalism in terms which do not belong to it. Such an explication would require a still larger conceptual framework which in turn could not be formalized in its own terms.

To repeat, by the conceptual net of science we shall understand the totality of concepts by means of which the content of science is expressed. Such a totality cannot be formalized. The conceptual net of a given science is thus not an easily definable entity because (a) it merges with the nets of other sciences, (b) because it merges with ordinary language, and (c) because it changes historically. We may attempt to formalize a part of the content of a science, either for aesthetic reasons or for the sake of logical clarity. We must be aware, however, that it is only a *part* of the content that has been formalized. The growth of knowledge does not follow any formalizable patterns. Epistemological inquiry concerned with this growth cannot be formalized either. If we focus our attention on formalizable parts and conceal non-formalizable ones, we distort the nature of our epistemological enterprise.

The development of the conceptual net with its complicated mesh of interrelationships is, as we have argued, an inseparable component in the growth of science. But this is only a part of the history of human cognition. This part may be called *external*. It is external because our knowledge, once formulated in language, could theoretically be assimilated by non-humans. We may imagine that, after a holocaust which annihilates all human beings, some extra-terrestrial intelligences visit our planet and discover our learned treatises. We may further imagine that they "learn" to read these treatises and thus assimilate our knowledge. In assimilating this knowledge they would have to decipher the conceptual net within which our knowledge is located and through which it is expressed. Briefly, if formulated by means of concepts and expressed through intersubjective language, knowledge becomes external to the mind. In theory non-humans could learn and assimilate it.

Now, the other part of human cognition is *internal*. It is internal because it takes place in the mind. We ought to distinguish here between cognitive *acts* and cognitive *results*. Cognitive acts occur in the mind. They are the internal part of the process of cognition. They represent the structure of the mind and the function of the mind. Cognitive results on the other hand are theories and statements—linguistic utterances or other symbolic representations—which express the contents of these acts; they are the external part. Expressed by means of intersubjective language, cognitive acts become externalized. Their content becomes independent of particular minds.

Now specific patterns of thought are not given to us *deus ex machina*. We have already suggested a hypothesis that there is an isomorphism between the net of science and the conceptual order of the mind, that is to say that the *conceptual development of science is parallelled by the conceptual development of the mind*. The conceptual arrangement of the mind with its specific patterns of thought thus mirrors the development of the conceptual net of science with its complicated mesh of concepts.

Science, that is the conceptual net of science, forms the scientific mind, but this mind does not remain passive. It interacts and thus transforms the net. These continuous interactions between the mind of a scientist and science are in actual fact interactions between the mind of an individual scientist and a particular science. But the result of these individual interactions constitutes a new stage in the development of science and a new stage in the conceptual development of the mind.

Every new hypothesis is an invention of a new possible world. The development of science in the last century has proved that so many of these inventions fit nature beautifully (although at first they appeared to be bizarre and impossible), that there seems to be no limit to imagination. Consequently, the human mind may come to organize the knowledge of the world in such fantastic and impossible units (patterns, categories, forms), that is to say, *fantastic* from our contemporary point of view, that whatever presuppositions we lay down as necessary for today, might be shattered tomorrow. Radically new ways of organizing our knowledge may bring a change in our understanding of such concepts as "rationality" and "intelligibility." After all, the concept of intelligibility is a function and a product of our knowledge. The only possible candidate for an absolute form of intelligibility seems to be the logical law of non-contradiction. Indeed, it is very hard to imagine any schemata or units or organizing knowledge in which this principle is not kept intact.

We have said that the conceptual net of science stands for the totality of concepts which express the entire content of science. This net determines the structure of our problems, their validity and even the meaning of particular concepts. The closest approximation to the idea of the conceptual net is Kuhn's idea of the paradigm. The scientific paradigm, as Kuhn uses the term, is a framework which is usually larger than particular scientific theories and which determines the nature of theories and the kinds of problems which are the legitimate problems within the paradigm.[24] The conceptual net is more comprehensive than the paradigm; it determines not only the nature of scientific problems but also the nature of scientific frameworks, of paradigms.

It may be asked how the changes of paradigms are to be viewed within the framework here proposed. Scientific revolutions are indeed significant changes in the conceptual net of science. Kuhn gives the impression, however, that after the revolution is completed, almost everything is changed—which might mean that the whole net is replaced. This cannot be so. In every scientific revolution the part of the conceptual net which is left untouched is larger than the part which is remolded. In every scientific revolution known so far, language, logic and mathematics were left essentially unchanged. Thus, paradigms can be described as easily recognizable parts of the conceptual net. To replace our entire conceptual net would mean to create an entirely different kind of knowledge that would not be accessible to the present human mind which is linked to the present kind of knowledge; we can ascend to new forms of knowledge and comprehend them only on the scaffolding of existing knowledge, only by moving like a spider through the web of existing knowledge. If this web is removed, like the spider we cannot move, we cannot comprehend the world. Our relationships with the world are annihilated, dissolved.

Kuhn ties the idea of scientific paradigms to the idea of scientific revolutions. But the idea of the revolution is not inherent in the idea of the paradigm; the former is not entailed in the latter. Kuhn's critics have been tearing him apart on the subject of normal science and of revolutions, some insisting (Stephen Toulmin, for instance)* that there are no revolutions in science, and consequently that the history of science is normal science *ad infinitum*; others arguing (J.W.N. Watkins, for example)* that there is no such thing as normal science, and consequently that the history of science is revolution in permanence. But none of these criticisms touches the idea of the paradigm with its insistence on discontinuous, non-homogeneous growth of scientific knowledge.

Crucial to the understanding of the growth of knowledge is the understanding of the nature of conceptual change. The nature of conceptual change cannot be fully comprehended without understanding the nature of the interaction between knowledge and the mind. The model of the mind outlined in this essay is a dynamic matrix which allows for the study of this interaction, and thus for the study of conceptual change. Latent in this model is the idea of evolutionary epistemology. Evolutionary epistemology is the discipline which provides a new perspective on the nature of our problems concerned with the growth of science and of conceptual change. This discipline when worked out in detail will be an alternative to the logico-empiricist epistemology that has been for too long dominant in twentieth-century philosophy.

---

*See I. Lakatos, *Criticism and the Growth of Knowledge* (Cambridge, 1970).

NOTES

[1] N. Chomsky, *Cartesian Linguistics* (New York: Harper and Row, 1966), p. 60.

[2] *Ibid.*

[3] D. O. Hebb, "Intelligence, Brain Function, and the Theory of Mind," *Brain*, Vol. 82, Part II (1959), p. 265.

[4] *Ibid.*

[5] *Ibid.*

[6] *Ibid.*

[7] *Ibid.*, p. 266.

[8] See in this respect Raziel Abelson's "A Spade Is a Spade, So Mind Your Language," in *Dimensions of Mind*, Sidney Hook, ed. (New York, 1966).

[9] F. T. Crosson and K. M. Sayre (eds.), *Philosophy and Cybernetics* (Notre Dame, 1967), p. 61.

[10] See John Lucas, "Minds, Machine and Gödel," in *Minds and Machines* (Englewood Cliffs, New York: A. Anderson, 1964).

[11] Hubert Dreyfus, "Cybernetics as the Last Stage of Metaphysics," *Proceedings of the XIVth International Congress of Philosophy*, Vol. II, pp. 497 and 498 (Vienna, 1968).

[12] Minsky is notorious in simplifying the phenomenon and the behavior of man so that it fits the pre-arranged categories characteristic of the structure of computers. This eagerness to "translate" man into the computer causes him to make outrageous claims such as: (in "I think, therefore I am," *Psychology Today* April 1969) ". . . the mind-body problem is not so much an elusive and difficult philosophical problem as it is an elusive and difficult engineering problem." About *will* or *spirit* or *conscious agent* he says: "Naturally, we can't say anything meaningful about it." With his philosophical sophistication, perhaps Minsky can't. But to suggest this as if it were a matter of fact is to ignore a philosophical tradition which has existed for some millenia. What is shocking about Minsky's attitude is not some specific philosophical issues which he tackles in the wrong way, but his total arrogance toward philosophy.

[13] Oliver G. Selfridge and Ulric Neisser, "Pattern Recognition by Machine," in *Computer and Thoughts,* ed. E. A. Feigenbaum and J. Feldman (New York: McGraw-Hill, 1963).

[14] N. S. Sutherland, "Machines Like Men," *Science Journal*, Vol. 4, No. 10 (October, 1968), p. 47.

[15] L. S. Vygotsky, *Thought and Language* (Cambridge: M.I.T. Press, 1962), p. 51.

[16] J. Bronowski, "Human and Animal Languages," in *To Honor Roman Jacobson: Essays on the Occasion of his Seventieth Birthday* (The Hague: Mouton, 1967), p. 387.

[17] Noam Chomsky, "Recent Contributions to the Theory of Innate Ideas," *Boston Studies in the Philosophy of Science*, Vol. III (1968), p. 81.

[18] Noam Chomsky, "Knowledge of Language," *Times Literary Supplement* (May 5, 1969), p. 523.

[19] Noam Chomsky, *Language and Mind* (1968), p. 81.

[20] The deep structure "is the underlining abstract structure of a sentence that determines its semantic interpretation." Put in other words, "it is the deep structure underlining the actual utterance, a structure that is purely mental, that conveys the semantic content of the sentence." "Deep structures," we are told, "are fundamentally the same across languages, although the means for their expression may be quite diverse." Quite consistently, Chomsky is unwilling to commit himself to say something more specific about these structures.

[21] Noam Chomsky, "Recent Contributions to the Theory of Innate Ideas," p. 88.

[22] *Ibid.,* p. 89.

[23] See, for instance, *The Concept of Matter in Greek and Medieval Philosophy*, ed. Ernan McMullin (1965).

[24] See Thomas Kuhn, *The Structure of Scientific Revolutions*, Foundations of the Unity of Science, Vol. II, No. 2 (1962).

CHAPTER XXII

Marginal Notes on Schrödinger

Wolfgang Yourgrau
University of Denver

Even the non-scientist among scholars is more or less acquainted with the fact that Erwin Schrödinger, who died in 1961, belongs to that galaxy of original thinkers in physics, which is distinguished by names like Planck, Einstein, Bohr, Heisenberg, Born, Dirac, Pauli, and so forth. And the physicist learns during his study of quantum theory that Schrödinger is the founder of wave mechanics, and thus the creator of one of the most impressive physical edifices of modern theoretical physics. The various versions of the famous equation called after him have become permanent features in current physical theory. I do not think that wave mechanics will ever reach the exceptional, prominent status of general theory of relativity. Anyhow, it is not custom to compare the relevance of one physical theory with another. In other words, we usually refrain from ranking, as it were, the accomplishments of physicists whose contributions have not only affected the domain of physics, but changed our general conception of the universe.

Yet, although he belongs to the small category of geniuses like a Planck or Einstein, one is prone to forget that Schrödinger's legacy to our culture displays by far more facets than the written work of his equally brilliant co-workers.

Like Planck, he received his training as an adolescent in a humanistically oriented gymnasium. In his delightful little book called *Nature and the Greeks,* he showed his wide and profound knowledge of the highlights of Greek culture in so far as the origin of scientific reasoning is concerned.

In four public lectures delivered in 1950 under the title of *Science as a Constituent of Humanism,* which were later published under the heading of *Science and Humanism,* he succeeded in demonstrating some of the most interesting results in various branches of physics and provided an informative insight into some modern discoveries in such a manner that the generally educated reader could become convinced that the physics in our time has become an indispensable ingredient of the culture of our age.

Like Planck, he was a universal mind, though one would hesitate to call him a universalist in the sense of Leonardo da Vinci or Goethe. The working physicist—be he attached to industry or to academic work—will be not only acquainted with his collection of essays in which he put down the substance of his wave mechanics, but also with his intriguing excursion into cosmology (*Expanding Universes*) and his beautifully written *Space-Time Structures* in which he discusses some fascinating aspects of Einstein's theory, conservation laws and variational principles in a highly original manner, though the ordinary student will complain about the very condensed and too succinct form of presentation. Schrödinger was an impatient teacher and truly believed that one could teach a subject like statistical thermodynamics in a booklet of 95 pages. Nevertheless, his *Statistical Thermodynamics* is a delight to read provided one has, prior to it, studied statistical mechanics and thermodynamics on a sophisticated level. Again, the average student—even if he has successfully gone through the assigned courses in those subjects—will have difficulty to do full justice to Schrödinger's exposition. While I was a student and later an assistant of his, I always felt that he treated even the most complicated and demanding concepts in contemporary theoretical physics more like an artist than a methodical pedagogue. Although his lectures, books, and papers prove beyond doubt that he could reason rigorously and cogently no less than any other of his illustrious colleagues, it is easy to point at passages in his total work that are full of artistic temperament and a certain predilection for subjective involvement.

In another book (*Science, Theory and Man,* formerly published under the title, *Science and the Human Temperament*) he discusses philosophic questions like, "What is a Law of Nature?", "Science, Art and Play", and finally attempts to instruct the uninitiated about the embarrassing question: "What is an elementary particle?" This book shows that he has thought in a historical and philosophical manner about those concepts and claims which the average physicist or scientist in general either entirely ignores or takes for granted because he is of the opinion that many of those topics treated by Schrödinger have been mentioned, though perfunctorily, in the courses of current academic training or in some textbooks. I think it was Holton who tried to prove that Einstein did use, once during his studies, *one* formal textbook, otherwise solely original papers or books, such as Newton's *Principia,* Maxwell's *A Treatise on Electricity and Magnetism,* and so forth.

After he retired from the Institute in Dublin, which he directed for 17 years, he published in the country of his origin, Austria, his so-called personal world view (*Meine Weltansicht*). This book which is not written in a very clear and simple style, contends that he found in Indian thought, especially in the

Upanishaden, his philosophic home; yet he also mentions somewhere that he feels a great intellectual affinity for the views of E. Cassirer whose philosophic work shows hardly any predilection for Eastern thinking. Again, Schrödinger read philosophy in a very unsystematic manner and I found many phrases so obscure that it amounted to almost physical labor to absorb the content of this book which contains some very original interesting observations, but also many statements which have been expressed by other authors before him. It is truly surprising to hear him enunciate that exact science is never really possible. In contrast to Born and Pauli, for instance, Schrödinger became more and more attracted to metaphysics and I found it very irritating that he resorted to physiology and other branches of science to justify his surrender to a certain type of Indian mysticism. Somehow, such an orientation is not disturbing when one deals with Schopenhauer, but it becomes rather vexing when these many references to Indian thought appear not only in his semi-philosophic papers, but even in strictly scientific essays.

Schrödinger had no patience for any form of positivism. Fortunately, some of his almost classic volumes dealing with wave mechanics, cosmology, astrophysics, statistical thermodynamics, and so forth, are not contaminated with his excursions into any abstruse philosophic speculations. That statement is not in contrast with the fact that his frequent references to ancient thought are worthwhile to be read even if they do only on occasion contain independent observations.

In *Meine Weltansicht,* he philosophizes about many purely philosophic or psychological subjects. It is almost impossible to place Schrödinger's philosophic creed in traditional categories. He pays lip service to some of the ideas introduced by naive realism, but in spite of his frequent references to physiology and neurology, his final views are closer to Fechner and Semon than to the research results of contemporary scientists who work in cognate fields.

According to him, or rather according to the views expressed in this little book that is supposed to reflect his world views, he pleads strongly for his particular brand of metaphysics. I find this work, not only because of its strong trend to Indian mysticism, a great disappointment, some of its occasional profound concepts notwithstanding. For instance, it is hardly permissible to claim that metaphysics transforms itself during the course that physics develops. His views on consciousness culminate in the judgment that modern rationalism is bankrupt. Hence, categories such as *numbers,* the *total* and its *parts* have no place in his metaphysic universe that reminds one not only of the unity which the mystics try to attain, but also of the Greek thinker Parmenides. It is strange to hear from a scientist of the magnitude of a Schrödinger that exact science is never truly possible. The claim that we are actually facets of a single being (which one may call God in Western terminology) is difficult to accept because Schrödinger does not even attempt to provide convincing arguments for his notion of metaphysical unity which he adopted uncritically from its Oriental sources. At the same time, he deems it possible that we may arrive at a scientific understanding of ethics in general, without resorting to any of the existing religions of the East or of the West. He pleads for the biological role of an ethical value judgment and regards the accomplishment of certain desirable state of affairs as the first step for the transformation of man to become an *animal sociale.*

In spite of his great admiration for Bertrand Russell, Schrödinger refuses to accept Russell's notion of the real external world. It appears that for Schrödinger all reality is fundamentally reducible to psychological, individual experience. Therefore, he arrived at the Eastern notion that there exists only *one* Mind in spite of the truism that we are surrounded by a multitude of bodies. He also gives more than one interpretation to the notion of causal relation, though he agrees with Hume's argument that causality in the ordinary usage of this term is not observable nor demonstrable.

He is obviously deeply impressed by the paradoxical situation that we as individuals assume, on the whole, the existence of a real objective world, a so-called external world and yet we are not usually aware that one's own body is part of that world too. To put it in more modern terminology: We are members of a class or category and at the same time we engender a metatheory—ignoring the fact that we are simultaneously members of the same class of which we have developed out metatheory. To belong to both categories or classes represents for Schrödinger a true antinomy.

All in all, this book which is intended to present the philosophic world view of Schrödinger is, according to my opinion, one of his weakest intellectual achievements and seems to be almost incompatible with the main work of that great scientist.

In the Tarner lectures (1956) which he delivered at Trinity College, Cambridge, he did not discuss any subject that was even remotely connected with his main contribution to theoretical physics, namely, wave mechanics. He did not refer at all to his current research projects as far as physics is concerned: in those lectures which were published in 1959 under the title of *Mind and Matter,* he deals with biological, psychological and religious topics. The physical basis of consciousness—one of the subjects—presents views which are hardly acceptable today by experimental psychologists or neurophysiologists. This particular chapter is a typical example of Schrödinger's manner of "philosophizing". His frequent references to Spinoza, and in particular, to Richard Semon whose book *Mneme* had caused quite a sensation at its time and also impressed Bertrand Russell very profoundly, would hardly satisfy the contemporary thinker who investigates the phenomenon of consciousness by invoking all the tools recent researchers could furnish most effectively. And yet, as occurs so often in those aspects of his work which are not devoted to physical science proper, one can discover some very stimulating and intriguing statements throughout that book, although it lacks any systematic treatment of the subjects mentioned and excels by rather sporadic, original ideas which literally invaded his mind incessantly.

Thus, when he reflects on the notion of "understanding," he approaches Darwin's theory in an unorthodox manner foreign to the professional evolutionists. He claims that Darwin's theory need not lead to a "gloomy" interpretation as far as the future is concerned. He reaches this verdict by attempting to reconcile Lamarckism with Darwinism! He asserts apodictically that behavior *does* influence selection. The question whether further biological development of man is likely, is approached in an interesting but not strictly scientific manner. Again, occasionally one encounters some trivially true observations like the remark that our biological future is nothing else but history on a large scale. Schrödinger is firmly convinced "that the increasing mechanization and 'stupidization' of most manufacturing processes involve the serious danger of a general degeneration of our organ of intelligence."[1]

Throughout this book one becomes aware of how much the physicist-scholar Schrödinger was influenced by Greek thought. The chapter on "objectivation" ends with a doctrine which Planck, his predecessor in the Chair of Theoretical Physics at Berlin, would have condemned as faulty reasoning: Schrödinger maintains that subject and object are *one* unity. But already earlier thinkers and some religions arrived at the same non-commonsensical dogma; that is, the barrier between subject and object, i.e. dualism, had been torn down. According to Schrödinger, recent results in the physical sciences too support this Eastern viewpoint.

Thus, one of his favorite subjects is the so-called "oneness of mind." He knows, of course, that Greek science—and the science of our age is certainly derived from Greek scientific reasoning—is incompatible with that kind of Eastern thought that upholds the unity of mind in contrast to the multitude of somatic individuals. This whole chapter, though it refers frequently to modern thinkers like Sherrington, is, according to my opinion, a hodge-podge of scientific viewpoints: Eastern mysticism, and highly bizarre claims which are beyond verification or falsification and barely escape—merely thanks to Schrödinger's competence as a scientific thinker—the danger of succumbing to pseudo-assertions.

Like Planck, Schrödinger too contemplated the ever-present question about the relation between science and religion. This chapter does not tell us much about the actual problem under discussion but it contains some valuable thoughts about certain philosophic and "mathematical" subjects!

The physicist will enjoy the way in which he deals with what one may call the foundation of statistical mechanics. Schrödinger agrees with Gibbs and

Boltzmann that—with some exceptions like fluctuations, etc.—"the course of events in nature is irreversible."[2] Planck had held that although the laws governing the behavior of gases, for instance, are definitely statistical, the behavior of each individual molecule or atom is strictly causally determined. The total aggregate, however, obeys statistical laws. In other words, he believed that one derives irreversible events from reversible elementary situations. On the other hand, one could maintain that the individual event is indeterministic—the apparent regularity of some natural laws makes only sense if one regards the single event as subject to chance; the law of large numbers alone allows us to deal with certain laws as if they were subject to causality and determinism. Schrödinger decided finally that the universe is neither big nor old enough to allow for reversibility on a large scale. Of course, he knew well that Eddington's "Arrow of Time" does not hold for Brownian movement, and similar phenomena. In contrast to Einstein, Schrödinger propounds a statistical theory of time which is in contradistinction to the generally accepted unidirectional flow of time. Although Schrödinger regarded himself as an acausalist and strict indeterminist, he seems not entirely to eliminate the unfounded hope of a Planck and an Einstein that causality and determinism will some time in the future rule our scientific reasoning again. But there cannot be any doubt that throughout his life work he shared the belief of men like Born that classical causality and determinism have at present become obsolete owing to our newly-gained knowledge in microphysics. Schrödinger even believed that one could demonstrate indeterminism in the macrocosmos.

If one were to express Schrödinger's genius in a few terse sentences, one would have to focus one's attention on his almost unique ability to see associations, links, and bridges, which escaped his fellow workers although the material was directly available. It is already classical history how he showed that the principle of least action in its original form, later expressed as Hamilton's principle and finally Fermat's principle (in de Broglie's terms), allowed him to draw the fruitful analogy between mechanics and optics. His wave mechanics was, with respect to classical mechanics, what wave optics is concerning geometrical optics. I think no physicist in the history of that subject has employed the technique of variational principles in a more comprehensive, efficient manner than Schrödinger.

The student who wishes to conquer wave mechanics and is accustomed to work with one form or other of Schrödinger's wave equation is not always aware of the fact that e.g. $\hat{H}\psi = E\psi$ can be derived from the variational principle $\delta\int\psi^*(H\text{-}E)\psi dq = 0$. No doubt, his wave equation surpasses, concerning generality, Newton's equations of laws of motion for a particle. Strangely enough, he never accepted the statistical interpretation of his wave equation by Born, and in the controversy corpuscle-wave he did not only side with the "wavists"—he tried to create a world picture in which *only* waves in one form or other constitute the fundamental substance of the universe: particles, he insisted, are no more than illusions. One can discover in his later writings occasional remarks that seem to indicate his reconciliation with the viewpoint generally accepted since Born. However, in his private correspondence and to his last breath he remained an incorrigible "wavist," his (rare and unconvinced) concession to the Copenhagen school notwithstanding.

In the history of physics one can find several examples for the fact that sometimes men trained in one profession made some momentous discoveries or developed valuable theories in domains which lay outside their own professional prowess. We all know that Julius Robert Mayer published in 1842 the first version of the First Law of Thermodynamics: The so-called energy principle. Mayer was a medical doctor, in fact a ship's doctor, whose discovery affects literally every branch of the physical and natural sciences. Analogously, Hermann Helmholtz was a well-trained medical doctor and became one of the greatest mathematical physicists of all time. It is amusing to observe how many members of the medical profession are unaware of the fact that one of their most frequently used instruments was invented by this theoretical physicist.

When the booklet entitled, *What is Life?* was first published in 1944, based on lectures given at the Institute at Trinity College, Dublin, February 1943, this very short and scientifically unorthodox book elicited sensational response among biologists—experimentalists and theorists alike. Most of the observations made in this booklet were not entirely unknown to some biologists who were of the opinion that not only present-day biochemistry, but also current physics would play an ever greater part in our understanding of living entities. I personally objected to the heading of the book, because I do not think that any scholar has succeeded in providing us with an exhaustive definition of "life." It is more reasonable to talk about living things or living matter as Szent-Györgyi does.

Following the concept of the physicist Delbrück, Schrödinger treated the most relevant part of a living cell, i.e. the chromosome fibre, as an aperiodic crystal. And since we deal in physics, on the whole, only with periodic crystals, the conception of the chromosome being an aperiodic crystal provided at least one criterion for living matter. For Schrödinger, the aperiodic crystal is "the material carrier of life."[3] It is very suggestive to compare the approach of Schrödinger to scientific issues with that of an "artist"—it often lacks dull, so-called systematic, rigidity. Thus, the book contains interesting remarks which are not immediately related to the subject which he promises in his title to discuss. For example, he tells us why atoms are as small as they are and he asked the question which goes beyond textbook knowledge by asking why human bodies are so enormously large compared with an atom.

Despite the fact that he was a layman as far as sophisticated physiology is concerned, he considered the human organism as a well-ordered organization and entirely subject to rigorous *physical* laws (to a very high degree of accuracy). But since physical laws rest on atomic statistics, they can be only approximate and all the physical and chemical laws that can be discovered when we examine living organisms, are no less of a statistical type than other aggregates, for instance, gases. As I mentioned above, this little, but extremely weighty book, is pervaded by purely physical notions often without containing any novel information. For instance, the pages devoted to *diffusion* do not increase in any way the knowledge of any trained biologist who is familiar with fundamental chemical or physical facts but becomes slightly impatient if he reads that he is once again compelled to interpret Brownian movement according to statistics. In other words, the non-biologist Schrödinger was obviously unaware of the amount of chemistry and physics which was demanded from an advanced biology student already in 1944.

The paragraph in which he discusses the $\sqrt{n}$ rule is presented in a very succinct and sensible manner, although the professional mathematician may not be quite satisfied with the form in which this rule is supposed to render it a general property of Nature. Nevertheless, it was relevant to show that the $\sqrt{n}$ rule plays an important part as far as organisms are concerned too.

In his discussion of the hereditary mechanism he admits that his knowledge of genetics displays in many respects a "dilettante character". True, he realized the relevance of the structure of the chromosome fibres and employs in this connection the term "codescript". This reminds us automatically that it was a physicist, namely, the late George Gamow, who introduced the concept of genetic code into biology. Today, after the revolutionary discovery by Crick and Watson, Schrödinger's treatment of mitosis, meiosis, and syngamy does not make any longer exciting reading.

Textbook knowledge is oddly mixed with some very felicitous expositions. Thus the two pages in which he treats Haploid individuals is expressed in an exemplary lucid and rational manner. It speaks for Schrödinger's extraordinary capacity for detecting ambiguities, if not naive errors, in scientific expositions which lie even outside his field of competence. This fact can be demonstrated, I think, in each book he ever wrote and his numerous papers testify to his rare gift for noting weak arguments or naive claims. To illustrate this point, I should like to draw the reader's attention to Schrödinger's unexpected but extreme caution with respect to the term "property." This term is perhaps one of the most important concepts in genetics, and geneticists and biologists in general should use this term in an adequately defined manner in order to avoid the ambiguities of past usage.

I wonder whether the average reader of his book, when it appeared and aroused sharp controversy among professional biologists, noticed that

Schrödinger considered the gene to be the "hypothetical material structure" located somewhere in the chromosome. That is, a certain "locus" situated in the chromosome is what we consider today as a *real* ostensible entity—the gene. In his estimate of the maximum size of a gene, Schrödinger follows the work of C.D. Darlington and agrees with the claim that the volume of a gene is equal to a cube of edge 300Å. I am referring to page 28-30 in which partly Schrödinger's knowledge of statistical physics paved the way, together with Delbrück, Gamow, and Crick, for the most exciting enrichment of contemporary science, namely, molecular biology.

The chapter on mutations is more or less based on historical sources and later upon the work of M. Delbrück, N. W. Timoféëf, and K. G. Zimmer. The very far-reaching and decisive question of what the human race can do in order to escape unwanted latent mutations was anticipated by Schrödinger but, unfortunately, not dealt with in any concrete manner. In other words, he saw the problem but did not make any positive efforts or proposals as far as this ever-present danger of mankind is concerned.

Biologists and physicists alike will profit from Schrödinger's quantum-mechanical treatment of genetics. In order to establish a link between quantum theory and the mechanism of heredity, he still accepts the validity (or truth) of what is usually designated as "quantum jumps." Several years later he published papers in which he uttered serious doubts as to the the plausibility of this "mysterious" behavior of the individual quantum. The mathematical treatment of this whole issue will only be of interest to the bio-mathematician and, to some extent, also to the molecular biologist. After all, the theoretical physicist Schrödinger whose life work was occupied with the solution of differential and partial differential equations, could not enter the science of biology without resorting, if possible, to what he called "mathematical interlude."

His analysis of Delbrück's celebrated model makes interesting reading and raises some intriguing questions, but it does not actually add any profound insight in our understanding of that model and its implications. None the less, once again he points out some misconceptions which arise if one does not study Delbrück's theory methodically and with patience.

The reader will recall that Schrödinger conceived of the gene as a hypothetical material structure located somewhere in the chromosome. On p. 61 he expresses his hypothesis by claiming that the gene (if not the whole chromosome fibre) is an aperiodic solid. In other words, he adopts without qualification Delbrück's model which emphasizes a molecular explanation of the genetic mechanism. The medical profession would certainly enjoy the (much too short) section which shows how X-rays produce mutation. Of course, the X-ray specialist is certainly acquainted with the content of this paragraph, but for the medical layman it is definitely of great informational value.

Boltzmann's statistical-mechanical explanation of the Second Law of Thermodynamics found in Schrödinger a tireless advocate. It is well known that this explanation is the basis for the belief that there are no causal, strict laws, so-called dynamic laws, as Planck called them. And hence it is only natural that Schrödinger in his enthusiasm for Boltzmann's interpretation considered entropy, energy, and also the law of the conservation of energy, as purely statistical concepts.

Perhaps the most sensational chapter of this little book by Schrödinger that deals with the physical aspect of the living cell is his treatment of the entropy function in its role in biology. "How would we express in terms of statistics the faculty of a living organism by which it delays the decay into thermodynamic equilibrium, (death)?" Well, answers Schrödinger, "It feeds upon negative entropy." The organism attracts a stream of negative entropy upon itself, so as to compensate the entropy increase which it produces by living. And thus it is able to maintain itself on a stationary and fairly low entropy level.[4]

The term "negative entropy" seems, at first glance, to be in contra-distinction to the Second Law. However, if one scrutinizes Boltzmann's equation carefully, the concept of negative entropy is therein contained, though the Second Law emphasizes that in all natural processes the entropy must increase within a closed system. We recall that Boltzmann suggested a statistical-mechanical definition for the entropy of a thermodynamic state: $S = k \log W + \text{const}$. Now Schrödinger uses a kind of mathematical trick since he introduces sharp definitions of "order" and "disorder." His reasoning is extremely simple and convincing and expressed in such a manner that one is prone to forget that the gist of his derivation is already entailed in Boltzmann's celebrated entropy function.

Thus, if D is a measure of disorder—reasons Schrödinger—$1/D$ may be looked at as an immediate measure of order. Now, we know that the logarithm of $1/D$ is simply minus the logarithm of D. Hence, the original Boltzmann equation can be written

$$\text{minus (entropy)} = k \log (1/D).$$

One may summarize his reasoning in the following manner: The device by which any organism maintains its stationary state at a "fairly high level of orderliness (equals fairly low level of entropy) really consists in continually sucking orderliness from its environment."[5]

Other authors too, for example, Brillouin, von Bertalanffy, Prigogine—to mention only a few known workers—have concentrated on the concept of negative entropy, which was hardly referred to in the physical literature before Schrödinger. In information theory this concept plays a relevant role quite apart from the fact that molecular and/or theoretical biologists have become nowadays thoroughly familiar with the notion of negative entropy. One could argue that the discussion of entropy is incomplete if it does not also consider its relation to the universe at large. In other words, any form of the Second Law must not only apply to micro- or macro- events, but to cosmological theory as well.

Schrödinger approaches the statistical meaning of entropy by resorting to the concepts of order and disorder. Thus, we have D and 1/D. Now, this dichotomy reminds one of Landé's concepts of quantum physics where he postulates the existence of fractional equality. This fractional equality introduces an intermediate state between two contradictory states. In contrast, Schrödinger's antithesis of order-disorder does not allow for different types of order as far as the living organism is concerned. I only want to point out that from a statistical point of view the dichotomy orderliness-disorder forbids the existence of different degrees of, or rankings of order. And we must not forget that negative entropy can be, on occasion, very positive. When Schrödinger asserts that an organism feeds upon negative entropy, it amounts to the fact that the organism can maintain its order (type of organization) by inducing "order" from its environment. (One may point out that plants receive their incessant and large supply of negative entropy in the light emitted by the sun).

The old controversy between mechanism and vitalism is treated by Schrödinger entirely from the viewpoint of a physicist. In other words, he discusses the resemblance between a clockwork and an organism.[6] Like Gamow, Schrödinger is satisfied that the new biology, which employs as much physics as it possibly can, suffices in order to account for the functioning of a living being.

When I studied that book, immediately after its publication in England, I received a shock in reading the epilogue. In a somewhat inexplicable manner this epilogue treats the philosophic issues of determinism and free will. According to him, a discussion of this philosophic subject is implied in the preceding physicobiological chapters. He arrives at the conclusions that body functions can be regarded as purely mechanical and are in concordance with the Laws of Nature. However, subjective experience lets us assume that we ourselves, as individuals, are "steering" these bodily motions. On the whole, we are able to predict the results which follow those deliberate motions. From these two assertions or suppositions, Schrödinger feels compelled to infer: "every conscious mind that has ever said or felt 'I'—am the person ... who controls the 'law of the atoms' according to the Laws of Nature."[7] He even seems to claim that the inference referred to is the most plausible one to which a biologist can resort in order to prove "God and immortality at one stroke."[8]

Again, this whole manner of contemplation—one can hardly call it rational reasoning—dates back to the Indian Upanishads. This means that Schrödinger followed the main doctrines of the Indian Vedanta. Nowhere does he argue or support his almost uncritical religious attitude toward Eastern thought and he also avoids any comparison with Western religious traditions like, for instance, the various interpretations of Christianity and so forth.

In this same chapter he compares the mystics (who, according to him, are in perfect harmony with each other) to the particles in an ideal gas! It is difficult to specify when Schrödinger gave up the main dogmas and messages of Western religious ideology but, if my memory does not deceive me, I think that he adopted his radical way of religious orientation long before he wrote this little book.

He starts from the premiss that we never experience consciousness as a multitude, but only as singular individuals. In contrast, there cannot be any doubt that we constitute a huge multitude of similar bodies. It would therefore be (seemingly) quite natural to suppose a plurality of minds (consciousnesses). As a matter of fact, Schrödinger is quite right when he claims that most Western thinkers have adopted the above hypothesis. Yet he arrives at the unorthodox view that the so-called plurality hypothesis, as far as the mind is concerned, cannot be upheld if we are inadvertently led to the question of mortality or immortality of our consciousness. And thus, he is perhaps the only contemporary physicist of that magnitude who seriously pleads for the oneness of the human mind or consciousness! It seems to me that his endeavor to answer the perennial question *"What is this 'I'?"*[9] has been the source of his Eastern commitment, so alien to the average occidental scientific scholar or even the enlightened layman.

The last page of that epilogue, which does not seem to be justifiably published in this volume, reads like the confession of a mystic who attempts in vain to express his subjective experiences in a manner that is intelligible, coherent, and definitely communicable. It is my firm conviction that this mystical out-pouring can never become intersubjective and convince any scientist, who otherwise enjoys Schrödinger's opinions, as long as he remains within the rigorous bounds of the physicist's approach to biology.

It is certainly not necessary to know Schrödinger's personality in order to do justice to his position among the outstanding few physicists of this age. As a matter of fact, many critical readers of the enormous advances made in almost all branches of physics during the last 70 years show a definite antagonism against all attempts to pay particular attention to the individual person, the original scholar who contributed in a unique manner to the progress in his particular domain.

I think that any serious student of Schrödinger's extensive output, and not merely in physics, can only gain by becoming acquainted with his rich and colorful personality. His style, his humor, his vast knowledge in the humanities, and his rare charm enable, in particular the young reader, to realize that he was not only one of the leading physicists of our time, but also one of the most fascinating and inspiring scholars whose lifework testifies to the claim that a great scientist need not, by necessity, be dry, dull, and afflicted with a narrow horizon, euphemistically called his "particular field of specialization."

NOTES

[1] *Mind and Matter,* 1959, p. 34.
[2] *Ibid.,* p. 83.
[3] *What is Life?,* 1944, p. 3.
[4] *Ibid.,* p. 74.
[5] *Ibid.,* pp. 74-75.
[6] *Ibid.,* p. 85.
[7] *Ibid.,* p. 88.
[8] *Ibid.,* p. 88.
[9] *Ibid.,* p. 90.

Index

Abelson, R., 329
Abercrombie, M., 60, 77
acausalist, 336
acceptor, 34
acquired adaptive modification, 83
actinomycin, 93
action, principles of, 256-257
Adams, H., 227-228, 231-232
Adams, H. B., 227
adaptation, 7, 10-11, 25, 200-201, 228, 234, 240
adenine, 91
Adey, W. R., 55, 77
adrenal steroids, 35
Alcmaeon, 113-114
Alfert, M., 104, 109
Allbut, C., 117
Ansevin, K. D., 61, 77
anamorphosis, 27
Anshen, R. N., 208
aperiodic crystal, 337, 339
apnea, 112
Arakilian, P., 45, 77
archeus, 191
Aristotle, 9, 18, 114-116, 140, 150, 181-182, 308-309
Armstrong, D. M., 300
arthropods, 41, 47-48, 50, 67
ascidia nigra, 104, 107
ascidians, 105, 107
Asclepiades, 183
asthma, 114
astronomy, 272, 279-280
gravitational, 265
Atman, 112, 120
ATP, 27, 81, 88
Austin, C. R., 104, 109
automaton, 20-21, 38, 291, 294, 296-297
axiomatization, 244-246, 248
Ayala, F. J., 1, 15, 207
Ayer, A. J., 205

Bacon, F., 249

bacteriophage T4, 84-86
Baglivi, G., 173, 188, 190-191
Bancroft, G., 225
Barghoorn, E. S., 215
Bargman, W., 77
Barnes, C. D., 43
Barthez, P. J., 192
Bates, J. W., 121
Baxter, R., 43
Bayes, T., 141
Beadle, G. & M., 80, 101
Beard, C. A., 230
Beermann, W., 104, 109
behavior, 311-312, 314, 318-320, 334
physiological, 294
protopsychological, 47
psychological, 294
random, 73-76
Beier, W., 29
Beling, I., 54, 71, 77
Bennett, E., 43
Benzer, S., 70
Berman, E., 121
Berlin, I., 231
Bernal, J. D., 96
Bernoulli, D., 175-177
Bernoulli, Jacques, 174
Bernoulli, Jean, 174-175
Bernoulli, Johann, 189
Bertalanffy, F. D., 26, 29
Bertalanffy, L. von, 17, 29, 340
Bessel, F. W., 279
Best, J. B., 37, 39, 41, 43, 47, 77-78
biocybernetics, 180
biological templates, 79, 100-101
biological system, 233-234, 236, 239-240
biology, 218, 313, 315, 319, 326
autonomous aspects of, 154
molecular, 339
organismic, 21
biomechanism, 183
biometrics, 180
biophysics, 193
biosphere, 210

345

biotonic laws, 28
Birch, L. C., 216
Block, D. P., 106, 109
Boerhaave, H., 192-193
Bohr, N., 128, 136-138, 162, 331
Boltzmann, L., 146, 148, 336, 340
Boltzmann's entropy function, 340
Bonner, J., 93
Borelli, G. A., 118, 173, 189-190
Born, M., 331, 333, 336-337
Boscovich, R., 280-281, 288
Boyle, R., 118
Brace, C. L., 232
Breck, A. D., 217
Bremermann, H. J., 155
Brentano, F., 205
Brillouin, L., 340
Brobeck, J. R., 121
Bronowski, J., 319, 330
Brown, D. D., 104, 109
Brown, F. A., Jr., 68, 78
Brownian movement, 336, 338
Bruce, H. M., 48
Bruno, G., 269
Buchsbaum, R., 61, 77
Buck, C. D., 120
Buffon, G. Le Comte de, 284, 286
Bunning, E., 70, 71, 77
Burke, E., 221
Bushi, 235
Bushido 234

Cabanis, P. J. G., 195
Calvin cycle, 63
Calvin, M., 95-96
Canguilheim, G., 190
Cannon. W. B., 55, 77
cartesianism, 189-190
Cassirer, E., 333
catastrophism, 230
causal chains, linear, 28
causality, 19, 210, 248, 334, 336
Cavafy, C., 112
Cavendish, H., 281, 288
cell-assembly, 302-303
cells, neurosecretory, 50
    goblet, 47, 57, 61
cellular biology, 103
Chambers, R., 225
charge transfer, 34
Charles I, 256-257
Chironomus, 104
chiton tuberculatum,
    trochophores, 105-106
choloroplast, 86
Chomsky, N., 320-322, 324, 329-330

Cicero, 115
ciona intestinalis, 107
Clairant, A. C., 265
Clark, R. B., 50, 77
classification, historical, 258
Clausius, R. J. E., 27
clavelina picta,, 107
clavis coelestis, 265-266, 269, 287-288
Cloudsley-Thompson, J. L., 41, 48, 70, 77
Clowes, R., 101
cluster of cistrons, 104
codescript, 338
coelenterates, 105
colinearity, 80
collagan, 86-87
Collingwood, R. G., 231
Columbo, R., 117
compartment theory 23-24
complementarity, 168
complexity, 211-214, 216
computer, 305-319
    universalism, 309
Comroe, J. H., 120
Comte, A., 218
concepts, *innate*, 322-324
    growth of, 322-323
conceptual net of science, 325-328
condition of connectability, 3, 5
condition of derivability, 3, 5
Condorcet, M. J. A., 218, 230
congruence, classes, 156-159
consciousness, 211-214, 216, 334, 342
    perceptual, 293-294, 296-298
conservation laws, 332, 340
conservation of angular momentum,
    principle of, 286
contradiction, 313-314
Coombs, R. R. A., 121
Copenhagen school, 337
Copernicus, 172, 208
Corning, W. C., 43
corpuscle-wave, 336
cosmo-genesis, 208
cosmogony, 212, 289
cosmology, 263-264, 266-267, 278-279,
    283-284, 332-333, 341
Cournot, A. -L., 179
Cowden, R. R., 103, 105-107, 109
Cox, R. T., 142, 162
Crick, F. H. C., 79, 338-339
Crosson, F. T., 329
Ctesibios, 182
Cura foremanii, 39
Curti, M. E., 232
cybernetics, 13, 18, 187, 292, 294-297
cytoplasm, 94-95, 104
cytosine, 91

D'Acquapendente, F., 188
D'Alembert, 175, 177, 263, 265, 287
Danery, A., 130, 133
Daniélou, J., 208
Danilevsky, N. Y., 224
Darlington, C. D., 339
Darwin, C., 7, 190, 195, 208, 217-229,
     230, 231, 335
  social, 226
Deboer, M., 43
deductive abstraction, 245-246
De human corporis fabrica, 183
Delbrück, M., 337, 339
Democritus, 114, 181, 248
Denbig, K. G., 29
Derham, W., 280, 285
De Robertis, E. D. T., 47
Descartes, 20, 172, 181, 184-190,
     192-193, 284, 293, 299
determinism, 211, 227, 253, 292,
     296, 336, 341
Dethier, V. G., 76-77
De Vries, H., 89, 229
De Witt, J., 174
diastole, 188
Diderot, 177, 263
differential reproduction, 11
Diocles, 116
Diogenes, the cynic, 116
Diogenes of Apollonia, 116
Dirac, P. A. M., 147, 331
disorder, 340-341
DNA, 27, 79-81, 83-85, 89-93, 96-98,
     100-101
  mitochondrial, plastid, 215
  nuclear, 83
  molecule, 19-20
donor, 34
Dobzhansky, T., 209, 232
double helix, 81, 90-91, 96
Doyon, E. -L., 194
Dray, W., 256
Dreyfus, H., 312, 329
Driesch, H., 24
Drosophila melanogaster, 70
Dryden, J., 269
Dugesia dorotocephala, 39, 41-42, 52,
     58, 72
Dunn, L. C., 39
Dürer, A., 172

Ebers, G. M., 113
ecological system, 168
ecosystem, 5-6
Ectienascidia turbinata, 107
Eddington, A. S., 336
Edgar, R. S., 84, 101

Edman, P., 77
Ehrensvärd, G., 215
Einstein, 331-332, 336
élan vital, 5
Elazar, A., 55, 77
El Mehairy, M. M., 121
Elsasser, W. M., 126, 162-163
Elshtain, E., 43
emergent properties, 6
Empedocles, 114, 116
endogenous oscillator, 70
energy bands, 33
energy principle 337
Engels, F., 222-223, 231
entelechy, 5, 18, 24
entropy, 24, 26, 145-148, 167, 340-341
  function, 27
    expanded, 23
  maximum, 26
  minimum production, 26
  negative, 23, 167, 241, 340-341
  negentropy, 24
  production, 23
ensembles, 145-146, 148-150, 152-155
  quantum-mechanical, 162
enzyme proteins, 87
Epicurus, 181
epiphenomenalism, 291, 296-298
epistemology, 299, 301, 304-305, 319,
     321-322, 325-326, 329
equality, fractional, 341
equifinality, 22, 24
Erasistratus, 115, 116, 182, 187
Erasmus, 193
ergodicity, 132
Escherichia coli, 93
Euler, L., 176, 265, 286
Eudendrium racemosum, 105, 106
evolution, 62, 79, 89-91, 94, 97-101,
     104, 108, 150, 208-215, 218-219,
     223, 225-226, 228-230, 234, 295,
     297, 312, 315, 318
  cosmic, 286
  social, 228
  societal, 253, 260
exclusion principle, Pauli, 32
extremum principle, 144-147, 149, 153

Falliers, C. J., 111, 121
Fange, R., 77
Farquar, I. E., 124, 133
Farrington, B., 114, 121
Fechner, G. I., 333
Feigenbaum, E. A., 330
Feldman, J., 330
Fenn, W. O., 121

Fermat, 174
Fermat's principle, 336
field biology, 103
finite classes, partial
    regularities of, 162
finite classes, 152
    principle of, 155
Fisher, R., 180
Fiske, J., 226
Fliessgleichgewicht, 21
Florey, E., 41, 48, 50, 77
Florkin, M., 101
Fontenelle, B. de, 193, 265
forces, supramolecular, 28
formula, 217, 219
Fox, S. W., 95, 101, 215
Fraenkel, G. C., 54, 77
Freeman, E. A., 221, 227
Freud, S., 241
Froude, J. A., 221
Fulton, J. F., 50
Funccius, J. N., 218

Galbraith, J. K., 37, 77
Galen, 115-117, 183
Galileo, 18, 172-174, 180, 185, 189, 208
Gallie, W. B., 250
Galton, F., 179-180
Gamow, G., 338-339, 341
Garrison, F. H., 121
Garstens, M.A., 123
Gasser, H. S., 69, 77
Gauss, 179
Gell, P. J. H., 121
gene-pool, 168
generalized complementarity, 136-139,
    143, 149, 156
generative grammar, 321
genetic code, 5, 20, 27, 82, 90, 92,
    168, 338
genetic information, 96
genetic recombination, 89
genetic variants (mutations), 10-11
genotypes, 27, 169
geosphere, 210
Gerard, R. W., 160-161, 163
Gesner, K. von, 188
Gibbs, J. W., 146, 148, 155, 335
Gilbert, M., 307
Gillispie, C. C., 179
Gilson, E., 187
Glisson, F., 188
glutamic acid, 80
glycolysis, 24, 25
Gnosiology, 300
Gödel, K., 310-311

Goethe, 24, 332
Goodfield, J., 232
Goodman, A. B., 41, 77
Goudge, T. A., 15
Granick, S., 96
Grant, U. S., 231
Graunt, J., 174
gravitation, 281
    laws of, 265, 279, 283
    universal, 280
    principle of, 285
    center of, 278
gravity, 323
Gray, W., 30
Grmek, M. D., 172-173, 181
Grodins, F. S., 121
Gross, J., 102
gaunine, 91
Gunn, D. L., 54, 77
Gurdon, J. B., 94, 104, 109

Hägerström, A., 205
Halas, E. S., 43
Haldane, J. B. S., 95
Hall, C. S., 39, 77
Haller, A. von, 188
Halley, E., 174, 176, 280
ham, 112
Hamilton's principle, 336
Ham–Sa, Ham–Sa, 112, 120
Hardin, C. L., 288
Hartmann, N., 205
Hartshorne, C., 216
Harvey, W., 116-117, 185, 188, 194
Hastie, W., 289
Hayes, C. J. H., 220
Hebb, D. O., 301-302, 329
Hegel, 218, 222, 249
Hegelianism, 225
Heisenberg, W., 331
helix, 107
Helix aspersa, 106
Helix pomatia, 106
Helvetius, C. A., 195
Helmholtz, H., 337
Helmont, J. B., von, 191
hemoglobin, 87, 90-91
Henderson, T., 279
Heraclitus, 115
Herbert of Cherbury, 301
Herlant–Meewis, H., 77
Heron, 182
Herophilus, 115
Herschel, J., 179
Herschel, W., 270, 280, 282-283,
    286, 289

heterogeneity, 161
  organized, 212
Hevelius, J., 280
Hew, H. C. Y., 106, 109
Hilbert space, 143
Hippocrates, 112-113, 115-116
Hippocratic humorism, 191
historian, 241-242, 246, 249-256,
  259, 261
historical processes, 249
  system, 233, 236-237, 239,
    241, 252
historiography, 217, 219, 225, 253, 255
history, 17, 211-212, 214, 217-219,
  221-222, 225-226, 229-230, 232
  243, 249, 255, 258, 335
history, comparative, 260
  scientific, 219-220, 226
  morphological, 224
  salvation, 209, 215
  cosmic, 209, 215
history of science, 323, 329
Hobbes, 222
Hodgkin, A. L., 68, 77
Hoffmann, F., 192
Hofstadter, R., 231
Holbach, P. H. D. Baron von, 195
Holmes, O. W., 221
Holton, G., 332
homeostasis, 9, 10, 112
Homer, 112, 113
homo economicus, 246-247
Hook, S., 329
Hooke, R., 118
Horowitz, N., 95
Hoskin, M., 287-288
Howell, W., 41
Howze, G. B., 60 78
Hoyle, F., 314
Hubl, H., 50, 77
Hudde, J., 174
Hughes, E., 287
Hughes, H. S., 232
Hume, 334
Huxley, J. S., 195
Huxley, T. H., 228
Huygens, C., 174, 265, 269
hydrogen bonding, 80, 88
hypothetico—dedcutive system, 17

iatrochemistry, 191, 193
iatrophysicists, 20
iatrophysics, 185, 189, 191
ideas,innate, 321-323
ideterminacy, 150-151
  principle, 17

indeterminism, 336
individuality, 158-162
induction, 140-141
  for universals, 141
  for probabilities, 141
  quantum—mechanical, 149
inductive method, 162
inductive probabilities, theory of,
  139-141
information, 167, 169
  theory, 28, 340
Ingram, V., 91
inoculation, 175-176
insufficient reason,principle of, 144
integration, biological, 160
intelligibility, 328
invariance above heterogeneity, 161
invertebrates, 67
irreducibility of biology, 5
irreversibility, 22-23, 26, 124
Irvine, W., 231
Irving, W., 225

Jacobson, A., 43
James, R. L., 43
Jarvik, M., 41-42
Jaynes, E. T., 126, 128, 147-149, 153,
  163
Jefferson, T., 225
Jeffreys, H., 142, 148, 163
Jenner, E., 178
Jordy, W. S., 232
John, E. R., 43
Jukes, T. H., 90, 102

Kant, 264, 270, 283-287, 289, 299, 301,
  325
Katzung, B. G., 43
Keeley, E., 120
Keill, J., 280
Kellenberger, E., 84
Kelsen, H., 205
Kenk, R., 41, 78
Kepler's laws, 173
Keynes, J. M., 141, 163
Kimble, D. P., 43
Kimmel, H. D., 43
Kirk, G. S., 121
knowledge, linguistic, 307
Kolmogoroff, A., 141
Konopka, R., 70
Kopp, R., 41-42
Kornberg, A., 81
Körner, S., 206, 243
Kotarbinski, T., 300

Koyré, A., 289
Krebs cycle, 20
Kropotkin, P., 224
Kuhn, T., 328, 330

Lacroix, S. F. de, 178
Lagrange, J. L., 118
Lakatos, I., 329
Lalande, J. J. J. de, 280
Lamarck, J. B. P. A., 83, 219, 228, 335
Lambert, J. H., 286
La Mettrie, J. O. de, 192-193, 195,
    299-300, 309
Landé, A., 341
Laplace, P. S. de, 141, 147-148, 178, 286
Lau, C., 29
Lavoisier, A. L., 118
least action, principle of, 336
Le Cat, C. N., 194
Leibnitzian–Wolffian system, 264
Leibniz, 174, 192-193
Lenin, 223, 240
Lentz, J. P., 60, 78
Leonardo da Vinci, 172, 183, 332
Lerner, D., 15-16
Leucippus, 181
Liaigre, L., 194
Lincoln, A., 227
limbic system, 55, 57
Limnaea saginalis, 106
linguistic aspect of mind, 305
Locke, 323
logicality, 309, 314, 318, 320
logical relativity, 137, 142
Loinger, A., 133
longitudinal transmission system, 57
Louis, P. C. A., 178
Lucas, J., 310, 312, 329
Lucretius, 181, 285, 287
Ludwig, G., 125, 133
Luther, 218, 240
Lyell, C., 219, 225

machine, 291-299, 305-306, 309-311, 314,
    316, 319
machine, *chemico–dynamic*, 20
    cybernetic, 20
    molecular, 20
macromolecules, 31
macrovariables, 139-141, 145-146,
    148-153, 156-159
Magnus, G., 118
Maimonides, 114
Maine, H., 221-222
Major, R. H., 121

Malebranche, N., 188
Malthus, T. R., 222
Markert, C. L., 107, 109
Marxism, 222-223
Marx, K., 222-223, 230-231, 240-241,
    252-255
materialism, semantic, 300
Mathan, D., 93
Maupertuis, P. L. M. de, 175-176, 280,
    285
Maxwell, 179, 322
May, H. F., 232
Mayer, J. R., 337
Mayr, E., 15
maze epistemology, 203-204
McConnell, J. V., 43
McCormmach, R., 288-289
McLaughlin, J., 33
McMullin, E., 330
measurement, 136-139, 150, 153
    process, 127, 130
    problem, 129
    theory, 123
mechanics, celestial, 175
    quantum, 292
    statistical, 292, 332
mechanism, 183, 341
mechanistic metaphysics, 152
Medawar, P., 102
Mehring, F., 231
Meixner, J., 21
Mendelian principles of inheritance,
    2, 6
mentalism, 321
mentality, interprelative, 294
Mergener, H., 105-106, 109
Mersenne, M., 188
metabolism, 21
meta–historians, 251-253, 255
metaphor, 217-219, 223, 225, 237
    biological, 224
metaphysical preconceptions, 4
Michell, J., 280-283, 287-289
microstates, 139-141, 145-146, 149, 158,
    161
microvariables, 148, 157
Milky Way, 269-270, 272, 279, 284-286
Mill, J. S., 179
Miller, S. L., 95
Milner, P., 55, 78
Milton, J., 269
mind, 299, 304-307, 310-311, 318-325,
    327-329
    oneness of, 335
mind-body dualism, 299, 304
Minsky, M., 314-315, 330
Mintz, B., 104, 109

mitochondria, 20, 83-84, 99
mitosis, 25
   rate of in red tissues, 26
mobility, 32
   electronic, 31-36
   of reactions, 31
model 20-21, 23-24
mode switching, 73
Moivre, A. de, 174
molecular biology, 5-6, 17, 19, 31, 36,
      90, 103, 180
molluscs, 105
Monad, J., 93, 102
monism, materialistic, 299
Montmort, A. de, 174
Moore, G. E., 205
Moore, J. A., 102
Moore, J. and B., 94
More, H., 186
Morgan, T. H., 89
Morgenstern, O., 75-76, 78
Morita, M., 47, 49, 76-78
morphogenetic pathway, 84-85
morphology, 61
Moscona, A. A., 60, 78
motion, laws of, 265
Motley, J. L., 225
Mulliken, R. S., 34
Mulry, R. C., 43
Munn, N. L., 39, 78
Muntner, S., 121
mutation, 89-94, 96-100
   lanceolate, 93
myoglobin, 91

$\sqrt{n}$ rule, 338
Naess, A., 197, 205-206
Nagel, E., 3, 12, 14-16
Nanney, D. L., 102
naturalism, 199
naturalistic objectivism, 205
natural sciences, 243-244, 249
natural selection, 7, 62, 66-67, 89,
      94, 99, 226, 235
nauropil, 57
Needham, J., 195
Neisser, U., 315, 330
neo—mechanism, 310
neosphere, 210
neurophysiology, 5, 300-307, 310, 320,
      334
neurosecretory cells, 61
neurosecretory granules, 50, 52
Newton, 141, 174, 223, 265, 269, 280,
      284, 288, 332, 336
Newtonian mechanics, 323, 325-326

Nichols, R., 232
Nietzsche, F., 206
Noel, J., 47, 57, 76-77
non—cognitivism, 205
non—constructive propositions, 157-158
normative ethics, 197-199, 201-202
Northrop, F. S. C., 205
nucleic acid, 80
   acids, 27
nucleolar organizer, 104, 108
nucleoli, 82
nucleolus formation, 105
nucleotides, 80-81, 89, 90, 97
nystagmns, 54

objectivism, 204
Odum, H. G., 232
Olds, J., 55, 78
Omega, 214
Ong, W. J., 231
Onsager, L., 21
   reciprocal relations, 29
Oparin, A. I., 215
Opatowski, I., 65, 78
optimal design of organisms, 63
oocytes, 106-108
oogenesis, 103-106
order, 340-341
order above heterogeneity, 160
Oresme, N., 172
organelles, 82, 85, 99
organic soup hypothesis, 95
organism, 152, 159-160, 166-167, 234,
      237, 258, 337-338, 340-341
organismic classes, 159-160
   relationships, 158, 161
organo genesis, 60-62
origin of life, 95
Ostlund, E., 77

Pacchioni, A., 191
papain, 98
Paracelsus, 191
paradigm, 328
paramecium, 85
Parkman, F., 225
Parmenides, 333
Parrott, D. M. V., 48
Pascal, 174
Pattee, H. H., 97
Pauli, W., 331, 333
Pavlov, 55, 57, 78
Pearson, K., 180
Peirce, C. S., 320
pennaria tiarella, 105-106

Pepys, S., 118
Pereira, G., 183-185
Petty, W., 174
pharmacodynamic action, 25
phenotypes, 169
Philistion, 116
Phillips, D. C., 102
phosphorylation, 81
photosynthesis, 25
phren, 113
physical determinism, 159
physical laws, 338
Pigon, A., 41, 77
planarian, 38-39, 41-43, 45-50, 52-55, 57-58, 61-62, 65, 72
Planck, M., 331-332, 335-336, 340
plastids, 84
Plato, 249, 281, 323
plethysmography, 118
plurality hypothesis, 342
pneuma zotikon, 114, 117
    psychikon, 117
Poisson, S. D., 179
Pollock, F., 222
Pope, A., 269, 285
positivism, 333
potential for interaction, 140
Praxagoras, N., 116
predictability, 292, 296
Prescott, W. H., 225
Prichard, H. A., 205
Priestley, J., 118
Prigogine, I., 21, 23, 27, 340
Prigogine's Theorem 26
probabilistic stellar distribution, 283
probability, 142-144
    algebra, 142
    set–theoretical, 148
    foundation of, 145
probabilities, deductive, 143, 147
    inductive, 143-144, 146-147
    inverse, 141
processes, reversible, 27
property, 338
Prosperi, G. M., 130, 133
Prosser, C. L., 68, 78
proteins, electrons in, 33
psychism, 213-214
psychologism, 321
Ptolomeus, 172
Purana, 112
Pythagoras, 115

quantification theory, 245
"quantum jumps", 339

quantum mechanics, 2, 35, 136, 138-140, 143, 145, 147, 149-151, 153-155, 159, 243
quantum theory, 123-124, 126, 339
    measurement of, 149
Quetelet, A., 178-179

Rabelais, 183
Rahn, H., 121
Raisman, G., 55, 78
random fluctuation, 98
Ranvier, nodes of, 69
Rappaport, C., 60, 78
Rapport, A., 29
rationalism, 333
rationality, 308-309, 312-316, 318-319, 328
Raven, C. P., 106, 109
Raven, J. E., 121
Raw, G., 30
Ray, J., 190
reduction, 2-6, 244
reductionism, 199-200, 291, 296, 300-301, 310, 321
regulation, 20
Reiner, R., 205
reism, 300
relativity, 2
    Einsteinian, 140
    Galilean, 140
    general theory of, 331,332
    geometrical, 140
    logical, 140, 146, 149-150
Rensch, B., 216
replication, 98
reproduction, 297
reproductive isolation, 89, 94
Rescigno, A., 29
resonance energy 35
respiration, 25
reversibility, 336
rhinencephalon, 53, 55, 57
ribosomal RNA synthesis, 103, 105, 107-108
ribosomes, 82, 99, 104-105
Riegel, V., 57, 77
Rizzo, N., 30
RNA, 80-83, 85, 93, 96-99, 105-107
    cytoplasmic, 106
    ooplasimic levels, 108
    ribosomal, 104, 106
Robinson, J. H., 220, 229
Roeder, R. E., 251
Rogers, J. A., 232
Romanticism, 225

Roosevelt, T., 227-229
Rosen, R., 29
Rosenfeld, L., 130, 133
Rosenvold, R., 57, 77
Ross, S., 205
Ross, W. D., 205
Rousseau, 313
Rubinstein, I., 39, 47, 77
Ruch, T. C., 50
Ruckert, H., 224
Russell, B., 334
Russell, E. S., 15
Ryle, G., 299-300

sa, 112
Saggiatore, 172
St. Paul, 214
St. Thomas Aquinas, 172
St. Augustine, 9, 249
saltatory transmission, 69
sampling, 157
Santorio, G. D., 173, 185
Sarton, G., 113, 121
Saturn, 272, 284
Saveth, E. N., 232
Sayre, K. M., 329
Scheler, M., 205
Schopenhauer, A., 333
Schopf, J. W., 215
Schrödinger, E., 241-242, 331-343
science, an historical system, 168
scientific knowledge, objectivity of,
    324
Segre, G., 29
self–identifying groups, 259
Selfridge, O., 315, 330
Semon, R., 333-334
Servetus, M., 201, 206
Shafer, B. C., 232
Shannon, C. E., 146
Sherrard, P., 120
Sherrington, C. S., 335
Siegel, R. E., 121
Simon, H., 28, 29
Simpson, G. G., 4, 15, 197
simulation, 24
Sirius, 279, 281
Skolimowski, H., 299
Smart, J. J. C., 300
Smith, G. E., 113
Smith, A., 230
Smith, K. D., 107, 109
Sobornost, 223
pre–Socratics, 248
Socrates, 200, 202

somatoplasm, 83
Sonneborn, T. M., 84, 102
species concept, 260
Spencer, H., 220-221, 225-227, 230-231
Spengler, O., 224, 230, 260
spinning electrons, 302-304
Spinoza, 202, 334
Stahl's animism, 190, 192
statistical matrix, 147
statistical mechanics, 2, 123-126, 128,
    145, 148, 155
    irreversible, 129
statistical physics, 339
steady state, 21-22, 24, 26
Stebbins, G. L., 79, 102
Steen, K., 76
Stein, L., 55
Steinberg, M. S., 60, 78
Stensen, N., 188, 190
stentor, 71-72
Stevenson, C. L., 205
Strato, 115
Stubbs, W., 227
Student, (W. S. Gosset), 180
subjectivism, 199, 203
Sun, 281-283
Sutherland, N. S., 318, 330
Swammerdam, J., 188
Swedenborg, E., 284
Sylvius, F., 191
symbiotic systems, 168
system, 18, 22
system, anterior motor, 50
    closed, 21, 23-24, 26-27
    conjugated double bond, 34
    general theory, 25, 29
    metabolic, 25
    open, 17, 19, 21-24
    organismic, 23
    somatrophic control, 50
systole, 188
Szent–Györgyi, A., 31, 337

Taton, R., 171
taxonomy, 251-255, 258, 260-261
    biological, 251, 258
    historical, 251, 255-261
Teilhard de Chardin, 17-19, 207-216
teleology, 6-15, 18, 222-223
    (internal and external), 12
    scientific, 7
    natural and artificial, 12
teleonomic systems, 9
template, 84, 215
    cultural, 100-101

template (cont'd)
  replicative, 84, 96-99
  positional, 84-88, 99-100
  self–activating, 88
testability, criterion of, 1
Themison, 183
theology, 272
theoretical biology, 135, 145, 155
theoretical physics, 135
thermodynamics, Second Law of, 145, 228,
  340-341
thermodynamics, 2, 20, 23, 26-27, 129,
  332
  irreversible, 21, 27-29
  second principle of, 23, 27
thinking, historical, 248
  scientific, 248
Thompson, R., 43
Thucydides, 217
thymine, 91
time, arrow of 26, 336
Timoféëf, N. W., 339
Tolman, R. C., 124, 133
Törnebohm, H., 165
Toulmin, S., 232, 329
Toynbee, A. J., 254
transcription, 98
transduction, 36
transition probability, 126
translation, 98
transmission of information, 68, 79
transplantation of nuclei, 94
Tribus, M., 147, 163
Trincher, K. S., 28-29
Trinkaus, J. P., 60, 78
tropism, 184
Truesdell, C., 289
tunicates, 107
Turing machine, 21
Turner, F. J., 225, 228
Tyndall, J., 228

Ubbels, G. A., 106-107, 109
uncertainty relations, 137
universal brain, 58-59
Upanishaden, 333, 342
Urey, H., 95
Urochordata, 107
utilitarianism, 199

Van der Merwe, A. J., 30
Van Kampen, N. G., 124, 133
variational principles, 332
Vaucanson, J., 193-194

Vartanian, A., 193
Venus, 281, 288
vertebrates, 38, 47-48, 50, 53, 57-58,
  61-62, 67
Vesalius, A., 117, 183, 188
vestibulo–ocular system, 53
Vico, J. B., 249
vitalism, 24, 31, 159, 182, 191, 341
vitalistic mechanism, 192
vital principle, 12
Voltaire, 176
Von Neumann, J., 75-76, 78, 145, 147,
  153-154, 163
Von Neumann's theorem, 161
Vygotsky, L., 319, 330

Warburg machine, 18
Watkins, J. W. N., 329
Watson–Crick model, 89
Watson, J. D., 79-80, 102, 338
wave equation, 336
wave function, 124, 127-128, 138
  collapse of, 143
wave mechanics, 331-334, 336
wave packet, collapse, 126-127
  reduction of, 124, 130-131
Weber, M., 253, 255
Weismann, A., 83
Weiss, J., 34
Weiss, P. A., 30, 161
Welsh, J. H., 57, 78
White, H. V., 198, 233
White, M., 231
Whitehead, A. N., 216
Whitrow, J. J., 289
Whyte, L. L., 30, 99
Wilberforce–Huxley debate, 221
Williams, L. D., 57, 78
Willis, T., 188, 191
Wilson, A. G., 30
Wilson, D., 30
Wilson, D. D., 43
Winsor, J., 225
Wisdom, J. O., 291, 298
Woese, C. R., 102
Wolfson, A., 41, 48
Wood, W. B., 84, 101
Woolf, H., 262, 288
Wright, S., 216
Wright, T., 264-266, 269-270, 272,
  278-280, 282-288

Xenopus, anucleolate embryo, 108
Xenopus laevis, 104

Yamamoto, W. S., 121
yang, 113
yin, 113
yoga, 112
yogi, 120
Young, E., 269
Yourgrau, W., 30, 331

Zapffe, P. W., 200, 206
Zimmer, H., 120
Zimmer, K. G., 339
zooids, 107
Zweig, S., 206
zygote, 9, 24